W9-ABP-251

Quality of Service

Delivering QoS on the Internet and in Corporate Networks

NOVA
SOUTHEASTERN
UNIVERSITY
SCIS

Quality of Service

Delivering QoS on the Internet and in Corporate Networks

Paul Ferguson

Geoff Huston

WILEY COMPUTER PUBLISHING

John Wiley & Sons, Inc.

New York • Chichester • Weinheim • Brisbane • Singapore • Toronto

Publisher: Robert Ipsen
Editor: Carol Long
Assistant Editor: Pam Sobotka
Managing Editor: Frank Grazioli
Electronic Products, Associate Editor: Mike Sosa
Text Design & Composition: Douglas & Gayle, Ltd.

Designations used by companies to distinguish their products are often claimed as trademarks. In all instances where John Wiley & Sons, Inc., is aware of a claim, the product names appear in initial capital or ALL CAPITAL LETTERS. Readers, however, should contact the appropriate companies for more complete information regarding trademarks and registration.

The opinions expressed within the book are solely the opinions of the authors and are not to be construed as those opinions represented by Cisco Systems, Inc., or Telstra.

This book is printed on acid-free paper. ♾

Copyright © 1998 by Paul Ferguson and Geoff Huston. All rights reserved.

Published by John Wiley & Sons, Inc.

Published simultaneously in Canada.

No part of this publication may be reproduced, stored in a retrieval system or transmitted in any form or by any means, electronic, mechanical, photocopying, recording, scanning or otherwise, except as permitted under Section 107 or 108 of the 1976 United States Copyright Act, without either the prior written permission of the Publisher, or authorization through payment of the appropriate per-copy fee to the Copyright Clearance Center, 222 Rosewood Drive, Danvers, MA 01923, (508) 750-8400, fax (508) 750-4744. Requests to the Publisher for permission should be addressed to the Permissions Department, John Wiley & Sons, Inc., 605 Third Avenue, New York, NY 10158-0012, (212) 850-6011, fax (212) 850-6008, E-Mail: PERMREQ@WILEY.COM.

This publication is designed to provide accurate and authoritative information in regard to the subject matter covered. It is sold with the understanding that the publisher is not engaged in rendering legal, accounting, or other professional services. If legal advice or other expert assistance is required, the services of a competent professional person should be sought.

Library of Congress Cataloging-in-Publication Data:

Ferguson, Paul,
Quality of service: delivering QoS on the Internet and in corporate networks / Paul Ferguson, Geoff Huston.
p. cm.
Includes bibliography references and index.
ISBN 0-471-24358-2 (pbk. : alk. paper)
1. Internet (Computer network)—Evaluation. 2. Telecommunication—Traffic—Management. 3. Intranets (Computer networks)—Evaluation.
I. Huston, Geoff. II. Title.
TK5105.875.I57F47 1998
004.6'5—dc21

97-41656
CIP

Printed in the United States of America.
10 9 8 7 6 5 4 3 2

This book is dedicated to network engineering flunkies everywhere. Keep the faith.

— pf & gh

Acknowledgments

First, I would like to express my appreciation to my management at Cisco Systems, Inc., for allowing me the time and latitude to undertake this project. While the views and opinions expressed in this book do not always align with those of my employer (and in fact sometimes are in direct conflict), the freedom and encouragement to undertake this project illustrates something I've known for a couple of years now—Cisco is a great place to work. I would also like to express my appreciation to several other individuals who have helped, either directly or indirectly, to stimulate my thought processes over the course of the past several months while I was writing and doing research for this book. Among them are Fred Baker, Bruce Davie, Craig Partridge, Van Jacobson, and Sally Floyd. I would also like to extend my gratitude to the active participants of the IRTF's End-to-End research group mailing list (you know who you are), which illustrates that there is still intelligent life in the Internet community. In my humble opinion, the technical discussions that have taken place on this mailing list are shining examples of perhaps some of the most beneficial dialogue which has taken place in the Internet research community. Last, and by no means least, I would like to express my sincere appreciation and genuine affection to my loving wife, Christina. Without her support, this project would have been tossed into the garbage can on several occasions.

— Paul Ferguson

I'd like to express my profound appreciation to my co-author Paul Ferguson for his tireless work on this project, which has been an inspiration to me. While Paul has helped me refine my understanding of this subject, it was a conversation with Dr. Marshall Rose one sunny Sydney morning in late 1996 which set me off into this world of Quality of Service. I'd like to thank him of heading me in this direction. I'd also like to thank Fred Baker, whose patient guidance on matters queuing has been invaluable to me, and to Randy Bush and Peter Lothberg for their pragmatic perspective of network operations. Of course much of this work would've been impossible without the many fine lunches and dinners at the IETF meetings, which acted as a catalyst for extraordinarily fine conversations on this subject. My thanks to my dining companions over the years! And finally I'd like to say a simple and heartfelt "thank you" to my wife Michele and my children Chris, Sam, and Alice, for their patient forbearance throughout.

— Geoff Huston

Contents

Introduction

"Any sufficiently advanced technology is indistinguishable from magic."

— Arthur C. Clarke

Quality of Service (QoS) is one of the most elusive, confounding, and confusing topics in data networking today. Why has such an apparently simple concept reached such dizzying heights of confusion? After all, it seems that the entire communications industry appears to be using the term with some apparent ease, and with such common usage, it is reasonable to expect a common level of understanding of the term.

Our research into this topic for this book has lead to the conclusion that such levels of confusion arise primarily because QoS means so many things to so many people. The trade press, hardware and software vendors, consumers, researchers, and industry pundits all seem to have their own ideas and definitions of what QoS actually is, the magic it will perform, and how to effectively deliver it. The unfortunate byproduct, however, is that these usually are conflicting concepts that often involve complex, impractical, incompatible, and non-complementary mechanisms for delivering the desired (and sometimes expected) results. Is QoS a mechanism to selectively allocate scarce resources to a certain class of network traffic at a higher level of precedence? Or is it a mechanism to ensure the highest possible delivery rates? Does QoS somehow ensure that some types of traffic experience less delay than other types of traffic? Is it a dessert topping, or is it a floor wax? In an effort to deliver QoS, you first must understand and define the problem. If you look deep enough into the QoS conundrum, this is the heart of the problem—there is seldom a single definition of QoS. The term QoS itself is an ambiguous acronym. And, of course, multiple definitions of a problem yield multiple possible solutions.

To illustrate the extent of the problem, we provide an observation made in early May 1997 while participating in the Next Generation Internet (NGI) workshop in northern

Virginia. The workshop formed several working groups that focused on specific technology areas, one of which was Quality of Service. The QoS group spent a significant amount of time spinning its wheels trying to define QoS, and of course, there was a significant amount of initial disagreement. During the course of two days, however, the group came to the conclusion that the best definition it could formulate for QoS was one that defined methods for differentiating traffic and services—a fairly broad brush stroke, and one that may be interpreted differently. The point here is that Quality of Service is clearly an area in which research is needed; tools must be developed; and the industry needs much improvement, less rhetoric, and more consensus.

TIP You can find details on the Workshop on Research Directions for the Next Generation Internet (NGI) held in May 1997 and sponsored by the Computer Research Association at www.cra.org/Policy/NGI/.

We have all experienced similar situations in virtually every facet of our everyday professional lives—although QoS seems to have catapulted into a position where it is an oft-used buzzword, there are wildly varying opinions on what it means and how to deliver it.

A Quality of Service Reality Check

This book is a project we strongly felt needed be undertaken to inject some semblance of equity into the understanding of QoS—a needed reality check, if you will. As a community, we need to do a much better job of articulating the problem we are trying to solve before we start thrusting multiple solutions onto the networking landscape. We also need to define the parameters of the QoS environment and then deliver an equitable panorama of available options to approach the QoS problem. This book attempts to provide the necessary tools to define QoS, examine the problems, and provide an overview of the possible solutions.

In the process of reviewing the QoS issues, we also attempt to provide an objective summary of the benefits and liabilities of the various technologies. It is important to draw the distinction between service quality and quality of service and to contemplate why QoS has been such an elusive and difficult issue to pursue.

QoS Is a Moving Target

Technology tends to change quickly, and QoS technologies are no exception. We will discuss emerging technologies, while recognizing that some of these still are undergoing revision and peer review and that many of these proposals still are somewhat half-baked, untested on a large scale, or not yet widely implemented. Of course, this provides an avenue for future editions of this book, but we will take a look at emerging technologies that are on the horizon and provide some thoughts on how these technologies might address particular problems that may exist today and tomorrow.

This book is geared toward all types of readers—the network manager trying to understand the technology, the network engineering staff pondering methods to provide differentiated services, and the consultant or systems integrator trying to verify a theory on why a particular approach may not be optimal.

Expectations of QoS

Although QoS means different things to different people in the population at large, it also means different things to individuals within an organization attempting to deploy it. Empirical evidence indicates that QoS mechanisms are in more demand in corporate, academic, and other private campus *intranets* (private networks that interconnect portions of a corporation or organization) than they are on the global *Internet* (the Internet proper, which is the global network of interconnected computer networks) and the ISP (Internet Service Provider) community. Soliciting comments and discussing the issue of QoS on the NANOG (North American Network Operators Group) mailing list, as well as within several Usenet newsgroups, reveals that unique requirements exist in each community. Not only do we see unique expectations within each of these communities, but we also see different expectations within each organization.

> **TIP** For more information on NANOG, see www.nanog.org.
>
> In the simplest of terms, Usenet news, or *netnews,* is a forum (or, rather, several thousand forums) for online discussion. Many computers around the world regularly exchange Usenet articles on almost every subject imaginable. These computers are not physically connected to the same network. Instead, they are connected logically in their capability to exchange data; they form the logical network referred to as *Usenet.* For more information on Usenet, see www.eff.org/papers/eegtti/eeg_68.html.

You don't need a bolt of lightning to strike you to realize that engineering and marketing organizations sometimes have conflicting views of reality. The marketing organization of an ISP, for example, may be persistent about implementing some type of QoS purely on the expectation that it will create new opportunities for additional revenue. Conversely, the engineering group may want a QoS implementation only to provide predictable behavior for a particular application or to reduce the resource contention within a particular portion of the network topology. Engineering purists are not driven by factors of economics, but instead by the intrinsic value of engineering elegance and stability—building a network that functions, scales, and outperforms inferior designs.

By the same token, unique expectations exist in the corporate and academic communities with regard to what QoS can provide to their organizations. In the academic community, it is easy to imagine the desire to give researchers preferential use of network resources for their meritorious applications and research traffic instead of forcing them to contend for network resources within the same environment where dormitory students are playing Quake across the campus network. Campus network administrators also may want to pay a slightly higher fee to their service provider to transport the research traffic with priority, while relegating student traffic to a lower-fee service perhaps associated with a lower best-effort service quality. Similarly, in the corporate network environment, there may be many reasons to provide preferential treatment to a particular application or protocol or to assign some application to a lower level of service quality.

These are the types of issues for which QoS conjures the imagination—automagic mechanisms that provide this differential service quality. It is not a simple issue that generates industry consensus, especially when agreement cannot be reached within the network engineering community or, worse, within a single organization. This also introduces a fair amount of controversy as a result of the various expectations of what the problem is that must be solved. Definitions often are crafted to deliver functionality to meet short-term goals. Is it technical? Or economic? Market driven? Or all of these things?

These are the types of issues that may become more clear; while focusing on the technical aspects, perhaps some of the political aspects will shake out.

How This Book Is Organized

This book is arranged in a manner that first examines the definitions of QoS and provides a historical perspective on why QoS is not pervasively deployed in networks today. To understand why QoS is surrounded by controversy and disagreement in the networking industry, you will examine the way in which networks are constructed currently, which methods of implementing QoS have been tried (and why they have succeeded or failed), and why implementing QoS is sometimes so difficult.

Chapter 1: What Is Quality of Service? In this chapter, you'll examine the elusive definitions of *Quality of Service*, as well as the perceptions of what QoS provides. This aspect is especially interesting and is intended for all audiences, both lay and technical. We discuss how QoS needs have evolved from a historical perspective, what mechanisms exist, how they are being used, and why QoS is important today as well as in the immediate and far-reaching future.

Chapter 2: Implementing Policy in Networks. In this chapter, you'll examine significant architectural principles that form the basis of initial QoS deployment—concepts that seem to have degenerated in practice. Because any QoS implementation is based on a particular policy or set of policies, it is important to understand where specific policies should be implemented in the network topology. There are certainly more appropriate places within a network to implement policies than others, and we will discuss why this is important, as well as why it may not be readily evident.

Chapter 2 also focuses on defining policies in the network, because policies are the fundamental building blocks with which you can begin to implement QoS in any particular network environment. The policy aspects of QoS are intended for both network managers and administrators, because understanding what tools are available to control traffic flow is paramount in successfully implementing policy control. We also provide a sanity check for network engineers, planners, and architects, so that these networking professionals can validate their engineering, design, and architectural assumptions against the principles that enable networks to be constructed with the properties necessary to implement QoS. Additionally, Chapter 2 examines measurement tools for quantifying QoS performance and looks behind the tools to the underlying framework of network performance measurement.

Chapter 3: QoS and Queuing Disciplines, Traffic Shaping, and Admission Control. Here, we provide an overview of the various queuing disciplines you can use to implement differentiated QoS, because management of queuing behavior is a central aspect of managing a network's performance. Additionally, we highlight the importance of traffic shaping and admission control, which are extraordinarily important tools used in controlling network resources. Network engineers will find Chapter 3 especially interesting, because we illustrate the intrinsic differences between each of these tools and exactly how you can use them to complement any QoS implementation.

Chapter 4: QoS and TCP/IP: Finding the Common Denominator. Next, we examine QoS from the perspective of a number of common transport protocols, discussing features of QoS implementations used in IP, Frame Relay, and ATM networks. We also examine why congestion avoidance and congestion control are increasingly important.

This chapter on TCP/IP and QoS examines the TCP/IP protocol suite and discusses which QoS mechanisms are possible at these layers of the protocol stack. TCP/IP does provide an extensive set of quality and differentiation hooks within the protocol, and you will see how you can apply these within the network.

Chapter 5: QoS and Frame Relay. This chapter provides a historical perspective of X.25, which is important in understanding how Frame Relay came into existence. We then move to Frame Relay and discuss how you can use Frame Relay mechanisms to differentiate traffic and provide QoS. You'll also see how Frame Relay may not be as effective as a QoS tool.

Chapter 6: QoS and ATM. In this chapter, you'll examine the QoS mechanisms within ATM as well as the pervasive misperception that ATM QoS mechanisms are "good enough" for implementing QoS. Both Chapters 5 and 6 are fairly technical and are geared toward the network engineering professional who has a firm understanding of Frame Relay and ATM technologies.

Chapter 7: The Integrated Services Architecture. It is important to closely examine RSVP (Resource ReSerVation Setup Protocol) as well as the Internet Engineering Task Force's (IETF) Integrated Services model to define and deliver QoS. In this chapter, you'll take an in-depth look at the Integrated Services architecture, RSVP, and ancillary mechanisms you can use to provide QoS. By the same token, however, we play devil's advocate and point out several existing issues you must consider before deciding whether this approach is appropriate for your network.

Chapter 8: QoS and Dial Access. This chapter examines QoS and dial-up services. Dial-up services encompass a uniquely different set of issues and problems, which you will examine in detail. Because a very large majority of network infrastructure, as least in the Internet, consists of dial-up services, we also provide an analytical overview of QoS-implementation possibilities that make sense in the dial-access marketplace. This chapter provides technical details of how you might implement QoS in the dial-access portion of the network. Therefore, product managers tasked with developing a dial-up QoS product will find this chapter particularly interesting.

Chapter 9: QoS and Future Possibilities. Here, you'll take a look at QoS possibilities in the future. Several promising technologies may deliver QoS in a method that makes it easy to understand and simple to implement. However, if current trends are any indication, the underlying mechanics of these technologies may prove to be extraordinarily complex and excessively difficult to troubleshoot when problems arise. This may not be the case; however, it is difficult at this point to look into a crystal ball and see the future. In any event, we do provide an overview of emerging technologies that might play a role in future QoS schemes, and we provide some editorial comments on the applicability of each scheme.

Among the possible future technologies that may have QoS significance are the IETF's emerging MPLS (Multi Protocol Label Switching) and QoSR (Quality of Service Routing) efforts, each of which is still in the early development stage, each with various submitted proposals and competing technologies. You'll also take a look at QoS possibilities with IPv6 (IP version 6) and several proposed extensions to RSVP (Resource ReSerVation Setup Protocol).

Chapter 10: QoS: Final Thoughts and Observations. In this final chapter, you'll review all the possible methods of implementing QoS and look at analyses on how each might fit into the overall QoS scheme. This chapter is particularly relevant, because we provide an overview of economic factors that play a role in the implementation of QoS. More important, we provide a simple summary of actions you can undertake within the network, as well as within host systems, to improve the service efficiency and service quality. You'll then examine a summary of actions that can result in levels of differentiation in an effort to effectively provide QoS in the network environment.

Despite the efforts of many, other networking protocols are in use today, and many networks support multiprotocol environments, so QoS within IP networks is perhaps not all of the QoS story for many readers.

Afterword: QoS and Multiprotocol Corporate Networks. Here, we discuss some of the more challenging aspects of providing QoS and differentiated services in multiprotocol networks. Although this element is not intended to be an extensive overview on QoS possibilities with each and every protocol found in corporate multiprotocol networks, we do provide a brief commentary on the issues encountered in implementing QoS in these environments, especially with regard to legacy SNA. This afterword is tailored specifically for corporate network managers who are struggling with scaling and stability issues and might be considering methods to migrate their existing legacy networks to more contemporary protocols, platforms, and technologies.

So What about QoS Is Important?

As mentioned at the start of this introduction, it is important to understand what Quality of Service means to different people. To some, it means introducing an element of predictability and consistency into the existing variability of best-effort network delivery systems. To others, it means obtaining higher transport efficiency from the network and attempting to increase the volume of data delivery while maintaining characteristically consistent behavior. And yet to others, QoS is simply a means of differentiating classes of data service—offering network resources to higher-precedence service classes at the expense of lower-precedence classes. QoS also may mean attempting to match the allocation of network resources to the characteristics of specific data flows. All these definitions fall within the generic area of QoS as we understand it today.

It is equally important to understand the past, present, and future QoS technologies and implementations that have been or have yet to be introduced. It doesn't pay to reinvent the wheel; it is much easier to learn from past mistakes (someone else's, preferably) than it is to foolishly repeat them. Also, network managers need to seriously consider the level of expertise required to implement, maintain, and debug specific QoS technologies that may be deployed in their networks. An underestimation of the proficiency level required to comprehend the technology ultimately could prove disastrous. It is important that the chosen QoS implementation function as expected; otherwise, all the theoretical knowledge in the world won't help you if your network is broken. It is no less important that the selected QoS implementation be somewhat cost-effective, because an economically prohibitive QoS model that may have a greater chance of success, and perhaps more technical merit, may be disregarded in favor of an inferior approach that is cheaper to implement. Also, sometimes brute force is better—buying more bandwidth and more raw-switching capability is a more effective solution than attempting a complex and convoluted QoS implementation.

It was with these thoughts in mind that we designed the approach in this book.

chapter 1

What Is Quality of Service?

What's in a name? In this case, not much, because *Quality of Service* (QoS) sometimes has wildly varying definitions. This is partly due to the ambiguity and vagueness of the words *quality* and *service*. With this in mind, you also should understand the distinction between *service quality* and *quality of service*.

Quality of Service: The Elusive Elephant

When considering the definition of QoS, it might be helpful to look at the old story of the three blind men who happen upon an elephant in their journeys. The first man touches the elephant's trunk and determines that he has stumbled upon a huge serpent. The second man touches one of the elephant's massive legs and determines that the object is a large tree. The third man touches one of the elephant's ears and determines that he has stumbled upon a huge bird. All three of the men envision different things, because each man is examining only a small portion of the elephant.

In this case, think of the elephant as the concept of QoS. Different people see QoS as different concepts, because various and ambiguous QoS problems exist. People just seem to have a natural tendency to adapt an ambiguous set of concepts to a single paradigm that encompasses just their particular set of problems. By the same token, this ambiguity within QoS yields different possible solutions to various problems, which leaves somewhat of a schism in the networking industry on the issue of QoS.

Another analogy often used when explaining QoS is that of the generic 12-step program for habitual human vices: The first step on the road to recovery is acknowledging and defining the problem. Of course, in this case, the immediate problem is the lack of consensus on a clear definition of QoS.

To examine the concept of QoS, first examine the two operative words: *quality* and *service*. Both words can be equally ambiguous. This chapter examines why this situation exists and provides a laundry list of available definitions.

What Is Quality?

Quality can encompass many properties in networking, but people generally use *quality* to describe the process of delivering data in a reliable manner or even somehow in a manner better than normal. This method includes the aspect of data loss, minimal (or no) induced delay or latency, consistent delay characteristics (also known as *jitter*), and the capability to determine the most efficient use of network resources (such as the shortest distance between two endpoints or maximum efficiency of circuit bandwidth). Quality also can mean a distinctive trait or distinguishing property, so people also use *quality* to define particular characteristics of specific networking applications or protocols.

What Is Service?

The term *service* also introduces ambiguity; depending on how an organization or business is structured, *service* may have several meanings. People generally use *service* to describe something offered to the end-users of any network, such as end-to-end communications or client-server applications. Services can cover a broad range of offerings, from electronic mail to desktop video, from Web browsing to chat rooms. In multiprotocol networks, a service also can have several other definitions. In a Novell NetWare network, each SAP (Service Advertisement Protocol) advertisement is considered an individual service. In other cases, services may be categorized according to the various protocol suites, such as SNA, DECnet, AppleTalk, and so forth. In this fashion, you can bring a finer level of granularity to the process of classifying services. It is not too difficult to imagine a more complex service-classification scheme in which you might classify services by protocol type (e.g., IPX, DECnet), and then classify them further to a more granular level within each protocol suite (e.g., SAP types within IPX).

Service Guarantees

Traditionally, network service providers have used a variety of methods to accommodate service *guarantees* to their subscribers, most of which are contractual. Network availability, for example, is one of the more prevalent and traditional measurements for a Service Level Agreement (SLA) between a provider and a subscriber. Here, access to the network is the basic service, and failure to provide this service is a failure to comply with the contractual obligation. If the network is inaccessible, the quality aspect of the service is clearly questionable. Occasionally, service providers have used additional criteria to define the capability to deliver on a service guarantee, such as the amount of traffic delivered. If a provider only delivers 98 percent of your traffic, for example, you might determine that the service is lacking in quality.

Understandably, a segment of the networking community abhors the use of the term *guarantee*, because it can be an indiscriminate, ambiguous, and misleading contradiction in terms. This may be especially true in an environment where virtually all traffic is packet-based, and structured packet loss may not be to the detriment of an application; indeed, it may be used as a signal to dynamically determine optimal data-flow rates. Offering a guaranteed service of some sort implies that not only will no loss occur, but that the performance of the network is consistent and predictable. In a world of packet-based networks, this is a major engineering challenge.

Quality of Service versus Classes of Service

The marriage of the terms *quality* and *service* produces a fairly straightforward definition: a measurement of how well the network behaves and an attempt to define the characteristics and properties of specific services. However, this definition still leaves a lot of latitude and creative interpretation, which is reflected in the confusion evident in the networking industry. One common thread does run through almost all the definitions of Quality of Service—the capability to differentiate between traffic or service types so that users can treat one or more classes of traffic differently than other types. The mechanisms to differentiate between traffic or service classes is discussed in more detail throughout this book. It is important to make a distinction between what is appropriately called *differentiated Classes of Service* (CoS) and the more ambiguous *quality of service*. Although some people may argue that these terms are synonymous, they have subtle distinctions. QoS has a broad and ambiguous connotation. CoS implies that services can be categorized into separate classes, which can, in turn, be treated individually. The *differentiation* is the operative concept in CoS.

Implementing Service Classes

Another equally important aspect of QoS is that, regardless of the mechanism used, some traffic must be given a predictable service class. In other words, it is not always important that some traffic be given preferential treatment over other types of traffic, but instead that the characteristics of the network remain predictable. These behavioral characteristics rely on several issues, such as end-to-end response time (also known as RTT or round-trip time), latency, queuing delay, available bandwidth, or other criteria. Some of these characteristics may be more predictable than others, depending on the application, the type of traffic, the queuing and buffering characteristics of the network devices, and the architectural design of the network.

Two forms of latency exist, for example: real and induced. *Real latency* is considered to be the physical, binding characteristics of the transport media (electronic signaling and clocked speed) and the RTT of data as it traverses the distance between two points as bound by the speed of electromagnetic radiation. This radiation often is referred to as the *speed of light problem*, because changing the speed at which light travels generally is considered to be an impossibility and is the ultimate boundary on how much real latency is inherently present in a network and how quickly data can be transmitted across any arbitrary distance.

The Speed of Light

The speed of light in a vacuum, or the physical sciences constant *c*, is probably the most investigated constant in all of science. According to electromatics theory, its value, when measured in a vacuum, should not depend on the wavelength of the radiation nor (according to Einstein's postulate about light propagation) on the observer's frame of reference. Estimates of the value of *c* have been progressively refined from Galileo's estimate in *Two New Sciences* of 1638—"If not instantaneous, it is extraordinarily rapid"—to a currently accepted value of 299,792.458 kilometers per second (with an uncertainty of some 4 millimeters per second). The speed of light in glass or fiber-optic cable is significantly slower at approximately 194,865 kilometers per second, whereas the speed of voltage propagation in copper is slightly faster at approximately 224,844 kilometers per second.

Induced latency is the delay introduced into the network by the queuing delay in the network devices, processing delay inherent to the end-systems, and any congestion present at intermediate points in the transit path of the data in flight. Queuing delay is not well ordered in large networks; queuing delay variation over time can be described only as highly chaotic, and this resultant induced latency is the major source of the uncertainty in protocol-level estimates of RTT.

Why is an accurate estimate of RTT important? For a packet protocol to use a network efficiently, the steady-state objective of the protocol is to inject a new data packet into the network at the same moment a packet in the same flow is removed from the network. The RTT is used to govern TCP's timeout and retransmission algorithms and the dynamic data rate management algorithms.

A third latency type can be introduced at this point: *remembered latency*. Craig Partridge makes a couple of interesting observations in *Gigabit Networking* [Partridge1994a] concerning latency in networks and its relationship to human perception. Partridge refers to this phenomenon as the "human-in-the-loop" principle—humans can absorb large amounts of information and are sensitive to delays in the delivery and presentation of the information. This principle describes some fascinating situations; for example, users of a network service have a tendency to remember infrequent failures of the network more than its overall success in delivering information in a consistent and reliable manner. This perception leaves users with the impression that the quality of service is poor, when the overall quality of the service actually is quite good. As you can see, predictable service quality is of critical importance in any QoS implementation, because human perception is the determining factor in the success or failure of a specific implementation.

This facet of data networking has been particularly troublesome in the search for QoS mechanisms because of the variety of issues that can affect the response time in a network. A perception also seems to exist that some sort of QoS technology will evolve that will provide predictable end-to-end behavior in the network, regardless of the architectural framework of the network—the construction and design of the underlying infrastructure, the interconnection model, and the routing system. Unfortunately (or fortunately, as the case may be), there is no substitute for prudent network design and adherence to the general principles of fundamental networking.

So what is quality of service? Significant discussions have been held on providing a fairness mechanism in the network that prohibits any one subscriber from consuming all available network resources in the presence of contention for network resources, or at the very least, prohibits any one subscriber from consuming enough resources to interfere with any other subscriber. This is a noble aspiration, but you should note that there is a direct relationship between the number of subscribers and the total amount of available bandwidth, which can be expressed as

Available Bandwidth / Number of Users = Notional Bandwidth Per User

Unless the network is overengineered to accommodate all possible users with no congestion in any portion of the network, strict adherence to the fair allocation of per-user bandwidth is not a very plausible scheme.

What is needed for QoS, and more important, differentiated CoS, is unfairness—for example, the capability to provide premium treatment for one class of traffic and best effort for another class. Several mechanisms attempt to accomplish this, and you may need to use some mechanisms with others to yield any degree of success. Among the tools that currently provide this functionality are

Traffic shaping. Using a "leaky bucket" to map traffic into separate output queues to provide some semblance of predictable behavior. This can be done at the IP layer or lower in the stack at the Asynchronous Transfer Mode (ATM) layer.

Admission control. Physically limiting link speed by clocking a circuit at a specific data rate or using a token bucket to throttle incoming traffic. The token-bucket implementation can be a stand-alone implementation or part of the IETF (Internet Engineering Task Force) Integrated Services architecture.

> **TIP** A token-bucket scheme is different from a leaky-bucket scheme. The token bucket holds tokens, with each token representing the capability to send a certain amount of data. When tokens are in the bucket, you can send data. If no tokens are in the bucket, corresponding packets (or cells) are discarded. Chapter 3, "QoS and Queuing Disciplines, Traffic Shaping, and Admission Control," discusses token buckets in more detail.

IP precedence. Using the IP (Internet Protocol) precedence bits in the IP header to provide up to eight (0 through 7) classes of traffic.

Differential congestion management. A congestion-management scheme that provides preferential treatment to certain classes of traffic in times of congestion.

There is a amusing maxim in the network engineering world that says "there are no perfect solutions, just different levels of pain." There is a smidgen of truth in this statement—it ultimately boils down to an exercise of finding the lowest pain threshold a particular solution provides.

You now can begin to formulate a straightforward definition of QoS: any mechanism that provides distinction of traffic types, which can be classified and administered differently throughout the network. This is exactly what is meant by *differentiated Classes of Service* (CoS). Should you simply refrain from using the term *QoS* in favor of *differentiated CoS*? Perhaps. In this book, you will examine several methods of differentiating traffic classes.

QoS: A Historical Perspective

In the earlier days of networking, the concept of QoS did not really exist; getting packets to their destination was the first and foremost concern. You might have once considered the objective of getting packets to their destinations successfully as a primitive binary form of QoS—the traffic was either transmitted and received successfully, or it wasn't. Part of the underlying mechanisms of the TCP/IP (Transmission Control Protocol/Internet Protocol) suite evolved to make the most efficient use of this paradigm. In fact, TCP [DARPA1981a] has evolved to be somewhat graceful in the face of packet loss by shrinking its transmission window and going into *slow start* when packet loss is detected, a congestion avoidance mechanism pioneered by Van Jacobson [Jacobson1988] at Lawrence Berkeley Laboratory. By the same token, virtually all of the IP [DARPA1981b] routing protocols use various timer-expiration mechanisms to detect packet loss or oscillation of connectivity between adjacent nodes in a network.

Prior to the transition of the NSFnet from the supervision of the National Science Foundation in the early 1990s to the commercial Internet Service Providers (ISPs) that now comprise the national Internet backbones in the United States, congestion management and differentiation of services was not nearly as critical an issue as it is now. The principal interest prior to that time was simply keeping the traffic flowing, the links up, and the routing system stable [and complying with the NSFnet Acceptable Use Policy (AUP), but we won't go into that]. Not much has changed since the transition in that regard, other than the fact that the same problems have been amplified significantly. The ISP community still is plagued with the same problems and, in fact, are facing many more issues of complex policy, scaling, and stability. Only recently has the interest in QoS built momentum.

Why Is QoS Compelling?

In the commercial Internet environment, QoS can be a competitive mechanism to provide a more distinguished service than those offered by competing ISPs. Many people believe that all ISPs are basically the same except for the services they offer. Many see QoS as a means to this end, a valuable service, and an additional source of revenue. If the mechanisms exist to provide differentiated CoS levels so that one customer's traffic can be treated differently than another customer's, it certainly is possible to develop different economic models on which to base these services.

> **TIP** For more background on the NSFnet, see www.cise.nsf.gov/ncri/nsfnet.html and www.isoc.org/internet-history/.

By the same logic, there are equally compelling reasons to provide differentiated CoS in the corporate, academic, and research network arenas. Although the competitive and economic incentives might not explicitly apply in these types of networks, providing different levels of CoS still may be desirable. Imagine a university campus where the network administrator would like to give preferential treatment to professors conducting research between computers that span the diameter of the campus network. It certainly would be ideal to give this type of traffic a higher priority when contending for network resources, while giving other types of best-effort traffic secondary resources.

Accommodating Subscriber Traffic

One tried-and-true method of providing ample network resources to every user and subscriber is to over-engineer the network. This basically implies that there will never be more aggregate traffic demand than the network can accommodate, and thus, no resulting congestion. Of course, it is hardly practical to believe that this can always be accomplished, but believe it or not, this is how many networks have traditionally been engineered. This does not always mean, however, that all link speeds have been engineered to accommodate more aggregate traffic than what may be destined for them. In other words, if there is traffic from five 56-Kbps (kilobits per second) circuits destined for a single link (or aggregation point) in a network, you would explicitly over-provision the aggregation point circuit to accommodate more than 100 percent of all aggregated traffic (in this example, greater than 280 Kbps; Figure 1.1).

Instead, there may be cases when the average utilization on certain links may have been factored into the equation when determining how much aggregate bandwidth actually is needed in the network. Suppose that you are aggregating five 56-Kbps links that are each averaging 10 percent link utilization. It stands to reason that the traffic from these five links can be aggregated into a single 56-Kbps link farther along in the network topology, and that this single link would use only, on the average, 50 percent of its available resources (Figure 1.2). This practice is called *oversubscription* and is fairly common in virtually all types of

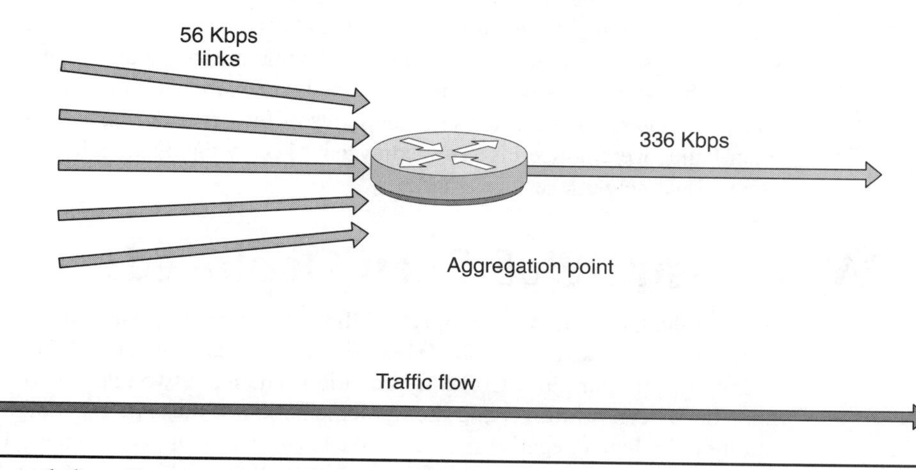

Figure 1.1: *Overengineered traffic aggregation.*

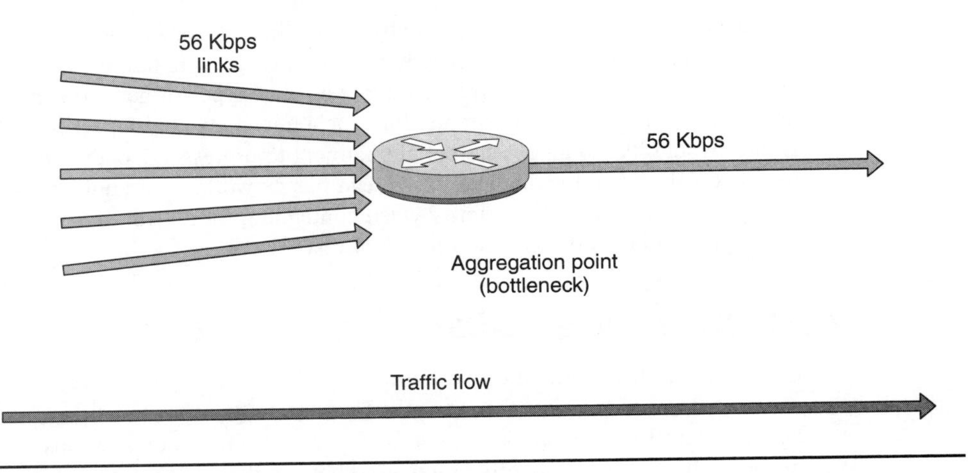

Figure 1.2: *Oversubscribed traffic aggregation.*

networks. It can be dangerous, however, if traffic volumes change dramatically before additional links can be provisioned, because any substantial increase in traffic can result in significant congestion. If each of the five links increase to an average 50 percent link utilization, for example, the one 56-Kbps link acting as the aggregation point suddenly becomes a bottleneck—it now is 250 percent oversubscribed.

For the most part, the practice of overengineering and oversubscription has been the standard method of ensuring that subscriber traffic can be accommodated throughout the network. However, oversubscribing often does not reflect sound engineering principles, and overengineering is not always economically feasible. Oversubscription is difficult to calculate; a sudden change in traffic volume or patterns or an innocent miscalculation can dramatically affect the performance of the overall network. Also, the behavior of TCP as a flow-management protocol can undermine oversubscription engineering. In this case, oversubscription is not simply a case of multiplexing constant rate sessions by examining the time and duration patterns of the sessions, but also a case of adding in adaptive rate behavior within each session that attempts to sense the maximum available bandwidth and then stabilize the data-transmission rates that this level. Oversubscription of the network inevitably leads to congestion at the aggregation points of the network that are oversubscribed. On the other hand, overengineering is an extravagant luxury that is unrealistic in today's competitive commercial network services market.

Why Hasn't QoS Been Deployed?

Most of the larger ISPs are simply peddling as fast as they can just to meet the growing demand for bandwidth; they have historically not been concerned with looking at ways to intelligently introduce CoS differentiation. There are two reasons for this. First, the tools have been primitive, and the implementation of these tools on high-speed links has traditionally had a negative impact on packet-forwarding performance. The complex queue-manipulation and packet-reordering algorithms that implement queuing-based CoS differentiation have, until recently, been a liability at higher link speeds. Second, reliable measurement tools have not been available to ensure that one class of traffic actually is being treated with preference over other classes of traffic. If providers cannot adequately demonstrate to their customers that a quantifiable amount of traffic is being treated with priority, and if the customer cannot independently verify these levels of differentiation, the QoS implementation is worthless.

There are also a few reasons why QoS has not historically been deployed in corporate and campus networks, and there are a couple of possible causes. One plausible reason is that on the campus network, bandwidth is relatively cheap; it sometimes is a trivial matter to increase the bandwidth between network segments when the network only spans several floors within the same building or several buildings on the same campus. If this is indeed the case, and because the traffic-engineering tools are primitive and may negatively impact performance, it sometimes is preferable to simply throw bandwidth at congestion problems. On a global scale, however, overengineering is considered an economically prohibitive luxury. Within a well-defined scope of deployment, overengineering can be a cost-effective alternative to QoS structures.

WAN Efficiency

The QoS problem is evident when trying to efficiently use wide-area links, which traditionally have a much lower bandwidth and much higher monthly costs than the one-time cost of much of the campus infrastructure. In a corporate network with geographically diverse locations connected by a Wide-Area Network (WAN), it may be imperative to give some mission-critical traffic higher priority than other types of application traffic.

Several years ago, it was not uncommon for the average Local-Area Network (LAN) to consist of a 10-Mbps (megabits per second) Ethernet segment or perhaps even several of them interconnected by a router or shared hub. During the same time frame, it was not uncommon for the average WAN link to consist of a 56-Kbps circuit that connected geographically diverse locations. Given the disparity in bandwidth, only a very small amount of LAN traffic could be accommodated on the WAN. If the WAN link was consistently congested and the performance proved to be unbearable, an organization would most likely get a

Moore's Law

In 1965, Intel co-founder Gordon Moore predicted that transistor density on microprocessors would double every two years. Moore was preparing a speech and made a memorable observation. When he started to graph data about the growth in memory-chip performance, he realized a striking trend. Each new chip contained roughly twice as much capacity as its predecessor, and each chip was released within 18 to 24 months of the preceding chip. If this trend continued, he reasoned, computing power would rise exponentially over relatively brief periods. Moore's observation, now known as *Moore's Law*, described a trend that has continued and is still remarkably accurate. It is the basis for many planners' performance forecasts.

faster link—unfortunately, it was also not uncommon for this to occur on a regular basis. If the organization obtained a T-1 (approximately 1.5 Mbps), and application traffic increased to consume this link as well, the process would be repeated. This paradigm follows a corollary to Moore's Law: As you increase the capacity of any system to accommodate user demand, user demand will increase to consume system capacity. On a humorous note, this is also known as a "vicious cycle."

Demand for Fiber

Recent years have brought a continuous cycle of large service providers ramping up their infrastructure to add bandwidth, customer traffic consuming it, adding additional bandwidth, traffic consuming it, and so forth. When a point is reached where the demand for additional infrastructure exceeds the supply of fiber, these networks must be carefully engineered to scale larger. If faster pipes are not readily available, these service providers do one of two things—oversubscribe capacity or run parallel, slower speed links to compensate for the bandwidth demand. After a network begins to build parallel paths into the network to compensate for the lack of availability of faster links, the issue of scale begins to creep into the equation.

This situation very much depends on the availability of fiber in specific geographic areas. Some portions of the planet, and even more specifically, some local exchange carriers within certain areas, seem to be better prepared to deal with the growing demand for fiber. In some areas of North America, for example, there may be a six-month lead time for an exorbitantly expensive DS3/T3 (45 Mbps) circuit, whereas in another area, an OC48 (2.4 Gb) fiber loop may be relatively inexpensive and can be delivered in a matter of days.

A Fiber-Optic Bottleneck

Interestingly, the fiber itself is not the bottleneck. Fiber-optic cable has three 200-nanometer-wide transmission windows at wavelengths of 850, 1300, and 1500 nanometers, and the sum of these cable-transmission windows could, in theory, be used to transmit between 50 and 75 terabits per second [Partridge1994c]. The electronics at either end of the fiber are the capacity bottleneck. Currently deployed transmission technology uses 2.5-Gbps (gigabits per second) transmission systems on each fiber-optic cable. Even with the deployment of Wave Division Multiplexing (WDM), the per-fiber transmission rates will rise to only between 50 and 100 Gbps per cable, which is still some two orders of magnititude below the theoretical fiber-transmssion capacity.

Availability of Delivery Devices

There is also a disparity in the availability of devices that can handle these higher speeds. A couple of years ago, it was common for a high-speed circuit to consist of a T3 point-to-point circuit between two routers in the Internet backbone. As bandwidth demands grew, a paradigm shift began to occur in which ATM was the only technology that could deliver OC3 (155 Mbps) speeds, and router platforms could not accommodate these speeds. After router vendors began to supply OC3 ATM interfaces for their platforms, the need for bandwidth increased yet again to OC12 (622 Mbps). This shift could seesaw back and forth like this for the foreseeable future, but another paradigm shift could certainly occur that levels the playing field—it is difficult to predict (Figure 1.3).

Changing Traffic Patterns

It is also difficult to gauge how traffic patterns will change over time; paradigm shifts happen at the most unexpected times. Until recently, network administrators engineered WAN networks under the *80/20 rule*, where 80 percent of the network traffic stays locally within the campus LAN and only 20 percent of local traffic is destined for other points beyond the local WAN gateway. Given this rule of thumb, network administrators usually could design a network with an ear to the ground and an eye on WAN-link utilization and be somewhat safe in their assumptions. However, remember that paradigm shifts happen at the most unexpected times. The Internet "gold rush" of the mid-1990s revealed that the 80/20 rule no longer held true—a larger percentage of traffic was destined for points beyond the LAN. Some parts of the world are now reporting a complete reversal of traffic dispersion along a 25/75 pattern, where 75 percent of all delivered traffic is not only non-locally originated but imported from an international source.

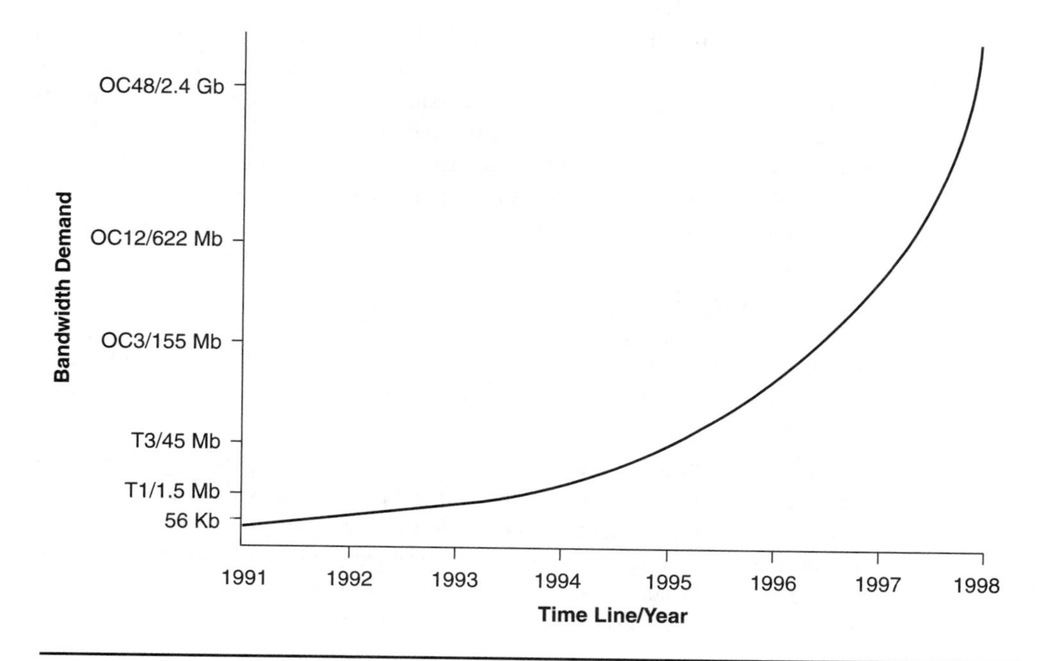

Figure 1.3: *Service provider backbone growth curve.*

Increased LAN Speed

Combine this paradigm shift with the fact that LAN media speeds have increased 100-fold in recent years and that a 1000-fold increase is right around the corner. Also consider that although WAN link speeds have also increased 100-fold, WAN circuits are sometimes economically prohibitive and may be unavailable in certain locations. Obviously, a WAN link will be congested if the aggregate LAN traffic destined for points beyond the gateway is greater than the WAN bandwidth capacity. It becomes increasingly apparent that the LAN-to-WAN aggregation points have become a bottleneck and that disparity in LAN and WAN link speeds still exists.

Other Changes Affecting QoS

Other methods of connecting to network services also have changed dramatically. Prior to the phenomenal growth and interest in the Internet in the early- to mid-1990s, providing dial-up services was generally a method to accommodate off-site staff members who needed to connect to network services. There wasn't a need for a huge commodity dial-up infrastructure. However, this changed inconceivably as the general public discovered the Internet and the number of ISPs offering dial-up access to the Internet skyrocketed. Of course, some interesting problems were introduced as ISPs tried to capitalize on the opportunity to make money and connect subscribers to the Internet; they soon found out that customers complained bitterly about network performance when the number of subscribers they had connected vastly outweighed what they had provisioned for network capacity. Many learned the hard way that you can't put 10 pounds of traffic in a 5-pound pipe.

The development and deployment of the World Wide Web (WWW or the Web) fueled the fire; as more content sites on the Internet began to appear, more people clamored to get connected. This trend has accelerated considerably over the last few years to the point where services once unimaginable now are offered on the Web. The original Web browser, *Mosaic*, which was developed at the National Center for Supercomputing Applications (NCSA), was the initial implementation of the *killer application* that has since driven the hunger for network bandwidth over the past few years.

Corporate networks have also experienced this trend, but not quite in the same vein. Telecommuting—employees dialing into the corporate network and working from home—enjoyed enormous popularity during this period and still is growing in popularity and practice. Corporate networks also discovered the value of the Web by deploying company information, documents, policies, and other corporate information assets on internal Web servers for easier employee access.

The magic bullet that network administrators now are seeking has fallen into this gray area called QoS in hopes of efficiently using network resources, especially in situations where the demand for resources could overwhelm the bandwidth supply. You might think that if there is no congestion, there is no problem, but eventually, a demand for more resources will surface, making congestion-control mechanisms increasingly important in heavily used or oversubscribed networks, which can play a key role in the QoS provided.

Class of Service Types

Before you can deliver QoS, you must define the methods for differentiating traffic into CoS types. You also must provide mechanisms for ordering the way in which traffic classes are handled. This is where preferential queuing and congestion-control mechanisms can play a role.

To understand the process of differentiated services, consider the airline industry's method of differentiating passenger seating. The manifest on any commercial passenger aircraft lists those passengers who have purchased "quality" seating in the first-class cabin, and it clearly shows that a greater number of passengers have purchased lower-priced, no-frills seating in coach class. Because the aircraft has only a certain number of seats, you can draw the analogy of the aircraft having a finite *bandwidth*.

Network administrators are looking for mechanisms to identify specific traffic types that can be treated as first class and other traffic types to be treated as coach. Of course, this is a simplistic comparison, but the analogy for the most part is accurate. Additional service classes may exist as well, such as second class, third class, and so forth. Overbooking of the aircraft, which results in someone being "bumped" and having to take another flight, is analogous to oversubscribing a portion of the network. In terms of networking, this means that an element's traffic is dropped when congestion is introduced.

Quality of service encompasses all these things: differentiation of CoS, admission control, preferential queuing (also known as class-based queuing or CBQ), and congestion management. It also can encompass other characteristics, such as deterministic path selection based on lowest latency, jitter control, and other end-to-end path metrics.

Selective Forwarding

Most of the methods of providing differentiated CoS until this point have been quite primitive. As outdated and simplistic as some methods may first appear, many still are used with a certain degree of success today. To determine whether some of these simpler approaches are adequate for CoS distinction in a network today, you need to examine the problem before deciding whether the solution is appropriate. In many cases, the benefits outweigh the liabilities; in other cases, the exact opposite is true.

In the simplest of implementations, and when parallel paths exist in the network, it would be ideal for traffic that can be classified as *priority* to be forwarded along a faster path and traffic that can be classified as *best-effort* to be forwarded along a slower path. As Figure 1.4 shows, higher-priority traffic uses the higher speed and a lower hop-count path, whereas best-effort traffic uses the lower-speed, higher hop-count path.

This simple scheme of using parallel paths is complicated by the traditional methods in which traffic is forwarded through a network. Traffic is forwarded based on its destination is called *destination-based forwarding*. This also has been referred to as *destination-based routing*, but this is a misnomer since a very important distinction must be made between routing and forwarding. Various perceptions and implementations of QoS/CoS hinge on each of these concepts.

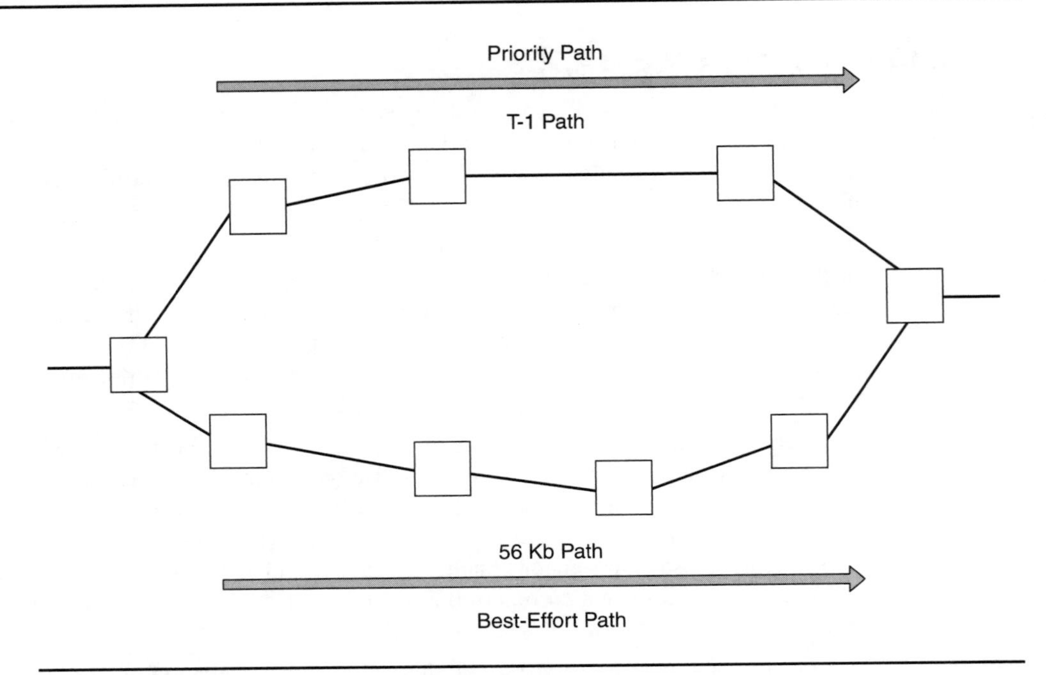

Figure 1.4: *Source-based separation of traffic over parallel paths.*

Routing versus Forwarding

Routing is, in essence, a mechanism in which a network device (usually a router) collects, maintains, and disseminates information about paths to various destinations on a network. Basically, two classes of routing protocols exist: distance-vector and link-state. This chapter is not intended to be a detailed overview of IP routing protocols, but it is important to illustrate the properties of basic IP routing protocols and packet-forwarding mechanisms at this point.

Most distance-vector routing protocols are based on the Bellman-Ford (also known as Ford-Fulkerson) algorithm that calculates the best, or shortest, path to a particular destination. A classic example of a routing protocol that falls into this category is RIP, or Routing Information Protocol [IETF1988], in which the shortest distance between two points is the one with the shortest *hop count* or number of routers along the transit path. Variations of modified distance-vector routing protocols also take into account certain other metrics in the path calculation, such as link utilization, but these are not germane to this discussion. With link-state routing protocols, such as OSPF (Open Shortest Path First) [IETF1994a], each node maintains a database with a topology map of all other nodes that reside within its portion or area of the network, in addition to a matrix of path weights on which to base a preference for a particular route. When a change in the topology occurs that is triggered by a link-state change, each node recalculates the topology information within its portion of the network, and if necessary, installs another preferred route if any particular path becomes unavailable. This recalculation process commonly is referred to as the SPF (Shortest Path First) calculation. Each time a link-state change occurs, each node in the proximity of the change must recalculate its topology database.

> **TIP** The SPF calculation is based on an algorithm developed by E.W. Dijkstra. This algorithm is described in many books, including the very thorough *Introduction to Algorithms*, by T. Cormen, C. Leiserson, and R. Rivest (McGraw Hill, 1990).

To complicate matters further, an underlying frame relay or ATM switched network also can provide routing, but not in the same sense that IP routing protocols provide routing for packet forwarding. Packet routing and forwarding protocols are connectionless, and determinations are made on a hop-by-hop basis on how to calculate topology information and where to forward packets. Because ATM is a connection-oriented technology, virtual-circuit establishment and call setup is analogous to the way in which calls are established in the switched telephone network. These technologies do not have any interaction with higher-layer protocols, such as TCP/IP, and do not explicitly forward packets. These lower-layer technologies are used to explicitly provide transport—to build an end-to-end link-layer path that high-layer protocols can use to forward packets. Frame Relay and ATM routing mechanisms, for example, are transparent to higher-level protocols and are designed primarily to accomplish VC (virtual circuit) rerouting in situations where segments of the inter-switch VC path may fail between arbitrary switches (Figure 1.5). Additionally, the ATM Forum's specification for the PNNI (Private Network-to-Network Interface) protocol also provides for QoS variables to be factored into path calculations, which you will examine in more detail in Chapter 6,

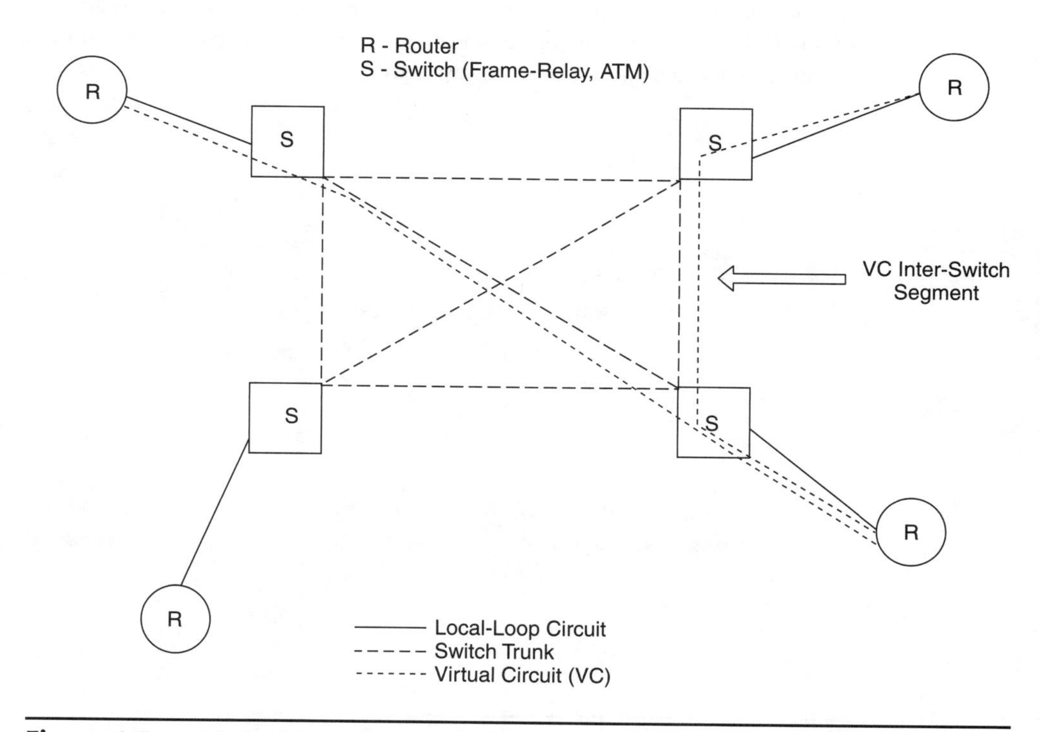

Figure 1.5: *A hybrid Layer 2/Layer 3 network.*

"QoS and ATM." Frame relay switching environments have similar routing mechanisms, but VC rerouting is vendor-specific because of the current lack of an industry standard.

> **TIP** Detailed PNNI specification documents are available from the ATM Forum at www.atmforum.com.

Packet Forwarding

Packet forwarding is exactly what it sounds like—the process of forwarding traffic through the network. Routing and forwarding are two distinct functions (Figure 1.6). Forwarding tables in a router are built by comparing incoming traffic to available routing information and then populating a forwarding cache if a route to the destination is present. This process generally is referred to as *on-demand* or *traffic-driven* route-cache population. In other words, when a packet is received by a router, it checks its forwarding cache to determine whether it has a cache entry for the destination specified in the packet. If no cache entry exists, the router checks the routing process to determine whether a route table entry exists for the specified destination. If a route table entry exists (to include a default route), the forwarding cache is populated with the destination entry, and the packet is sent on its way. If no route table entry exists, an ICMP (Internet Control Message Protocol) [DARPA1981c] unreachable message is sent to the originator, and the packet is dropped. An existing forwarding cache entry would have been created by previous packets bound for the same destination. If there is an existing forwarding cache entry, the packet is sent to the appropriate outbound interface and sent on its way to the next hop.

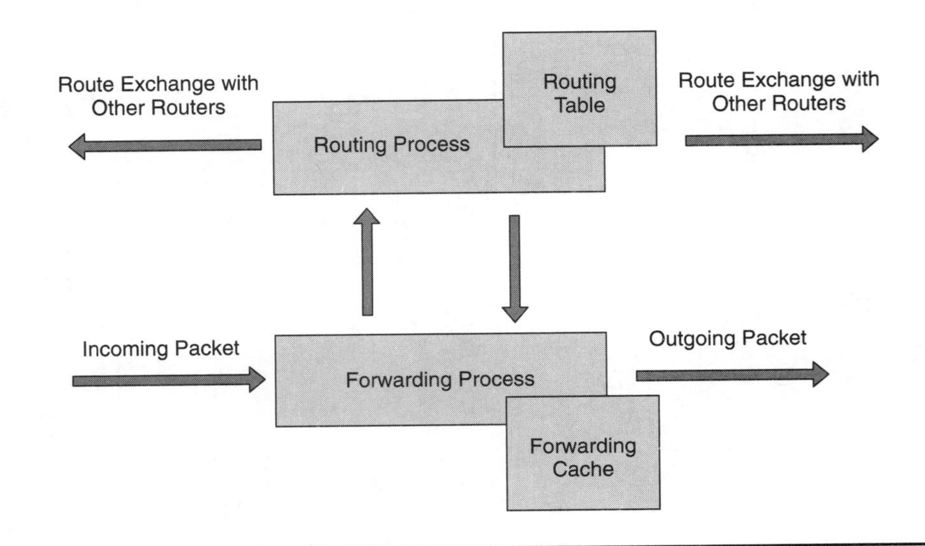

Figure 1.6: *Routing and forwarding components.*

Topology-Driven Cache Population

Another method of building forwarding cache entries is *topology-driven cache population*. Instead of populating the forwarding cache when the first packet bound for a unique destination is received, the forwarding cache is pre-populated with all the routing information from the onset. This method has long been championed as possessing better scaling properties at faster speeds. Topology-driven cache population has been or is being adopted and implemented by router vendors as the forwarding cache-population mechanism of choice.

As mentioned previously, traditional packet forwarding is based on the destination instead of the source of the traffic. Therefore, any current effort to forward traffic based on its origin is problematic at best. A few router vendors have implemented various methods to attempt to accomplish source-based packet forwarding, also known as *policy-based routing*, but because of undesirable (and historical) forwarding performance issues that occur when these mechanisms are used, this type of forwarding generally has been largely avoided.

> **TIP** Currently, several router vendors are working to develop policy-based forwarding in their products that has a negligible impact on forwarding performance.

Some source-based forwarding implementations do not provide the granularity required, require excessive manual configuration, or become too complex to manage and configure as

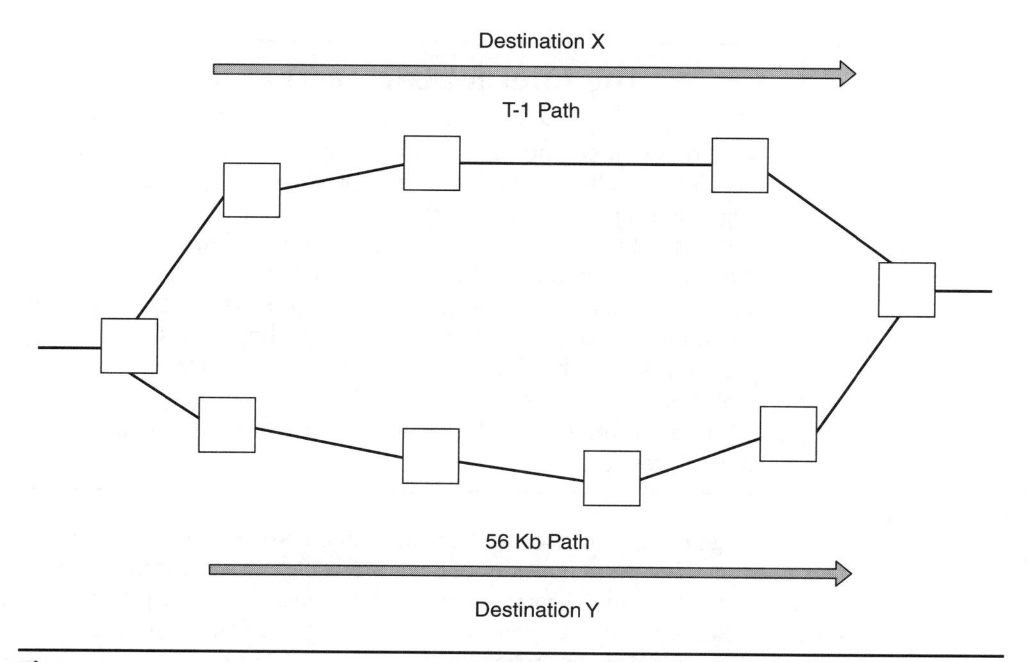

Figure 1.7: *Destination-based separation of traffic over parallel paths.*

the forwarding policies become complicated. As traffic policies become more convoluted, the need to provide this functionality in a straightforward and simple mechanism grows. It stands to reason that over time, these mechanisms will mature and evolve into viable alternatives.

As mentioned previously, the case of forwarding traffic along parallel paths becomes problematic, because no practical mechanism exists to forward traffic based on origin. As Figure 1.7 shows, it is trivial to advertise a particular destination or group of destinations along a particular path and another destination along another path. Forwarding traffic along parallel paths is easily accomplished by using static routing or filtering dynamic route propagation. Unfortunately, this method deviates from the desire to forward traffic based on source.

More elaborate variations on source-based routing are discussed later, but first you should examine a fundamental problem in using forwarding strategies as a method for implementing differentiation.

Insurmountable Complexity?

It has been suggested that most of the methods of implementing a QoS scheme are much too complex and that until the issues surrounding QoS are simplified, a large majority of network engineering folks won't even consider the more complicated mechanisms currently available. Not to cast a shadow on any particular technology, but all you need to do to understand this trepidation is to explore the IETF Web site to learn more about RSVP (Resource ReSerVation Setup Protocol) and the Integrated Services working group efforts within the IETF.

The Internet Engineering Task Force

The IETF is a loose international collection of technical geeks consisting of network engineers, protocol designers, researchers, and so forth. The IETF develops the technologies and protocols that keep the Internet running. The majority of the work produced by the various working groups (and individuals) within the IETF is documented in *Internet Drafts* (I-Ds). After the I-Ds are refined to a final document, they generally are advanced and published as RFCs (Requests For Comments). Virtually all the core networking protocols in the Internet are published as RFCs, and more recently, even a few technologies that are not native to the Internet have been published as RFCs. For more information on the IETF, see www.ietf.org and for more information on RSVP, see www.ietf.org/html.charters/rsvp-charter.html.

Although most of the technical documentation produced within the IETF has been complex to begin with, the complexity of older protocols and other technical documentation pales in comparison to the complexity of the documents recently produced by the RSVP and Integrated Services working groups. It is quite understandable why some people consider some of this technology overwhelming.

And there is a good reason why these topics are complex: The technology is difficult. It is difficult to envision, it is difficult to develop, it is difficult to implement, and it is difficult to manage. And just as with any sufficiently new technology, it will take some time to evolve and mature. And it also has its share of naysayers who believe that simpler methods are available to accomplish approximately the same goals.

You will examine the IETF RSVP and Integrated Services architecture in Chapter 7, but it is noteworthy to mention at this point that because of the perceived complexity involved, some people are shying away from attempting to implement QoS using RSVP and Integrated Services today. Whether the excessive complexity involved in these technologies is real or imagined is not important. For the sake of discussion, you might consider that perception *is* reality.

Layer 2 Switching Integration

Another chink in the armor of the QoS deployment momentum is the integration of Layer 2 switching into new portions of the network topology. With the introduction of Frame Relay and ATM switching in the WAN, isolating network faults and providing adequate network management have become more difficult. Layer 2 WAN switching is unique in a world accustomed to point-to-point circuits and sometimes is considered a mixed blessing. Frame Relay and ATM are considered a godsend by some, because they provide the flexibility of provisioning VCs between any two end-points in the switched network, regardless of the number of Layer 2 switches in the end-to-end path. Also, both Frame Relay and ATM help reduce the amount of money spent on a monthly basis for local loop circuit charges, because many virtual circuits can be provisioned on a single physical circuit. In the past, this would have required several point-to-point circuits to achieve the same connectivity requirements (consider a one-to-one relationship between a point-to-point circuit and a virtual circuit) and subsequently, additional monthly circuit charges.

One of the more unfortunate aspects of both ATM and Frame Relay is that although both provide a great deal of flexibility, they also make it very easy for people to build sloppy networks. In very large networks, it is unwise to attempt to make all Layer 3 devices (routers) only one hop away from one another. Depending on the routing protocol being used, this may introduce instability into the routing system, as the network grows larger, because of the excessive computational overhead required to maintain a large number of adjacencies. You'll look at scaling and architectural issues in Chapter 2, "Implementing Policy in Networks," because these issues may have a serious impact on the ability to provide QoS.

Problems of this nature are due to the "ships in the night" paradigm that exists when devices in the lower layers of the OSI (Open Systems Interconnection) reference model stack (in this case, the link layer) also provide buffering and queuing and in some cases, routing intelligence (e.g., PNNI). This is problematic, because traditional network management tools are based on SNMP (Simple Network Management Protocol) [IETF1990a], which is reliant on UDP (User Datagram Protocol) [DARPA1980] and the underlying presence of IP routing. Each of these protocols (SNMP, UDP, and IP) are at different layers in

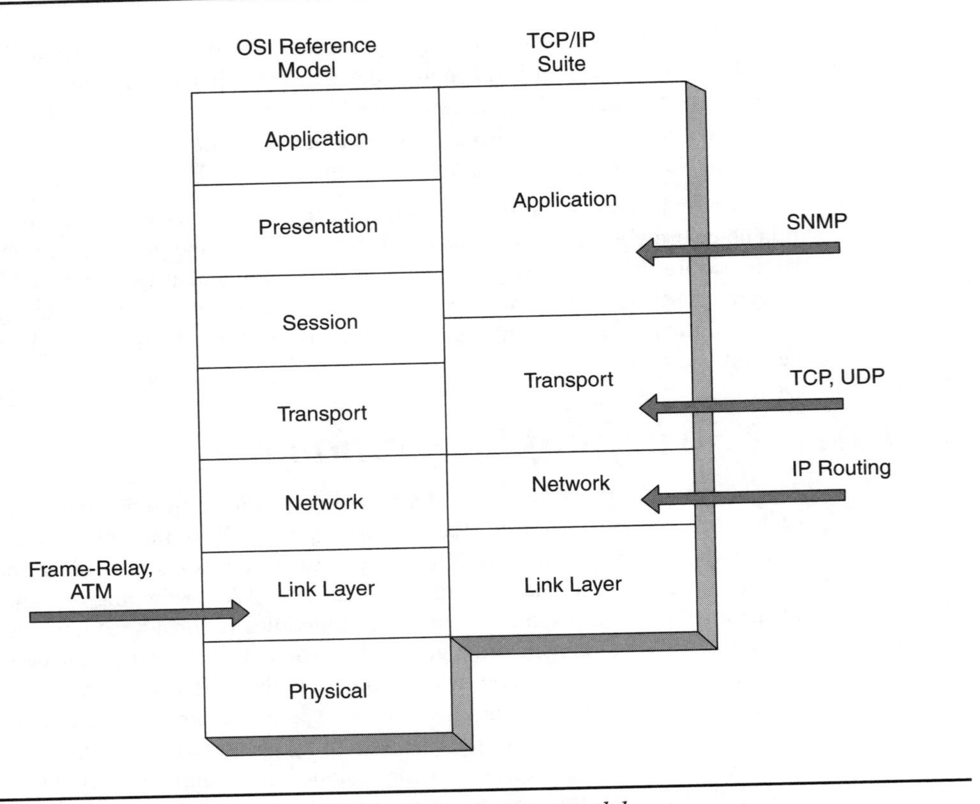

Figure 1.8: *The OSI and TCP/IP communications models.*

the TCP/IP protocol stack, yet there is an explicit relationship between them; SNMP relies on UDP for transport and an IP host address for object identification, and UDP relies on IP for routing. Also, a clean, black-and-white comparison cannot be made of the relationship between the TCP/IP protocol suite and the layers in the OSI reference model (Figure 1.8).

The "Ships in the Night" Problem

The problem represented here is one of no interlayer communication between the link-layer (ATM, frame relay, et al.) and the network-layer and transport-layer protocols (e.g., IP, UDP, and TCP)—thus the "ships in the night" comparison. The higher-layer protocols have no explicit knowledge of the lower link-layer protocols, and the lower link-layer protocols have no explicit knowledge of higher-layer protocols. Thus, for an organization to manage both a TCP/IP-based network and an underlying ATM (link-layer) switched network, it must maintain two discrete management platforms. Also, it is not too difficult to imagine situations in which problems may surface within the link-layer network that are not discernible to troubleshooting and debugging tools that are native to the upper-layer protocols. Although link-layer-specific MIB (SNMP management information base) variables may exist to help isolate link-layer problems, it is common to be unable to recognize specific link-layer problems with SNMP. This can be especially frustrating in situations in which ATM VC routing problems may exist, and the Network Operations Center (NOC) staff has to quickly isolate a problem using several tools. This is clearly a situation in which too much diversity can be a liability, depending on the variety and complexity of the network technologies, the quality of the troubleshooting tools, and the expertise of the NOC staff.

This problem becomes even more complicated when the network in question is predominantly multiprotocol, or even when a small amount of multiprotocol traffic exists in the network. The more protocols and applications present in the network, the more expertise required of the support staff to adequately isolate and troubleshoot problems in the network.

Differentiated Classes of Service (CoS)

It is possible that when most people refer to *quality of service*, they usually are referring to *differentiated classes of service*, whether or not they realize it, coupled with perhaps a few additional mechanisms that provide traffic policing, admission control, and administration. *Differentiation* is the operative word here, because before you can provide a higher quality of service to any particular customer, application, or protocol, you must classify the traffic into classes and then determine the way in which to handle the various traffic classes as traffic moves throughout the network. This brings up several important concepts.

When differentiation is performed, it is done to identify traffic by some unique criteria and to classify the incoming traffic into classes, each of which can be recognized distinctly by the classification mechanisms at the network ingress point, as well as farther along in the network topology. The differentiation can be done in any particular manner, but some of the more common methods of initially differentiating traffic consist of identifying and classifying traffic by

Protocol. Network and transport protocols such as IP, TCP, UDP, IPX, and so on.

Source protocol port. Application-specific protocols such as Telnet, IPX SAP's, and so on, dependent on their source host address.

Destination protocol port. Application-specific protocols such as Telnet, IPX SAP's, and so on, dependent on their destination host address.

Source host address. Protocol-specific host address indicating the originator of the traffic.

Destination host address. Protocol-specific host address indicating the destination of the traffic.

Source device interface. Interface on which the traffic entered a particular device, otherwise known as an *ingress interface*.

Flow. A combination of the source and destination host address, as well as the source and destination port.

Differentiation usually is performed as a method of identifying traffic as it enters the network or ensuring that traffic is classified appropriately so that, as it enters the network, it is subject to a mechanism that will force it to conform with whatever user-defined policy is desired. This entry-point enforcement is called *active policing* (Figure 1.9) and can be done at the network entry points to ensure that traffic presented to the network has not been arbitrarily categorized or classified before being presented to the network. Enforcement of another sort is also performed at intermediate nodes along the transit path, but instead of resource-intensive enforcement, such as various admission-control schemes, it is a less resource-intensive practice of administering what already has been classified, so that it can queued, forwarded, or otherwise handled accordingly.

The alternative approach to active differentiation is known as *passive admission*; here, the traffic is accepted as is from a downstream subscriber and transited across the network (Figure 1.10). The distinction is made here between *initial* differentiation through the use of selective policing and allowing differentiation to be done outside the administrative domain of your network. In other words, you would allow downstream connections outside your administrative domain to define their traffic and simply allow the traffic to move unadulterated. There are certain dangers in providing a passive-admission service—for example, downstream subscribers may try to mark their traffic as a higher priority than entitled and

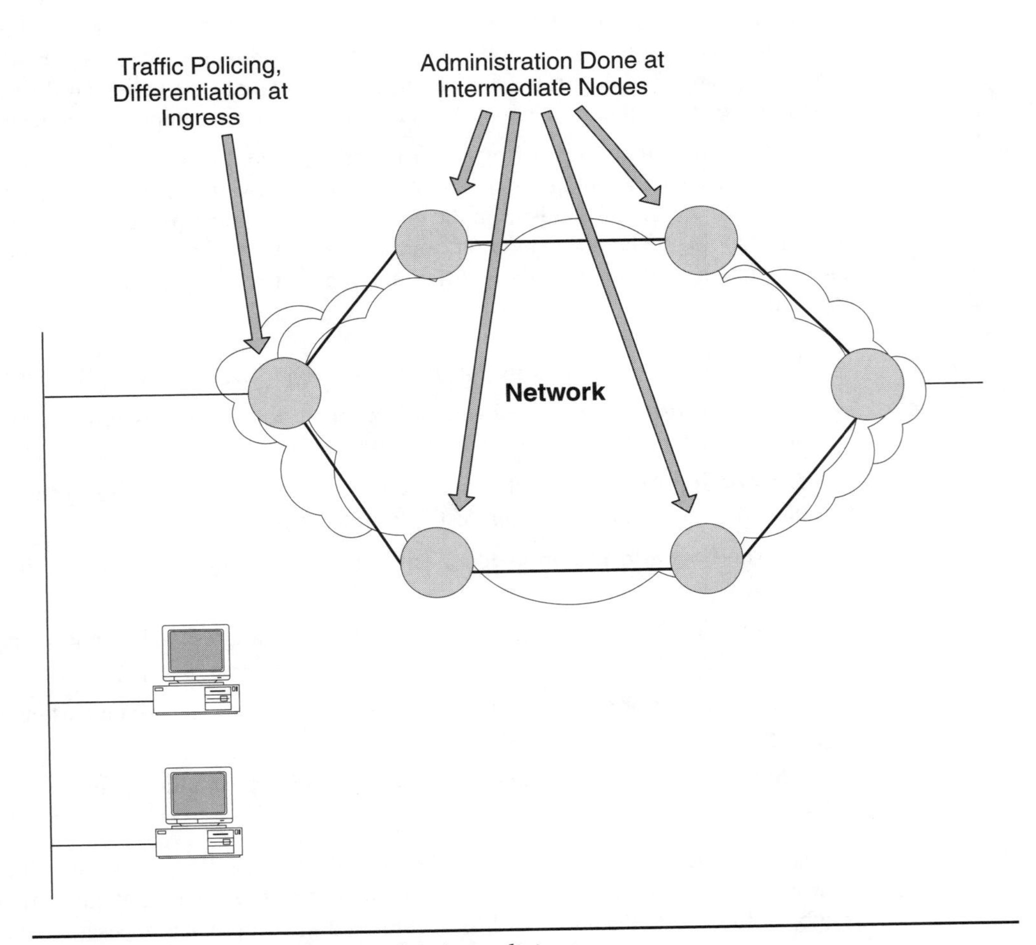

Figure 1.9: *Differentiation and active policing.*

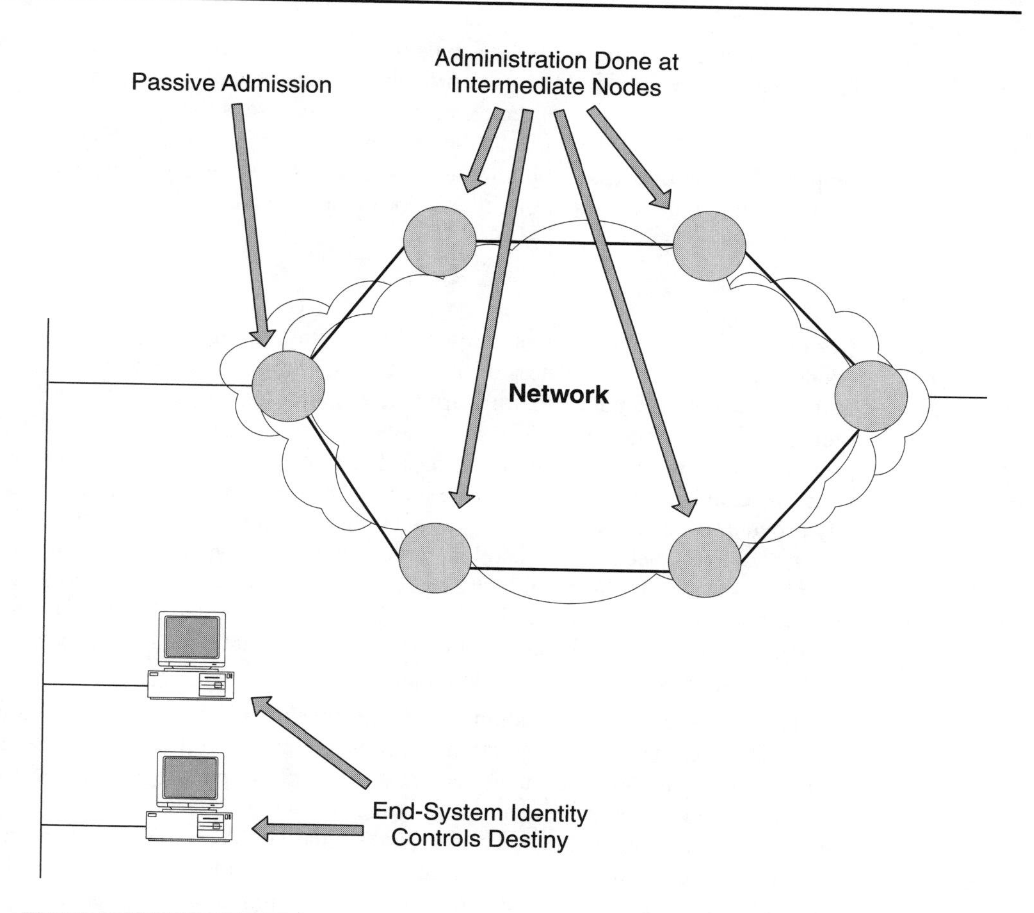

Figure 1.10: *Differentiation and passive admission.*

subsequently consume more network resources than entitled. The *marking* of the traffic may take different forms; it may be RSVP reservation requests or IP precedence to indicate preference, or it may be a Frame Relay DE (discard eligible) bit or an ATM CLP (cell loss priority) bit set to indicate the preference of the traffic in a congestion situation. The difficulty in policing traffic at the edges of a network depends on how and at what layer of the protocol stack it is done.

Based on the preceding criteria, a network edge device can take several courses of action after the traffic identification and classification is accomplished. One of the simplest courses of action to take is to queue the various classes of traffic differently to provide diverse servicing classes. Several other choices are available, such as selectively forwarding specific classes of traffic along different paths in the network. You could do this by traditional (or nontraditional) packet-forwarding schemes or by mapping traffic classes to specific Layer 2 (frame relay or ATM VCs) switched paths in the network cloud. A variation of mapping traffic classes to multiple Layer 2 switched paths is to also provide different traffic-shaping schemes or congestion thresholds for each virtual circuit in the end-to-end path. In fact, there are probably more ways to provide differentiated service through the network than are outlined here, but the important aspect of this exercise is that the identification, classification, and marking of the traffic is the fundamental concept in providing differentiated CoS support. Without these basic building blocks, chances are that any effort to provide any type of QoS will not provide the desired behavior in the network.

❖ ❖ ❖

Quality of Service (QoS) is an ambiguous term with multiple interpretations, many of which depend on the uniqueness of the problems facing the networking community, which in turn is faced with managing infrastructure and providing services within its particular organizations. QoS generally is used to describe a situation in which the network provides preferential treatment to certain types of traffic but is disingenuous in describing exactly which mechanisms are used to provide these types of services or what these services actually provide. By contrast, it may be more appropriate to use the term *differentiated Classes of Service* (CoS), because at least this implies that traffic types can be distributed into separate classes, each of which can be treated differently as it travels through the network. Of course, several methods are available for delivering differentiated CoS—and equally as many opinions as to which approach is better, provides predictable behavior, achieves the desired goal, or is more elegant than another.

The first issue a network administrator must tackle, however, is understanding the problem to be solved. Understanding the problem is paramount in determining the appropriate solution. There are almost as many reasons for providing and integrating QoS/CoS support into the network as there are methods of doing so—among them, capacity management and value-added services. On the other hand, understanding the benefits and liabilities of each approach can be equally important. Implementing a particular technology to achieve QoS differentiation without completely understanding the limitations of a particular implementation can yield unpredictable and perhaps disastrous results.

chapter 2

Implementing Policy in Networks

Quality of Service is almost synonymous with policy—how certain types of traffic are treated in the network. Humans define the policy and implement the various tools to enact them. Implementing policy in a large network can be complicated; depending on exactly where you implement these policies, they can have a direct impact on the capability of the network to scale, maintain peak performance, and provide predictable behavior.

An appropriate analogy for the way in which policy implementation affects a network is the importance of a blueprint to building a house. Building without a blueprint can lead to a haphazardly constructed network, and one in which service quality is severely lacking. Developing a traffic-management policy is paramount to integrating stability into the network, and without stability, service quality is unachievable.

QoS Architectures

What architectural choices are available to support QoS environments? To understand the architecture of the application of QoS in an Internet environment, you first should understand the generic architecture of an Internet environment and then examine where and how QoS mechanisms can be applied to the architecture.

Generic Internet Carrier Architectures

Figure 2.1 shows a generic carrier architecture for the Internet environment. This architecture separates the high-capacity systems that drive the high-speed trunking from the high-functionality systems that terminate customer-access links.

The key to scaling very large networks is to maintain strict levels of hierarchy, commonly referred to as *core, distribution*, and *access* levels (Figure 2.2). This limits the degree of meshing among nodes. The core portion of the hierarchy generally is considered the central portion, or backbone, of the network. The access level represents the outermost portion of the hierarchy, and the distribution level represents the aggregation points and transit portions between the core and access levels of the hierarchy.

The core routers can be interconnected by using dedicated point-to-point links. Core systems also can use high-speed switched systems such as ATM as the transport mechanism. The core router's primary task is to achieve packet switching rates that are as fast as possible. To achieve this, the core routers assume that all security filtering already has been performed and that the packet to be switched is being presented to the core router because it conforms to the access and transmission policy of the network. Accordingly, the core router has the task of determining which interface to pass the packet to, and by removing all other access-related tasks from the router, higher switching speeds can be achieved. Core routers also have to actively participate in routing environments, passing connectivity information

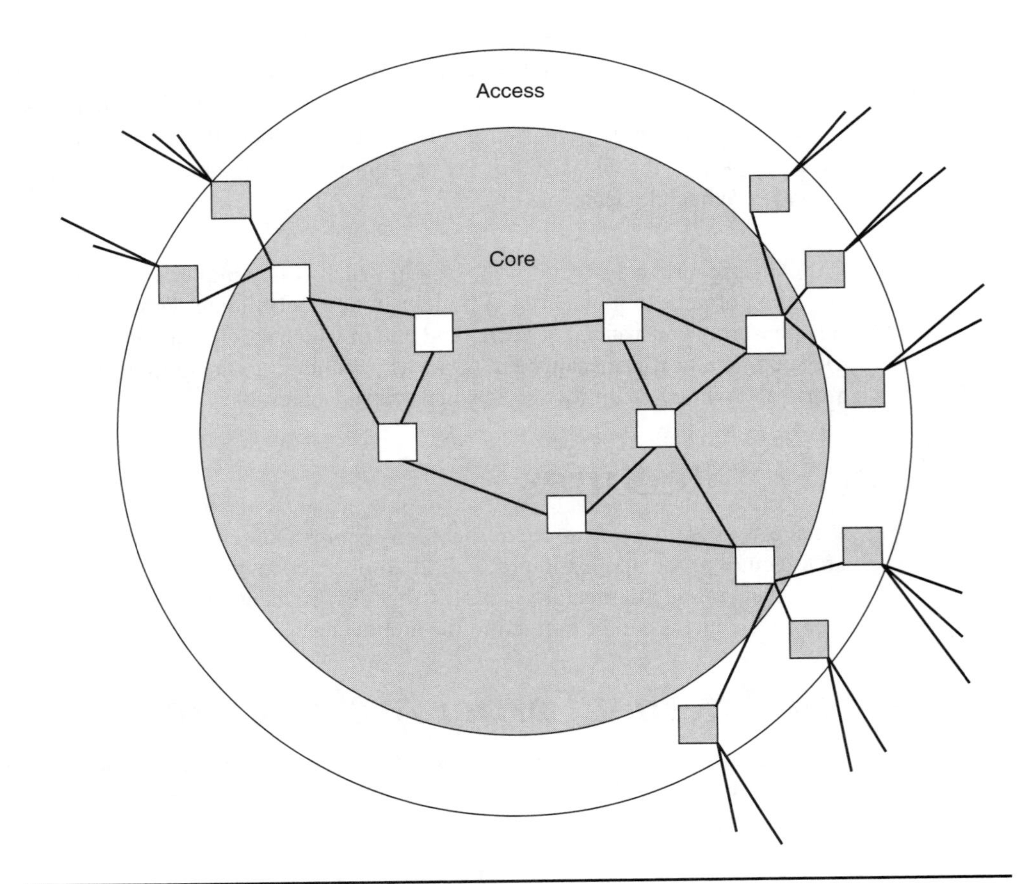

Figure 2.1: *Access/core router paradigm.*

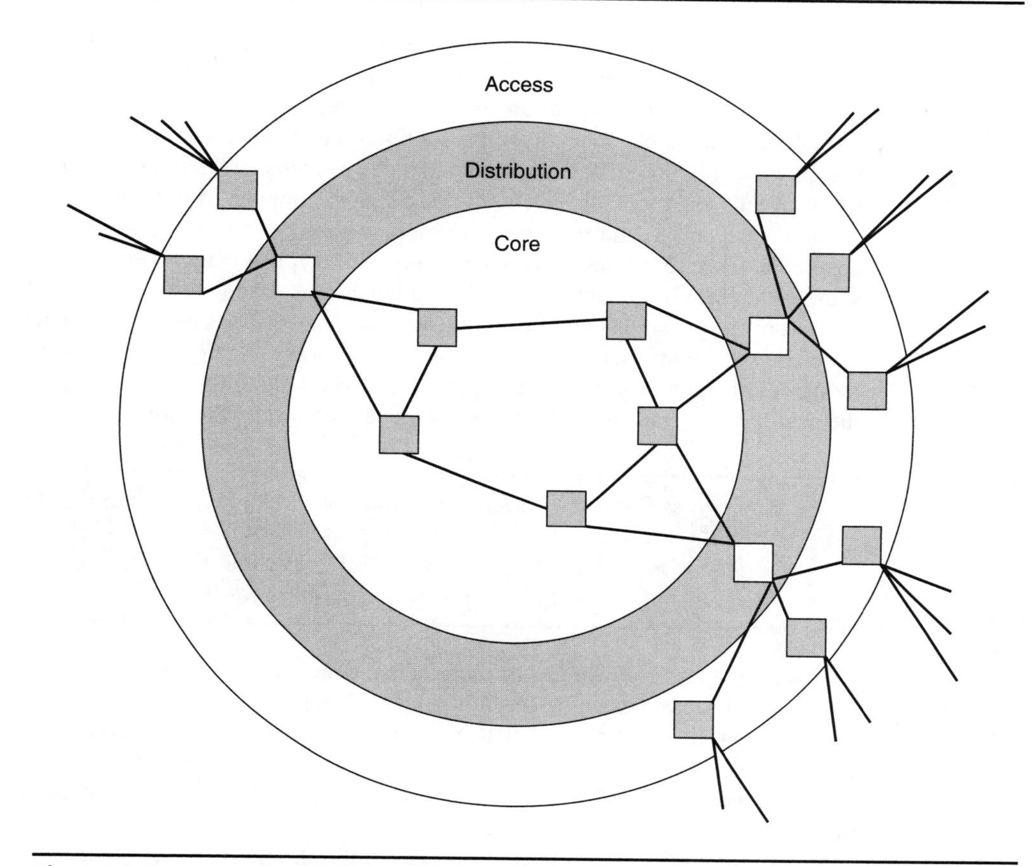

Figure 2.2: *Structured hierarchy: core, distribution, and access.*

along within the routing system, and populating the forwarding tables with routing information based on the status of the routing information.

Clocking Rates and Point-to-Point Links

Point-to-point links come in a variety of flavors. In countries that use a base telephony signaling rate of 56 Kbps, the carrier hierarchy typically provides point-to-point bearer circuits with a data clocking rate of 1.544 Mbps (T1) and 45 Mbps (T3). In countries that use a base telephony signaling rate of 64 Kbps, the data clocking rates are 2.048 Mbps (E1) and 34 Mbps (E3). In recent years, the higher-speed systems are working to a common Synchronous Digital Hierarchy (SDH) architecture, where data rates of 155 Mbps (STM-1) and 622 Mbps (STM-4) are supported.

Access routers use a different design philosophy. Access routers have to terminate a number of customer connections, typically of lower speed than the core routing systems. Access routers do not need to run complete routing systems and can be designed to run with a smaller routing table and a default route pointing toward a next-hop core router. Access routers generally do not have to handle massive packet switching rates driving high-speed trunk circuits and typically have to manage a more complex traffic-filtering environment where the packets to and from the customer are passed though a packet-filter environment to ensure the integrity of the network environment. A typical filtering environment checks the source address of each packet passed from the customer and checks to ensure that the source address of the packet is one of the addresses the customer has declared an intention to advertise via the upstream provider (to prevent a form of source address spoofing). The filtering environment also may check packets passed along to the customer, generally checking the source address to ensure that it is not one advertised by the customer.

TIP Ingress filtering is documented in "Network Ingress Filtering: Defeating Denial of Service Attacks Which Employ IP Source Address Spoofing," (P. Ferguson and D. Senie, October 1997). You can find it at ftp://ds.internic.net/internet-drafts/draft-ferguson-ingress-filtering-03.txt.

Scaling a large network is also a principal concern, especially at high speeds, and it is an issue you need to examine carefully in the conceptual design and planning stages. Striking the right hierarchical balance can make the difference between a high-performance network and one on the verge of congestion collapse. It is important to recognize the architectural concepts of applying policy control in a large network. This is necessary to minimize the performance overhead associated with different mechanisms that provide traffic and route filtering, bandwidth control (rate-limiting), routing policy, congestion management, and differentiated classes of service. Scaling a network properly also plays a role in which quality is maintained in the network. If the network is unstable because of sloppy design, architecture, or implementation, attempting to introduce a QoS scheme into the network is an effort in futility.

Another important aspect in maintaining a sensible hierarchy in the network is to provision link speeds in the network with regard to *topological significance* to avoid oversubscription. Link speeds should be slower farther out in the network topology and faster in the core of the network. This allows higher-speed links in the higher levels of the network hierarchy to accommodate the aggregate volume of traffic from all lower-speed links farther down in the network hierarchy.

Capacity Oversubscription

In Chapter 1, you looked at the concepts of overengineering and oversubscription. Without a complete review of these concepts, overengineering can be described as designing a network so that the overall demand for bandwidth will never surpass the amount of bandwidth available. You also learned in Chapter 1 that if a network constantly has excess bandwidth, as is the case in overengineered networks, it may never need QoS differentiation of traffic. For matters of practicality and economics, however, this is rarely the case. Instead, sensible oversubscription usually is done based on a back-of-the-envelope calculation that takes into account the average utilization on each link in the hierarchy; it also is based on the assumption that the network has been designed with relatively sound hierarchical principles and regard for the topological significance of each provisioned link. Based on this calculation, you might decide that it is safe to oversubscribe the network at specific points in the hierarchy and avoid chronic congestion situations. Of course, oversubscription is far more prevalent than overengineering in today's networks.

Hierarchical Application of Policy

It is critical to understand the impact of implementing policy in various points in the network topology, because mechanisms used to achieve a specific policy may have an adverse effect on the overall network performance. Because high-speed traffic forwarding usually is performed in the network core, and lower-speed forwarding generally is done lower in the network hierarchy, varying degrees of performance impact can result, depending on where these types of policies are implemented. In most cases, a much larger percentage of traffic transits the network core than transits any particular access point, so implementing policy in the network core has a higher degree of impact on a larger percentage of the traffic. Therefore, policy implementation in large networks should be done at the lowest possible levels of the hierarchy to avoid performance degradation that impacts the entire network (noting that in this architectural model, the access level occupies the lowest level in the hierarchy). Traffic and route filtering, for example, are types of policies that may fall into this category.

Implementation of policy lower in the hierarchy has a nominal impact on the overall performance of the network for several reasons compared to the impact the same mechanism may have when implemented in the core. Some mechanisms, such as traffic filtering, have less of an impact on forwarding performance on lower-speed lines. Because the speed of internode connectivity generally gets faster as you go higher in the network hierarchy, the impact of implementing policy in the higher levels of the network hierarchy increases.

The same principle holds true for traffic accounting, access control, bandwidth management, preference routing, and other types of policy implementations. These mechanisms are more appropriately applied to nodes that reside in the lower portions of the network hierarchy, where processor and memory budgets are less critical.

If traffic congestion is a principal concern in the distribution and core levels of the network hierarchy, you should consider the implementation of an intelligent congestion-management mechanism, such as Random Early Detection (RED) [Floyd1993], in these portions of the network topology. You'll examine RED in more detail in Chapter 3, "QoS and Queuing Disciplines, Traffic Shaping, and Admission Control."

Another compelling reason to push policy out toward the network periphery is to maintain stability in the core of the network. In the case of BGP (Border Gateway Protocol) [IETF1995a] peering with exterior networks, periods of route instability may exist between exterior routing domains, which can destabilize a router's capability to forward traffic at an optimal rate. Pushing exterior peering out to the distribution or access levels of the network hierarchy protects the core network and minimizes the destabilizing effect on the network's capability to provide maximum performance. This fundamental approach allows the overall network to achieve maximum performance and maintain stability while accommodating the capability to scale the network to a much larger size and higher speeds, thus injecting a factor of stability and service quality.

Approaches to Attaining a QoS Environment

QoS environments can be implemented in a number of modes. The first of these is a stateless mode, in which QoS is implemented by each router within the network. This normally takes the form of a number of local rules based on a pattern match of the packet header tied to a specific router action. Suppose that the environment to be implemented is one of service class differentiation, where the NNTP (Network News Transfer Protocol), which is used to transfer Usenet news [IETF1986], is placed into a lower priority, and interactive Telnet sessions are to be placed into a high priority. You could implement this service class by using multiple output queues and associated local rule sets on each router to detect the nominated service classes of packets and place them in the appropriate output queue with a scheduling algorithm that implements queue precedence. Note that this stateless queuing-based action is relevant only when the incoming packet rate exceeds the egress line capacity. When the network exhibits lower load levels, such QoS mechanisms will have little or no impact on traffic performance.

This stateless approach produces a number of drawbacks. First, the universal adoption of rule sets in every router in the network, as well as the consequent requirement to check each packet against the rule sets for a match and then modification of the standard FIFO (First In, First Out) queuing algorithm to allow multiple properties, is highly inefficient. For this QoS structure to be effective, the core routers may need to be loaded with the same rule sets as the access routers, and if the rule sets include a number of specific cases with corresponding queuing instructions, the switching load for each packet increases dramatically, causing the core network to require higher cost-switching hardware or to increase the switching time (induce latency). Neither course of action scales well. Additionally, the operator must maintain consistent rule sets and associated queuing instructions on all routers on the network, because queuing delay is accumulated across all routers in the end-to-end path.

Thus, stateless locally configured QoS architectures are prone to two major drawbacks: the architecture does not scale well into the core of the network, and operational management overhead can be prone to inconsistencies.

The second approach is a modification to the first, where the access routers modify a field in the packet header to reflect the QoS service level to be allocated to this packet. This marking of the packet header can be undertaken by the application of a rule set similar to that described earlier, where the packet header matching rules can be applied to each packet entering the network. The consequent action at this stage is to set the IP precedence header field to the desired QoS level in accordance with the rule set. Within the core network, the rule set now is based on the IP precedence field value (QoS value) in every packet header. Queuing instructions or discard directives then can be expressed in terms of this QoS value in the header instead of attempting to undertake a potentially lengthy matching of the packet header against the large statically configured rule set.

To extend this example, you can use the 3-bit IP precedence field to represent eight discrete QoS levels. All access routers could apply a common rule set to incoming packets so that Usenet packets (matching the TCP—Transmission Control Protocol—field against the decimal value 119) may have their IP precedence field set to 6, whereas Telnet and RLOGIN packets have their IP precedence field set to 1. All other packets have their QoS header field cleared to the value 0, which resets any values that may have been applied by the customer or a neighbor transit network. The core routers then can be configured to use a relatively simple switching or queuing algorithm in which all packets with IP precedence values of 6 are placed in the low-priority queue, all packets with values of 1 are placed in the high-priority queue, and all other packets are placed within the normal queuing process. The router's scheduler may use a relative weighting of scheduling four packets from the high-precedence queue for every two from the "normal" queue to every single packet from the low-precedence queue. This architectural approach attempts to define the QoS level of every packet on entry to the network, and thereby defines the routers' scheduling of the packet on its transit through the network. The task of the core routers can be implemented very efficiently, because the lookup is to a fixed field in the packet and the potential actions are limited by the size of the field.

The third approach to QoS architecture is to define a flow state within a sequence of routers, and then mark packets as being a component of the defined flow. The implementation of flows is similar to switched-circuit establishment, in which an initial discovery packet containing characteristics of the subsequent flow is passed through the network, and the response acknowledgment commits each router in the path to support the flow.

The major architectural issues behind QoS networks that maintain flow-state information are predominantly concerns about the control over the setting of QoS behavior and the transitivity of QoS across networks, as well as the computational and system-resource impact on maintaining state information on thousands of flows established across a diverse network.

Triggering QoS in Your Environment

Who can trigger QoS behavior within the network? Of the models discussed in the preceding section, the first two implicitly assume that the network operator sets the QoS behavior and that the customer has no control over the setting. This may not be the desired behavior of a QoS environment. Instead of having the network determine how it will selectively degrade service performance under load, the network operator may allow the customer to determine which packets should be handled preferentially (at some premium fee, presumably). Of course, if customers can trigger QoS behavior by generating QoS labeled packets, they may also want to probe the network to determine whether setting the QoS bits would make any appreciable difference in the performance of the end-to-end application.

If the network path is relatively uncongested, there probably will be no appreciable difference in triggering a QoS mechanism for the application; otherwise, if the network path does encounter points of high congestion and extended queuing delay, QoS mechanisms may produce substantial differences in data throughput rates. This particular approach leads to the conclusion that QoS becomes an end-to-end discovery problem similar to that of end-to-end MTU (Maximum Transmission Unit) discovery, where the customer can use probe mechanisms to determine whether every network operator in the end-to-end path will honor QoS settings, and whether setting the QoS headers will make any appreciable difference in the subsequent end-to-end service session.

QoS across Diverse Networks

Another architectural consideration is the transitivity of QoS across networks. The models discussed here implicitly assume that any two customers who want to use QoS in a service session are both customers of the same network operator and are, in effect, customers of the same QoS environment. In today's Internet environment, such an assumption is a very weak one, and the issue of transitivity of QoS is critical. In the example described earlier, the network operator has elected to use the IP precedence field as a QoS value and has chosen to take packets with field values of 6 as low-priority packets and packets with field values of 1 as high-priority packets.

Obviously, there is an implicit expectation that if the packet is passed to a neighbor network, the packet priority field will have a consistent interpretation, or at worst, the field will not be interpreted at all. If the neighbor network had chosen an opposite interpretation of field values, with 1 as low priority and 6 as high priority, obviously, the resultant traffic flow would exaggerate degradation levels. Because packets passed quickly through the first network will have a very high probably of delay—and even discard—through the second network, packets passed with high delay through the first network without discard then are passed quickly through the second network.

This type of problem has a number of solutions. One is that the network operator must translate the QoS field values of all packets being passed to a neighbor network into corresponding QoS field values that preserve the original QoS semantics as specified by the originating customer. Obviously, this type of manipulation of the packet headers involves some per-packet processing cost in the router. Accordingly, this approach does not scale easily in very high-speed network interconnect environments. A second solution is that a number of network operators reach agreement that certain QoS field values have a fixed QoS interpretation, so that the originally set QoS values are honored consistently by all networks in the end-to-end path.

In summary, stateless QoS architectures that rely on distributed filter lists and associated QoS actions on each router do not scale well in large networks. QoS architectures that set packet attributes on entry to the network or have QoS attributes set by the original customer are more scaleable. QoS architectures need to be consistently applied across networks, by common agreement on the interpretation of QoS attribute values, if QoS is going to be used in a scaleable fashion. Finally, if QoS is a customer-configured attribute, customers will be using some form of end-to-end QoS discovery mechanism to determine whether any form of QoS is required for any particular application.

Routing and Policy

There are basically two types of routing policy: intradomain and interdomain. *Intradomain policy* is implemented within a network that is a single administrative domain, such as within a single AS (Autonomous System) in the Internet or a single corporate network. *Interdomain policy* deals with more than a single administrative network domain, which is much more difficult, because different organizations usually have diverse policies and, even when their policies may be similar, they rarely can agree on similar methods of implementing them. When dealing with policy issues, it is much easier to implement policy when working within a single organization, where you can control the end-to-end network paths. When traffic leaves your administrative domain, you lose control over the interaction of policy and technology.

It is imperative to understand the relationship between the capability to control the forwarding paths in the network and the impact this has on the capability to introduce service quality into the overall equation. If multiple paths exist between the source and destination, or ingress and egress, it stands to reason that the capability to prefer one particular forwarding path over another can make the difference between poor service quality and service that is above average or perhaps even superior.

Mechanisms within both interior (intradomain) and exterior (interdomain) routing protocols can provide the capability to prefer one particular path over another. It is the capability to adequately understand these mechanisms and control the forwarding path that is important as a tool in implementing policy, and policy is a critical tool in beginning to understand and implement a QoS strategy.

As an illustration of intradomain routing control, Figure 2.3 shows that from node A, there are three possible paths to node 10.1.1.15, whereas one path (path B) is a much higher speed (T3) than the other two (paths A and C), yet a lower-speed link (path A) has a lower hop count. The path-selection process in this case would very much depend on the interior routing protocol being used. If RIP (Routing Information Protocol) was the routing protocol being used, router Z would use path A to forward traffic toward 10.1.1.15, because the hop

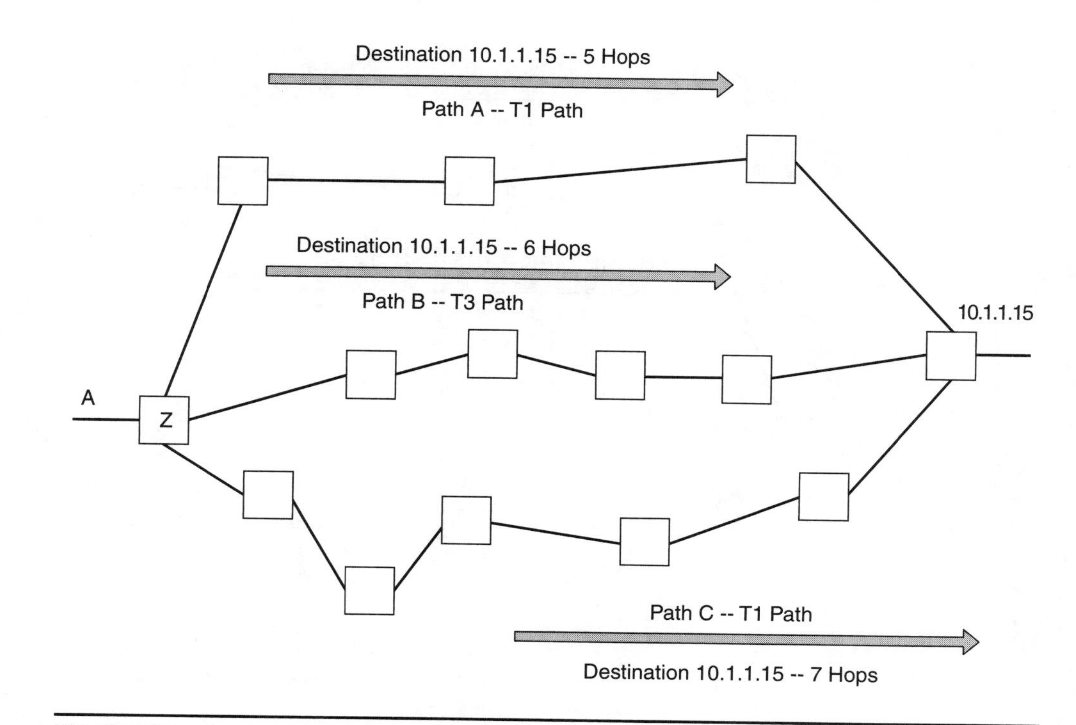

Figure 2.3: *Controlling routing policy within a single administrative domain (RIP).*

count is 5. However, an administrator who maintained the routers in this network could artificially increase the hop count at any point within this network to force the path-selection process to favor one path over another.

If OSPF (Open Shortest Path First) is the interior routing protocol being used, each link in the network can be assigned a cost, whereas OSPF will chose the least-cost path based on the sum of the end-to-end path cost of each link in the transit path to the appropriate destination. In this particular example (Figure 2.4), router Z would chose the lowest-cost path (20), because it is consists of a lower end-to-end path cost than the other two paths. This is fairly straightforward, because this path is also one that is higher bandwidth. For policy reasons, though, an administrator could alternatively cost each path differently so that a lower-bandwidth path is preferred over a higher-bandwidth path.

Similar methods of using other interior routing protocol metrics to manually prefer one path over another are available. It is only important to understand that these mechanisms can be used to implement an alternative forwarding policy in place of what the native mechanisms of a dynamic routing protocol may choose without human intervention (static configuration).

It also is important to understand the shortcomings in traditional hop-by-hop, destination-based routing protocols. When a "best" route is installed for a particular destination at any location, all packets that reach that location are forwarded to the appropriate next-hop router for that particular destination. This means that special handling situations, as mentioned in the case of policy-based forwarding, need to be considered and implemented separately.

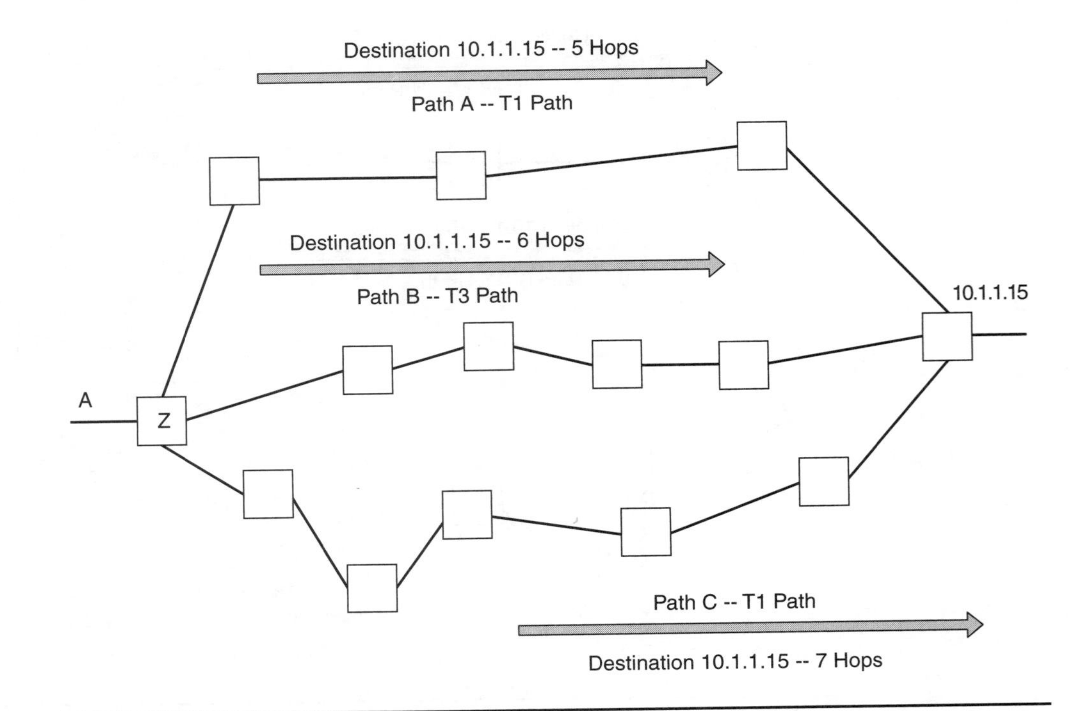

Figure 2.4: *Controlling routing policy within a single administrative domain (OSPF).*

Predictive path selection and control of routing information between different ASs also is important to understand; more important, it is vital to understand what is possible and what is *not* possible to control. As mentioned earlier, it is rarely possible to control path selection for traffic coming *into* your AS because of the routing policies of exterior neighboring ASs. There are several ways to control how traffic *exits* an interior routing domain, and there are a couple of ways to control how traffic *enters* your domain that may prove feasible in specific scenarios. You will briefly look at a few of these scenarios in a moment.

TIP For a more comprehensive overview of Internet routing, refer to Bassam Halabi, *Internet Routing Architectures,* Cisco Press, New Riders Publishing, 1997, ISBN 1-56205-652-2.

It is important to revisit the fundamental aspect of how traffic is forwarded within a network. When specific destinations (or an aggregate) are advertised neighbor-by-neighbor, hop-by-hop throughout the network, traffic is forwarded based on several factors, namely the best path to a specific destination. Therefore, the way in which destinations are advertised along certain paths usually dictates how traffic is forwarded. With BGP (Border Gateway Protocol), as with most other routing protocols, the most specific advertisement to a particular destination always is preferred. This commonly is referred to as *longest match*, because the *longest* or most specific prefix (route) always is installed in the forwarding table.

CIDR's Influence

A fundamental concept in route aggregation was introduced by CIDR (Classless InterDomain Routing), which allows for contiguous blocks of address space to be aggregated and advertised as singular routes instead of individual entries in a routing table. This helps keep the size of the routing tables smaller than if each address were advertised individually and should be done whenever possible. The Internet community generally agrees that aggregation is a good thing.

TIP The concepts of CIDR are discussed in RFC1517, RFC1518, RFC1519, and RFC1520. You can find the CIDR Frequently Asked Questions (FAQ) at www.rain.net/faqs/cidr.faq.html.

CIDR provides a mechanism in which IP addresses are no longer bound to network masking being done only at the byte level of granularity, which provides legacy addresses that are considered *classful*—class A, class B, and class C addresses. With CIDR, entire blocks of contiguous address space can be advertised by a single *classless* prefix with an appended representation of the bit mask that is applied across the entire 32-bit address. A traditional class C network address of 199.1.1.0 with a network mask of 255.255.255.0, for example, would be represented as 199.1.1.0/24 in classless address notation, indicating that the upper 24 bits of the address consists of network bits, leaving the lower 8 bits of the address for host addresses (Figure 2.5).

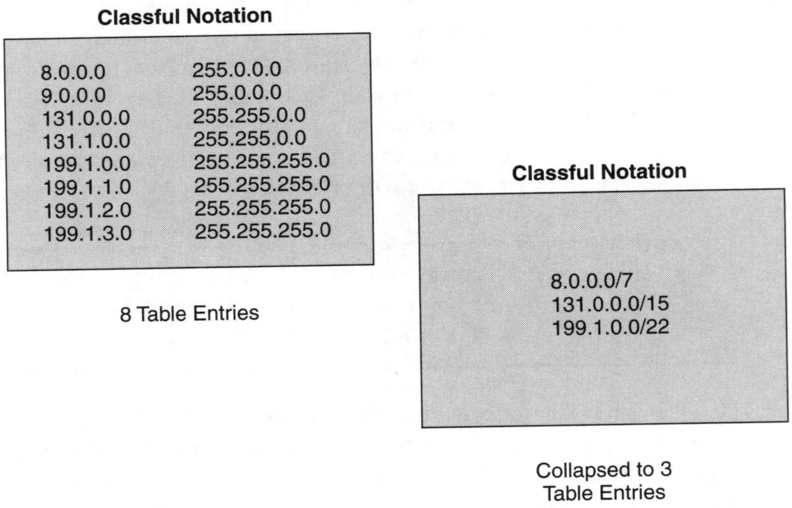

Classful Notation

8.0.0.0	255.0.0.0
9.0.0.0	255.0.0.0
131.0.0.0	255.255.0.0
131.1.0.0	255.255.0.0
199.1.0.0	255.255.255.0
199.1.1.0	255.255.255.0
199.1.2.0	255.255.255.0
199.1.3.0	255.255.255.0

8 Table Entries

Classful Notation

8.0.0.0/7
131.0.0.0/15
199.1.0.0/22

Collapsed to 3
Table Entries

Figure 2.5: *Comparison of classful and classless addressing.*

> **TIP** For a comparison of classful and classless address notation, see RFC1878, "Variable Length Subnet Table For IPv4" (T. Pummill and B. Manning, December 1995).

It also is important to understand the importance of route aggregation in relationship to scaling a large network to include the Internet. It is easy to understand how decreasing the number of entries in the routing table is beneficial—the fewer the number of entries, the less amount of computational power required to parse them and recalculate a new path when the topology changes. The fewer the number of entries, the more scaleable the network.

However, CIDR does reduce the granularity of routing decisions. In aggregating a number of destinations into a single routing advertisement, routing policy then can be applied only to the aggregated routing entry. Specific exceptions for components of the aggregated routing block can be generated by advertising the exception in addition to the aggregate, but the cost of this disaggregation is directly related to the extent by which this differentiation of policy should be advertised. A local differentiation within a single provider would impose a far lower incremental cost than a global differentiation that was promulgated across the entire Internet, but in both cases the cost is higher than no differentiation at all. This conflict between the desire for fine-level granularity of differentiated routing policy and the desire to deploy very coarse granularity to preserve functionality in the face of massive scaling is very common within today's Internet.

The Border Gateway Protocol (BGP)

The current implementation of routing policy between different administrative routing domains is accomplished with the use of the BGP protocol, which is the current de facto method of providing interdomain routing in the Internet. BGP can be used in any situation where routing information needs to be exchanged between different administrative routing domains, whether in the Internet or in the case of multiple private networks with dissimilar administrative policies.

There are several methods of influencing the BGP path-selection process, each of which has topological significance. In other words, these tools have very explicit purposes and are helpful only when used in specific places in the network topology. These tools revolve around the manipulation of certain BGP protocol attributes. Certain BGP attribute tools have very specific uses in some situations and are inappropriate in others. Also, some of these tools are nothing more than simply tie-breakers to be used when the same prefix (route) is advertised across parallel paths. A path for which a more specific prefix is announced generally will always be preferred over a path that advertises shorter aggregate. This is a fundamental aspect of longest-match, destination-based routing. The tools that can provide path-selection influence are:

- Filtering based on the AS path attributes (AS origin, et al.)
- AS path prepending
- BGP communities
- Local preference
- Multi-Exit Discriminator (MED)

BGP AS Path Filtering

A prevalent method used in controlling the method in which traffic is forwarded along specific paths is the capability to filter routing announcements to or from a neighboring AS based on the AS path attributes. This is based on the concept that if a route is not seen for a particular destination from a particular peer, traffic will not be forwarded along that path. Although there are far too many scenarios in which route filtering can be done that could adequately be discussed here, an example of this mechanism follows.

As depicted in Figure 2.6, a route for the prefix 199.1.1.0/24 that has traversed AS200 to reach AS100 will have an AS path of 200 300 when it reaches AS100. This is because the route originated in AS300. If multiple prefixes were originated within AS300, the route propagation of any particular prefix could be blocked, or filtered, virtually anywhere along the transit path based on the AS, in order to deny a forwarding path for a particular prefix or set of prefixes. The filtering could be done on an inbound (accept) or outbound (propagate) basis, across multiple links, based on any variation of the data found in the AS path attribute. Filtering can be done on the origin AS, any single AS, or any set of ASs found in the AS path information. This provides a great deal of granularity for selectively accepting and propagating routing information in BGP. By the same token, however, if an organization begins to get exotic with its routing policies, the level of complexity increases dramatically,

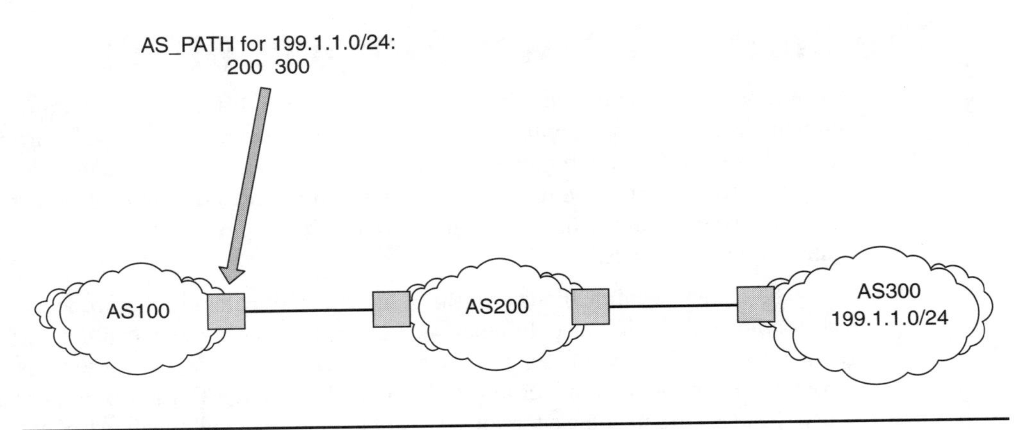

Figure 2.6: *BGP AS path.*

especially when multiple entry and exit points exist. The downside to filtering routes based on information found in the AS path attribute is that it only really provides binary selection criteria; you either accept the route and propagate it, or you deny it and do not propagate it. This does not provide a mechanism to define a primary and backup path.

AS Path Prepend

When comparing two advertisements of the same prefix with differing AS paths, the default action of BGP is to prefer the path with the lowest number of transit AS hops, or in other words, the preference is for the shorter AS path length. To complicate matters further, it also possible to manipulate the AS path attributes in an attempt to influence this form of downstream route selection. This process is called *AS path prepend* and is adopted from the practice of inserting additional instances of the originating AS into the beginning of the AS path prior to announcing the route to an exterior neighbor. However, this method of attempting to influence downstream path selection is feasible only when comparing prefixes of the same length, because an instance of a more specific prefix always is preferable. In other words, in Figure 2.7, if AS400 automatically aggregates 199.1.1.0/24 into a larger announcement, say 199.1.1.0/23, and advertises it to its neighbor in AS100, while 199.1.1.0/24 is advertised from AS200, then AS200 continues to be the preferred path because it is a more specific route to the destination 199.1.1.0/24. The AS path length is examined only if identical prefixes are advertised from multiple external peers.

Similar to AS path filtering, AS path prepending is a negative biasing of the BGP path-selection process. The lengthening of the AS path is an attempt to make the path less desirable than would otherwise be the case. It commonly is used as a mechanism of defining candidate backup paths.

BGP Communities

Another popular method of making policy decisions using BGP is to use the *BGP community* attribute to group destinations into communities and apply policy decisions based on this attribute instead of directly on the prefixes. Recently, this mechanism has grown in

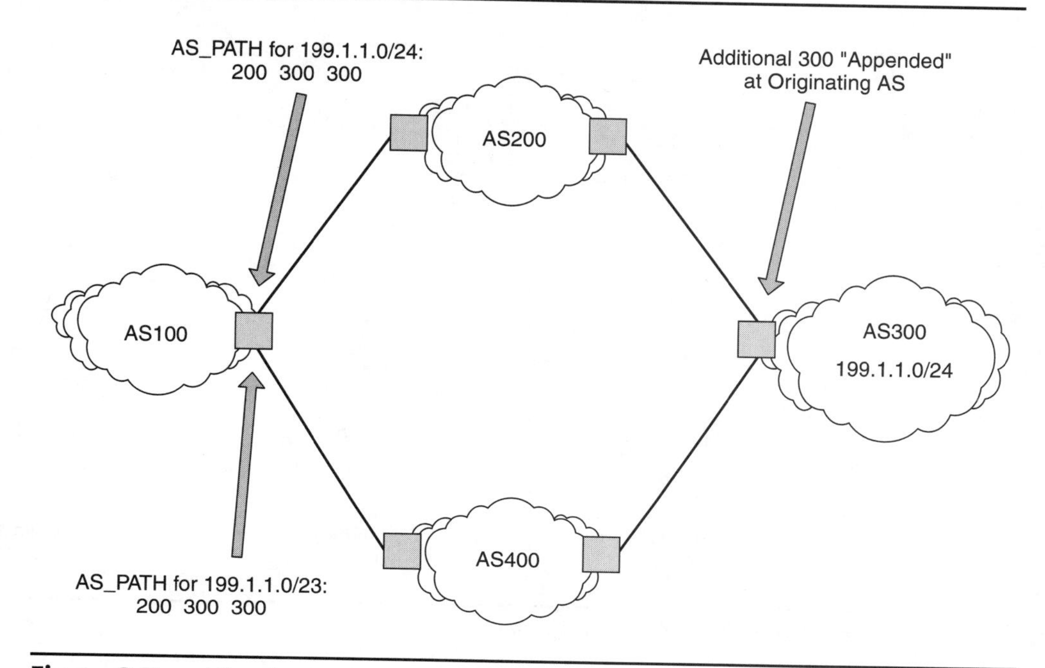

Figure 2.7: *AS path prepend.*

popularity, especially in the Internet Service Provider (ISP) arena, because it provides a simple and straightforward method with which to apply policy decisions. As depicted in Figure 2.8, AS100 has classified its downstream peers into BGP communities 100:10 and 100:20. These values need only have meaning to the administrator of AS100, because he is the one using these community values for making policy decisions. He may have a scheme for classifying all his peers into these two communities, and depending on a prearranged agreement between AS100 and all its peers, these two communities may have various meanings.

One of the more common applications is one in which community 100:10 could represent all peers who want to receive full BGP routing from AS100, and community 100:20 could represent all peers who want only partial routing. The difference in semantics here is that *full* routing includes all routes advertised by all neighbors (generally considered the entire default-free portion of the global Internet routing table), and *partial* includes only prefixes originated by AS100, including prefixes originated by directly attached customer networks. The latter could be a significantly smaller number of prefixes, depending on the size and scope of AS100's customer base.

BGP communities are very flexible tools, allowing groups of routing advertisements to be grouped by a common attribute value specific to a given AS. The community can be set locally within the AS or remotely to trigger specific actions within the target AS. You then can use the community attribute value in a variety of contexts, such as defining backup or primary paths, defining the scope of further advertisement of the path, or even triggering other router actions, such as using the community attribute as an entry condition for rewrit-

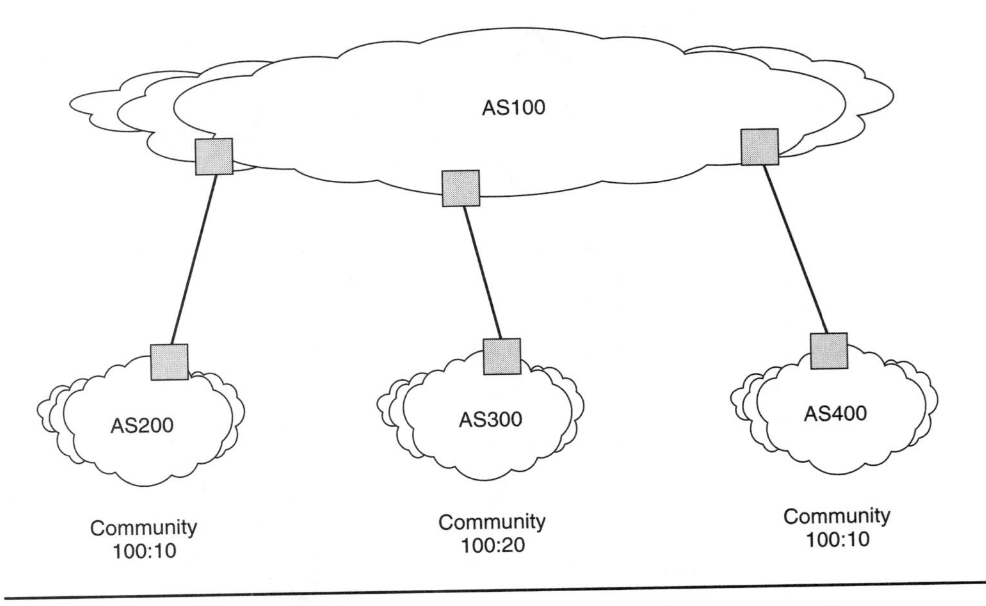

Figure 2.8: *BGP communities.*

ing the TOS (Type of Service) or precedence field of IP packets, which consequently allows a model of remote triggering of QoS mechanisms. This use of BGP communities extends beyond manipulation of the default forwarding behavior into an extended set of actions that can modify any of the actions undertaken by the router.

BGP Local Preference

Another commonly used BGP attribute is called *BGP Local Preference*. The local preference attribute is local to the immediate AS only. It is considered nontransitive, or rather, not passed along to neighboring ASs, because its principal use is to designate a preference within an AS as to which gateway to prefer when more than one exists.

As an example, look at Figure 2.9. Here, the AS border routers B and C would propagate a local preference of 100 and 150, respectively, to other iBGP (Internal BGP) peers within AS100. Because the local preference with the highest value is preferred when multiple paths for the same prefix exist, router A would prefer routes passed along from router B when considering how to forward traffic. This is especially helpful when an administrator prefers to send the majority of the traffic exiting the routing domain along a higher speed link, as shown in Figure 2.9, when prefixes of the same length are being advertised internally from both gateway routers. The downside to this approach, however, is that when an administrator wants equitable use among parallel paths, a single path will be used heavily, and others with a lower local preference will get very little, if any, use. Depending on what type of links these are (local or wide area), this may be an acceptable situation. The primary consideration observed here may be one of economics—exorbitant recurring monthly costs for a wide-area circuit may force an administrator to consider other alternatives. However, when network administrators want to get traffic out of their network as quickly as possible

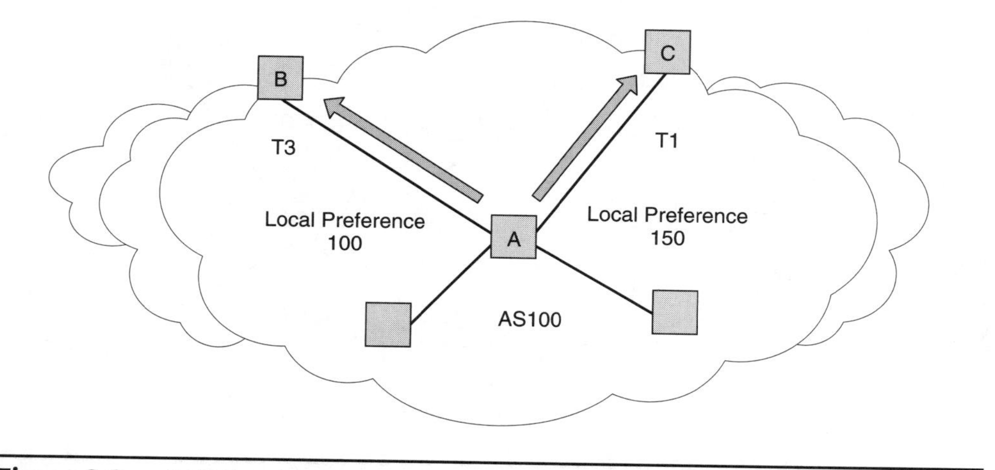

Figure 2.9: *BGP Local Preference.*

(also known as *hot-potato routing*), this may indeed be a satisfactory approach. The operative word in this approach is *local.* As mentioned previously, this attribute is not passed along to exterior ASs and only affects the path in which traffic is forwarded as it exits the local routing domain. Once again, it is important to consider the level of service quality you are attempting to deliver.

The concept of iBGP is important to understand at this point. In a network in which multiple exit points exit to multiple neighboring ASs, it becomes necessary to understand how to obtain optimum utilization from each external gateway for redundancy and closest-exit routing. A mistake network administrators commonly make is to redistribute exterior routing information (BGP) into their interior routing protocol. This is a very dangerous practice and should be avoided whenever possible. Although this book does not intend to get too far in-depth into how to properly dual- or multihome end-system networks, it is appropriate to discuss the concept of iBGP, because it is an integral part of understanding the merits of allowing BGP to make intelligent decisions about which paths are more appropriate for forwarding traffic than others.

iBGP is not a different protocol from what is traditionally known simply as BGP. It is just a specific application of the BGP protocol *within* an AS. In this model, the network interior routing (e.g., OSPF, dual IS-IS) carries next-hop and interior-routing information, and iBGP carries all exterior-routing information.

Using iBGP

Two basic principles exist for using iBGP. The first is to use BGP as a mechanism to carry exterior-routing information *along the transit path between exit peers* within the same AS, so that internal traffic can use the exterior-routing information to make a more intelligent decision on how to forward traffic bound for external destinations. The second principle is that this allows a higher degree of routing stability to be maintained in the interior network. In other words, it obviates the need to redistribute BGP into interior routing. As depicted in

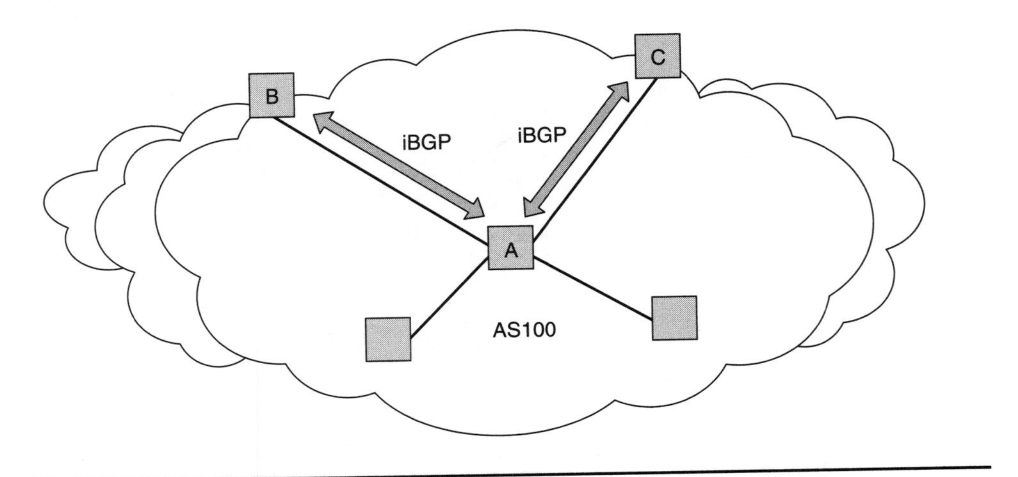

Figure 2.10: *Internal BGP (iBGP).*

Figure 2.10, exterior-routing information is carried along the transit path, between the AS exit points, via iBGP. A default route could be injected into the interior routing system from the backbone transit-path routers and redistributed to the lower portions of the hierarchy, or it could be statically defined on a hop-by-hop basis so that packets bound for destinations not found in the interior-routing table (e.g., OSPF) are forwarded automatically to the backbone, where a recursive route lookup would reveal the appropriate destination (found in the iBGP table) and forwarding path for the destination.

It is worthwhile to mention that it is usually a good idea to filter (block) announcements of a default route (0.0.0.0) to external peers, because this can create situations in which the AS is seen as a transit path by an external AS. The practice of leaking a default route via BGP generally is frowned upon and should be avoided, except in rare circumstances when the recipient of the default route is fully aware of the default-route propagation.

Scaling iBGP becomes a concern if the number of iBGP peers becomes large, because iBGP routers must be fully meshed from a peering perspective due to iBGP readvertisement restrictions to prevent route looping. Quantifying *large* can be difficult and can vary depending on the characteristics of the network. It is recommended to monitor the resources on routers on an ongoing basis to determine whether scaling the iBGP mesh is starting to become a problem. Before critical mass is reached with the single-level iBGP mesh, and router meltdown occurs, it is necessary to introduce iBGP route reflectors to create a small iBGP core mesh and iBGP peers that are local reflector clients from a core mesh iBGP participant.

Manipulating Traffic Using the Multi-Exit Discriminator (MED)

Another tool useful in manipulating traffic-forwarding paths between ASs is the Multi-Exit Discriminator (MED) attribute in BGP. The major difference between local preference and MED is that whereas local preference provides a mechanism to prefer a particular exit gateway from within a single AS (outbound), the MED attribute is used for providing a mechanism for an administrator to tell a neighboring AS how to return traffic to him across a preferred link (inbound). The MED attribute is propagated only to the neighboring AS and no further, however, so its usefulness is somewhat limited.

The primary utility of the MED is to provide a mechanism with which a network administrator can indicate to a neighboring AS which link he prefers for traffic entering his administrative domain when multiple links exist between the two ASs, each advertising the same length prefix. Again, MED is a tie-breaker. As illustrated in Figure 2.11, for example, if the administrator in AS100 wants to influence the path selection of AS200 in relation to which link it prefers to use for forwarding traffic for the same prefix, he passes a lower MED to AS200 on the link that connects routers A and B. It is important to ensure that the neighboring AS is actually honoring MEDs, because they may have a routing policy in conflict with a local administrator's (AS100, in this case).

Each of these tools provides unique methods of influencing route and path selection at different points in the network. By using these tools with one another, a network administrator can have a higher degree of control over local and neighbor path selection than if left completely to a dynamic routing protocol. However, these BGP knobs do not provide a great deal of functionality, including definitive levels of QoS. Path-selection criteria can be important in providing QoS, but it is only a part of the big picture. The importance of this path-selection criteria should not be underestimated, however, because more intelligent methods of integrating path selection based on QoS metrics are not available today.

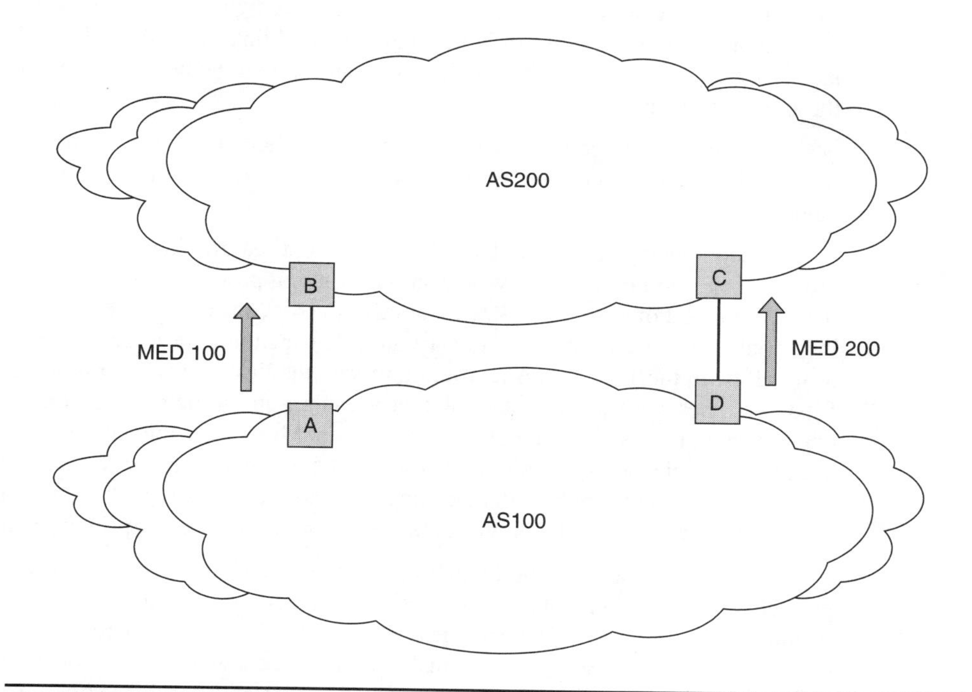

Figure 2.11: *BGP Multi-eExit Discriminator (MED).*

Asymmetric Routing

A brief note is appropriate at this point on the problems with asymmetric routing. Traffic in the Internet (and other networks, for that matter) is bi-directional—a source has a path to its destination and a destination to the source. Due to the fact that different organizations may have radically different routing policies, it is not uncommon for traffic to traverse a path in its journey back to its source entirely different from the path it traversed on its initial journey to the destination. This type of asymmetry may cause application problems, not explicitly because of the asymmetry aspects, but sometimes because of vastly different characteristics of the paths. One path may be low latency, whereas another may be extraordinarily high. One path may have traffic filters in place that restrict certain types of traffic, whereas another path may not. This is not always a problem, and most applications used today are unaffected. However, as new and emerging multimedia applications, such as real-time audio and video, become more reliant on predictable network behavior, this becomes increasingly problematic. The same holds true for IP-level signaling protocols, such as RSVP (Resource ReSerVation Setup Protocol), that require symmetry to function.

Policy, Routing, and QoS

A predictive routing system is integral to QoS—so much so that currently resources are being expended on examining ways to integrate QoS metrics and path-selection criteria into the routing protocols themselves. Several proposals are examined in Chapter 9, "QoS and Future Possibilities," but this fact alone is a clear indication that traditional methods of routing lack the characteristics necessary for advanced QoS functionality. These characteristics include monitoring of available bandwidth and link use on parallel end-to-end paths, stability metrics, measured jitter and latency, and instantaneous path selection based on these measurements.

This can be distilled into the two questions of whether link cost metrics should be fixed or variable, and whether a link cost is represented as a vector of costs or within a single value.

In theory, using variable link costs results in optimal routing that attempts to route around congestion in the same way as current routing protocols attempt to route around damage. Instead of seeing a link with an infinite cost (down) or a predetermined cost (up), the variable cost model attempts to vary the cost depending on some form of calculation of availability. In the 1977 ARPAnet routing model, the link cost was set to the outcome of a PING, in other words, setting the link metric to the sum of the propagation delay plus the queuing and processing delays. This approach does have very immediate feedback loops, such that the selection of an alternate route as the primary path due to transient congestion of the current primary path shifts the congestion to the alternate path, which during the next cycle of link metric calculation yields the selection of the original primary path—and soon.

Not unsurprisingly, the problem of introducing such feedback systems into the routing protocol results in highly unstable oscillation and lack of convergence of the routing system. Additionally, you need to factor in the cost of traffic disruption where the network topology is in a state of flux because of continual route metric changes. In this unconverged state, the probability of poor routing decisions increases significantly, and the consequent result may well be a reduced level of network availability. The consequent engineering problem in this scenario is one of defining a methodology of damping the link cost variation to reduce the level of route oscillation and instability.

To date, no scaleable routing tools have exhibited a stable balance between route stability and dynamic adaptation to changing traffic-flow profiles. The real question is perhaps not the viability of this approach but one simply of return on effort. As Radia Perlman points out, "I believe that the difference in network capacity between the simplest form of link cost assignment (fixed values depending on delay and total bandwidth of the link) and 'perfect routing' (in which a deity routes every packet optimally, taking into account traffic currently in the network as well as traffic that will enter the network in the near future) is not very great" [Perlman1992].

The other essential change in the route-selection process is that of using a vector of link metrics, such as that defined in the IS-IS protocol. A vector metric could be defined as the 4-tuple of (default, delay, cost, bandwidth), for example, and a packet could be routed according to a converged topology that uses the delay metric if the header TOS field of the packet specifies minimum delay. Of course, it does admit the less well-defined situation in which a combination of metrics is selected, and effectively if there are n specific metrics in the link metric vector, this defines $2n + 1$ distinct network topologies that must be computed. Obviously, this is not a highly scaleable approach in terms of expansion of the metric set, nor does it apply readily to large networks where convergence of a single metric is a performance issue, and increasing the convergence calculation by $2^{02286321n}$. Again, Radia's comments are relevant here, where she notes that, "Personally, I dislike the added complexity of multiple metrics. . . . I think that any potential gains in terms of better routes will be more than offset by the overhead and the likelihood of extreme sub-optimality due to misconfiguration or misunderstanding of the way multiple metrics work" [Perlman1992].

Regardless of the perceived need for the integration of QoS and routing, routing as a policy tool is more easily implemented and tenable when approached from an intradomain perspective; from an interdomain perspective, however, there are many barriers that may prevent successful policy implementation. This is because, within a single routing domain, a network administrator can control the end-to-end network characteristics. Once the traffic leaves your administrative domain, all bets are off—there is no guarantee that traffic will behave as you might expect and, in most cases, will not behave in any predictable manner.

Measuring QoS

Given that there are a wide variety of ways to implement QoS within a network, the critical issue is how to measure the effectiveness of the QoS implementation chosen. This section provides an overview of the techniques and potential approaches to measuring QoS in the network.

Who Is Measuring What?

The first question here is to determine the agent performing the measurement and what the agent is attempting to measure. An *agent* is any application that can be used for monitoring or measuring a component of a network. A relevant factor in determining what measurements are available to the agent is the level of access that agent has to the various components of the network. Additionally, the question of what measurements can be obtained directly from the various components of the network and what measurements can be derived from these primary data points is at the forefront of the QoS measurement problem.

A customer of a QoS network service may want to measure the difference in service levels between a QoS-enabled transaction and a non-QoS-enabled transaction. The customer also might want to measure the levels of variability of a particular transaction and derive a metric of the constancy of the QoS environment. If the QoS environment is based on a reservation of resources, the customer might want to measure the capability to consume network resources up to the maximum reservation level to determine the effectiveness of the reservation scheme.

One of the major factors behind the motivations for measurement of the QoS environment is to determine how well it can deliver services that correspond to the particular needs of the customer. In effect, this is measuring the level of consistency between the QoS environment and the subscriber needs. Second, the customer is interested in measuring the success or failure of the network provider in meeting the agreed QoS parameters. Therefore, the customer is highly motivated to perform measurements that correspond to the metrics of the particular QoS environment provided.

A customer of a remote network also might be interested in measuring the transitivity of a QoS environment. Will a transaction across the network that spans multiple service provider networks be supported at a QoS level reflecting the QoS of the originating network? Are the QoS mechanisms used in each network compatible?

The network operator also is keenly interested in measuring the QoS environment, for perhaps slightly different reasons than the customer. The operator is interested in measuring the level of network resources allocated to the QoS environment and the impact this allocation has on non-QoS resource allocations. The operator also is keenly interested in measuring the performance of the QoS environment as delivered to the customer, to ensure that the QoS service levels are within the parameters of a service contract. Similarly, the network operator is strongly motivated to undertake remote measurements of peer-networked environments to determine the transitivity of the QoS architecture. If there are QoS peering contracts with financial settlements, the network operator is strongly motivated to measure the QoS exchange to ensure that the level of resources consumed by the QoS transit is matched by the level of financial settlement between the providers.

Not all measurements can be undertaken by all agents. Although a network operator has direct access to the performance data generated by the networking switching elements within the network, the customers generally do not enjoy the same level of access to this data. The customers are limited to measuring the effects of the network's QoS architecture by measuring the performance data on the end-systems that generate network transactions and cannot see directly the behavior of the network's internal switches. The end-system measurement can be undertaken by measuring the behavior of particular transactions or by generating probes into the network and measuring the responses to these probes. For remote network measurements, the agent, whether a remote operator or a remote customer, generally is limited to these end-system measurement techniques of transaction measurement or probe measurement.

Accordingly, there is no single means of measuring QoS and no resulting single QoS metric.

Measuring Internet Networks

Measuring QoS environments can be regarded as a special case of measuring performance in an Internet environment, and some background in this subject area is appropriate. Measurement tools must be based on a level of understanding of the architecture being measured if it is to be considered an effective tool.

> **TIP** There has been a recent effort within the IETF (Internet Engineering Task Force) to look at the issues of measurement of Internet networks and their performance as they relate to customer performance, as well as from the perspective of the network operator. The IPPM (IP Provider Metric) working group is active; its charter is available at www.ietf.org. Activities of this working group are archived at www.advanced.org/IPPM. A taxonomy of network measurement tools is located at www.caida.org/Tools/taxonomy.html.

In its simplest form, an Internet environment can be viewed as an interconnected collection of data packet switches. Each switch operates in a very straightforward fashion, where a datagram is received and the header is examined for forwarding directives. If the switch can forward the packet, the packet is mapped immediately for transmission onto the appropriate output interface; if the interface already is transmitting a packet, the datagram is queued for later transmission. If the queue length is exceeded, the datagram may be discarded immediately, or it may be discarded after some interval spent waiting for transmission in the queue. Sustained queuing and switch-initiated packet discards occur when the sustained packet-arrival rate exceeds the available link capacity.

The Effect of Queue Length

A more precise statement can be made about the mechanics of switch-initiated discards. The transmission capacity and the queue length are related. If the transmission bandwidth is 1 Mbps and the packet arrival rate is 1.5 Mbps, for example, the queue will grow at a rate of 0.5 Mbps (or 64 Kbps). If the queue length is Q bytes, the transmission speed is T bits per second, and the packet arrival rate is A bits per second, packet discard occurs after (8 * Q) / (A - T) seconds. In this state, each queued packet is delayed by (8 * Q) / T seconds. Increasing the queue length as a fraction of the transmission bandwidth decreases the probability of dropped packets because of transient bursts of traffic. It also increases the variability of the packet-transmission time, however, which in turn desensitizes the TCP retransmission timers. This increased RTT variance causes TCP to respond more slowly to dynamic availability of network resources. Thus, although switch queues are good, too much queue space in the switch has an adverse impact on end-to-end dynamic performance.

The interaction of end-system TCP protocol behavior and the queue-management algorithm within the switch (or router) is such that if the switch determines to discard a packet, the end-system should note the packet drop and adjust its transmission rate downward. This action is intended to relieve the congestion levels on the switch while also optimizing the throughput of data from end-to-end in the network.

The effectively random nature of packet drops in a multiuser environment and the relatively indeterminate nature of cumulative queuing hold times across a series of switches imply that the measurement of the performance of any Internet network does not conform to a precise and well-defined procedure. The behavior of the network is based on the complex interaction of the end-systems generating the data transactions according to TCP behavior or UDP (User Datagram Protocol) application-based traffic shaping, as well as the behavior of the packet-forwarding devices in the interior of the network.

Customer Tools to Measure an Internet Network

Customer tools to measure the network are limited to two classes of measurement techniques: the generation of probes into the network, which generates a response back to the probe agent, and the transmission of traffic through the network, coupled with measuring the characteristics of the traffic flow.

The simplest of the probe tools is the PING probe. *PING* is an ICMP (Internet Control Message Protocol) echo request packet directed at a particular destination, which in turn responds with an ICMP echo reply. (You can find an annotated listing of the PING utility in W. Richard Stevens' *UNIX Network Programming*, Prentice Hall, 1990.) The system generating the packet can determine immediately whether the destination system is reachable. Repeated measurement of the delay between the sending of the echo request packet can infer the level of congestion along the path, given that the variability in ping transaction times is assumed to be caused by switch queuing rather than route instability. Additionally, the measurement of dropped packets can illustrate the extent to which the congestion level has been sustained over queue thresholds.

A Refinement of PING Is the TRACEROUTE Tool

TRACEROUTE uses the mechanism of generating UDP packets within a series of increasing TTL (Time to Live) values. *TRACEROUTE* measures the delay between the generation of the packet and the reception of the ICMP TTL exhausted return packet and notes which system generates the ICMP TTL error. *TRACEROUTE* can provide a hop-by-hop display of the path selected by the routing, as well as the round-trip time between the emission of the probe packet and the return of the corresponding IMCP error message.

The manual page for TRACEROUTE on UNIX distributions credits authorship of TRACEROUTE as "Implemented by Van Jacobson from a suggestion by Steve Deering. Debugged by a cast of thousands with particularly cogent suggestions or fixes from C. Philip Wood, Tim Seaver and Ken Adelman."

One variation of the PING technique is to extend the probe from the simple model of measuring the probe and its matching response to a model of emulating the behavior of the TCP algorithm, sending a sequence of probes in accordance with TCP flow-control algorithms. *TRENO* (www.psc.edu/~pscnoc/treno.html) is one such tool which, by emulating a TCP session in its control of the emission of probes, provides an indication of the available path capacity at a particular point in time.

A variation of the TRACEROUTE approach is *PATHCHAR* (also authored by Van Jacobson). Here, the probes are staggered in a number of sizes as well as by TTL. The resultant differential measurements for differing packet sizes can provide an indication of the provisioned bandwidth on each link in the end-to-end path, although the reliability of such calculations depends heavily on the quality of the network path with respect to queuing delays.

In general, probe measurements require a second level of processing to determine QoS metrics, and the outcome of such processing is weakened by the nature of the assumptions that must be embedded in the post processing of the subsequent probe data. The other weakness of this approach is that a probe is typically not part of a normal TCP traffic flow, and if the QoS mechanism is based on tuning the switch performance for TCP flow behavior, the probe will not readily detect the QoS mechanism in operation in real time. If the QoS mechanism being deployed is a Random Early Detection (RED) algorithm in which the QoS mechanism is a weighting of the discard algorithm, for example, network probes will not readily detect the QoS operation.

The other approach is to measure repetitions of a known TCP flow transaction. This could be the measurement of a file transfer, for example—measuring the elapsed time to complete the transfer or a WWW (World Wide Web) download. Such a measurement could be a simple measurement of elapsed time for the transaction; or it could undertake a deeper analysis by looking at the number of retransmissions, the variation of the TCP Round-Ttrip Timers (RTT) across the duration of the flow, the interpacket arrival times, and similar levels of detail. Such measurements can indicate the amount of distortion of the original transmission sequence the network imposed on the transaction. From this data, a relatively accurate portrayal of the effectiveness of a QoS mechanism can be depicted. Of course, such a technique does impose a heavier load on the total system than simple short probes, and the risk in undertaking large-scale measurements based on this approach is that the act of measuring QoS has a greater impact on the total network system and therefore does not provide an accurate measurement.

Operator Tools to Measure an Internet Network

The network operator has a larger array of measurement points available for use. By querying the switches using SNMP (Simple Network Management Protocol) query tools, the operator can continuously monitor the state of the switch and the effectiveness of the QoS behavior within each switch. Each transmission link can be monitored to see utilization rates, queue sizes, and discard rates. Each can be monitored to gauge the extent of congestion-based queuing and, if QoS differential switching behavior is deployed, the metrics of latency and discard rates for classes of traffic also can be monitored directly.

Although such an approach can produce a set of metrics that can determine availability of network resources, there is still a disconnect with these measurements and the user level performance because the operator's measurements cannot directly measure end-to-end data performance.

The Measurement Problem

When the complete system is simple and unstressed (where the data-traffic rates are well within the capacity of each of the components of the system), the network system behaves in a generally reproducible fashion. However, although such systems are highly amenable to repeatable measurement of system performance, such systems operate without imposed degradation of performance and, from a QoS measurement perspective, are not of significant interest. As a system becomes sufficiently complex in an effort to admit differential levels of imposed degradation, the task of measuring the level of degradation of the performance and the relative level of performance obtained by the introduction of QoS mechanisms also becomes highly complicated.

From a network router perspective, the introduction of QoS behavior within the network may be implemented as an adjustment of queue-management behavior so that data packets are potentially reordered within the queue and potentially dropped completely. The intention is to reduce (or possibly increase) the latency of various predetermined classes of packets by adjusting their positions in the output buffer queue or to signal a request to slow down the end-to-end protocol transmission rate by using packet discard.

When examining this approach further, it is apparent that QoS is apparent only when the network exhibits load. In the absence of queuing in any of the network switches, there is no switch-initiated reordering of queued packets or any requirement for packet discard. As a consequence, no measurable change is attributable to QoS mechanisms. In such an environment, end-system behavior dictates the behavior of the protocol across the network. When segments of the network are placed under load, the affected switches move to a QoS enforcement mode of behavior. In this model of QoS implementation, the entire network does not switch operating modes. Congestion is not systemic but instead is a local phenomenon that may occur on a single switch in isolation. As a result, end-to-end paths that do not traverse a loaded network may never have QoS control imposed anywhere in the transit path.

Although quality of Service mechanisms typically are focused on an adjustment of a switch's queue-management processes, the QoS function also may impose a number of discrete routing topologies on the underlying transmission system, routing differing QoS traffic types on different paths.

Of course, the measurement of a QoS structure is dependent on the precise nature of the QoS mechanism used. Some forms of QoS are absolute precedence structures, in which packets that match the QoS filters are given absolute priority over non-QoS traffic. Other mechanisms are self-limited in a manner similar to Frame Relay Committed Information Rate (CIR), where precedence is given to traffic that conforms to a QoS filter up to a defined level and excess traffic is handled on a best-effort basis. The methodology of measuring the effectiveness of QoS in these cases is dependent on the QoS structure; the measurements may be completely irrelevant or, worse, misleading when applied out of context.

How Can You Measure QoS?

An issue of obvious interest is how QoS is measured in a network. One of the most apparent reasons for measuring QoS traffic is to provide billing services for traffic that receives preferential treatment in the network compared to traffic traditionally billed as best-effort. Another equally important reason is the fact that an organization that provides the service must be cognizant of capacity issues in its backbone, so it will want to understand the disparity between best-effort and QoS-designated traffic.

There are two primary methods of measuring QoS traffic in a network: intrusive and nonintrusive. Each of these approaches is discussed in this section.

Nonintrusive Measurements

The *nonintrusive measurement* method measures the network behavior by observing the packet-arrival rate at an end-system and making some deduction on the state of the network, thereby deducing the effectiveness of QoS on the basis of these observations.

In general, this practice requires intimate knowledge of the application generating the data flows being observed so that the measurement tool can distinguish between remote-application behavior and moderation of this remote behavior by a network-imposed state. Simply monitoring one side of an arbitrary data exchange does not yield much information of value and can result in wildly varying interpretations of the results. For this reason, single-ended monitoring as the basis of measurement of network performance and, by inference, QoS performance is not recommended as an effective approach to the problem. However, you can expect to see further deployment host-based monitoring tools that look at the behavior of TCP flows and the timing of packets within the flow. Careful interpretation of the matching of packets sent with arriving packets can offer some indication as to the extent of queuing-induced distortion within the network path to the remote system. This interpretation also can provide an approximate indication of the available data capacity of the path, although the caveat here is that the interpretation must be done carefully and the results must be reported with considerable caution.

Intrusive Measurements

Intrusive measurements refer to the controlled injection of packets into the network and the subsequent collection of packets (the same packet or a packet remotely generated as a result of reception of the original packet). PING—the ICMP echo request and echo reply—packet exchange is a very simple example of this measurement methodology. By sending sequences of PING packets at regular intervals, a measurement station can measure parameters such as reachability, the transmission RTT to a remote location, and the expectation of packet loss on the round-trip path. By making a number of secondary assumptions regarding queuing behavior within the switches, combined with measurement of packet loss and imposed jitter, you can make some assumptions about the available bandwidth between the two points and the congestion level. Such measurements are not measurements of the effectiveness with regard to QoS, however. To measure the effectiveness of a QoS structure, you must undertake a data transaction typical of the network load and measure its performance under controlled conditions. You can do this by using a remotely triggered data generator. The requirement methodology is to measure the effective sustained data rate, the retransmission rate, the stability of the RTT estimates, and the overall transaction time. With the introduction of QoS measures on the network, a comparison of the level of degradation of these metrics from an unloaded network to one that is under load should yield a metric of the effectiveness of a QoS network.

The problem is that QoS mechanisms are visible only when parts of the network are under resource contention, and the introduction of intrusive measurements into the system further exacerbates the overload condition. Accordingly, attempts to observe the dynamic state of a network disturbs that state, and there is a limit to the accuracy of such network state measurements. Oddly enough, this is a re-expression of a well-understood physical principle: Within a quantum physical system, there is an absolute limit to the accuracy of a simultaneous measurement of position and momentum of an electron.

The Heisenberg Uncertainty Principle

Werner Heisenberg described the general principle in 1927 with *The Heisenberg Uncertainty Principle,* which relates the limit of accuracy of simultaneous measurement of a body's position and velocity to Planck's constant, *h*.

Werner (Karl) Heisenberg (1901–1976) was a theoretical physicist born in Würzburg, Germany. He studied at Munich and Göttingen. After a brief period working with Max Born (1923) and Niels Bohr (1924–1927), he became professor of physics at Leipzig (1927–1941), director of the Max Planck Institute in Berlin (1941–1945), and director of the Max Planck Institute at Göttingen. He developed a method of expressing quantum mechanics in matrices (1925) and formulated his revolutionary principle of indeterminacy (the uncertainty principle) in 1927. He was awarded the 1932 Nobel Prize for Physics.

This observation is certainly an appropriate comparison for the case of networks in which intrusive measurements are used. No single consistent measurement methodology is available that will produce deterministic measurements of QoS performance, so effectively measuring QoS remains a relatively imprecise area and one in which popular mythology continues to dominate.

Why, then, is measurement so important? The major reason appears to be that the implementation of QoS within a network is not without cost to the operator of the network. This cost is expressed both in terms of the cost of deployment of QoS technology within the routers of the network and in terms of cost of displacement of traffic, where conventional best-effort traffic is displaced by some form of QoS-mediated preemption of network resources. The network operator probably will want to measure this cost and apply some premium to the service fee to recoup this additional cost. Because customers of the QoS network service probably will be charged some premium for use of a QoS service, they undoubtedly will want to measure the cost effectiveness of this investment in quality. Accordingly, the customer will be well motivated to undertake comparative measurements that will indicate the level of service differentiation obtained by use of the QoS access structures.

Despite this wealth of motivation for good measurement tools, such tools and techniques are still in their infancy, and it appears that further efforts must be made to understand the protocol-directed interaction between the host and the network, the internal interaction between individual elements of the network, and the moderation of this interaction through the use of QoS mechanisms before a suite of tools can be devised to address these needs of the provider and the consumer.

chapter 3

QoS and Queuing Disciplines, Traffic Shaping, and Admission Control

In the context of network engineering, *queuing* is the act of storing packets or cells where they are held for subsequent processing. Typically, queuing occurs within a router when packets are received by a device's interface processor (input queue), and queuing also may occur prior to transmitting the packets to another interface (output queue) on the same device. Queuing is the central component of the architecture of the router, where a number of asynchronous processes are bound together to switch packets via queues. A basic router is a collection of input processes that assembles packets as they are received, checking the integrity of the basic packet framing, one or more forwarding processes that determine the destination interface to which a packet should be passed, and output processors that frame and transmit packets on to their next hop. Each process operates on one packet at a time and works asynchronously from all other processes. The binding of the input processes to the forwarding processes and then to the output processes is undertaken by a packet queue-management process, shown in Figure 3.1.

It is important to understand the role queuing strategies have played, and continue to play, in the evolution of providing differentiated services. For better or worse, networks are operated in a store-and-forward paradigm. This is a non-connection-oriented networking world, as far as packet technology is concerned (the Internet is a principal example), where all end-points in the network are not one hop away from one another. Therefore, you must

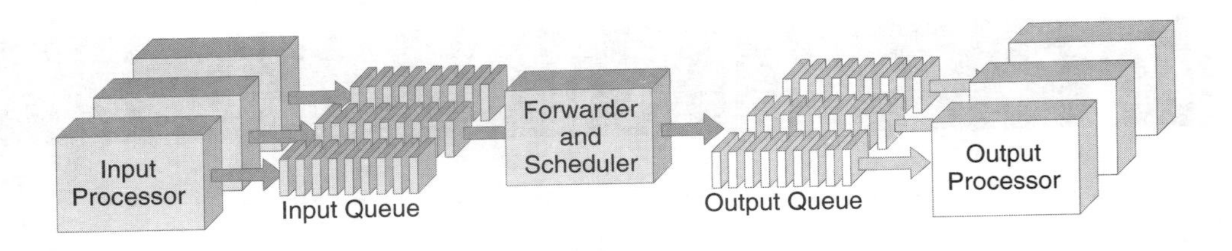

Figure 3.1: *Router queuing and scheduling: a block diagram.*

understand the intricacies of queuing traffic, the effect queuing has on the routing system and the applications, and why managing queuing mechanisms appropriately is crucial to QoS, with emphasis on *service quality*.

Managing Queues

The choices of the algorithm for placing packets into a queue and the maximal length of the queue itself may at first blush appear to be relatively simple and indeed trivial choices. However, you should not take these choices lightly, because queue management is one of the fundamental mechanisms for providing the underlying quality of the service and one of the fundamental mechanisms for differentiating service levels. The correct configuration choice for queue length can be extraordinarily difficult because of the apparently random traffic patterns present on a network. If you impose too deep a queue, you introduce an unacceptable amount of latency and RTT (Round-Trip Time) jitter, which can break applications and end-to-end transport protocols.

Not to mention how unhappy some users may become if the performance of the network degenerates considerably. If the queues are too shallow, which sometimes can be the case, you may run into the problem of trying to dump data into the network faster than the network can accept it, thus resulting in a significant amount of discarded packets or cells. In a reliable traffic flow, such discarded packets have to be identified and retransmitted in a sequence of end-to-end protocol exchanges. In a real-time unreliable flow (such as audio or video), such packet loss is manifested as signal degradation.

Queuing occurs as both a natural artifact of TCP (Transmission Control Protocol) rate growth (where the data packets are placed onto the wire in a burst configuration during slow start and two packets are transmitted within the ACK timing of a single packet's reception) and as a natural artifact of a dynamic network of multiple flows (where the imposition of a new flow on a fully occupied link will cause queuing while the flow rates all adapt to the increased load). Queuing also can happen when a Layer 3 device (a router) queues and switches traffic toward the next-hop Layer 3 device faster than the underlying Layer 2 network devices (e.g., frame relay or ATM switches) in the transit path can accommodate. Some intricacies in the realm of queue management, as well as their effect on network traffic, are very difficult to understand.

TIP Len Kleinrock is arguably one of the most prolific theorists on traffic queuing and buffering in networks. You can find a bibliography of his papers at millennium.cs.ucla.edu/LK/Bib/.

In any event, the following descriptions of the queuing disciplines focus on output queuing strategies, because this is the predominate strategic location for store-and-forward traffic management and QoS-related queuing.

FIFO Queuing

FIFO (First In, First Out) queuing is considered to be the standard method for store-and-forward handling of traffic from an incoming interface to an outgoing interface. For the sake of this discussion, however, you can consider anything more elaborate than FIFO queuing to be exotic or "abnormal." This is not to say that non-FIFO queuing mechanisms are inappropriate—quite the contrary. Non-FIFO queuing techniques certainly have their merit and usefulness. It is more an issue of knowing what their limitations are, when they should be considered, and perhaps more important, understanding when they should be avoided.

As Figure 3.2 shows, as packets enter the input interface queue, they are placed into the appropriate output interface queue in the order in which they are received—thus the name *first in, first out*.

FIFO queuing usually is considered default behavior, and many router vendors have highly optimized forwarding performances that make this standard behavior as fast as possible. In fact, when coupled with topology-driven forwarding cache population, this particular combination of technologies quite possibly could be considered the fastest of technology implementation available today as far as packets-per-second forwarding is concerned. This is because, over time, developers have learned how to optimize the software to take advantage of simple queuing technologies. When more elaborate queuing strategies are implemented

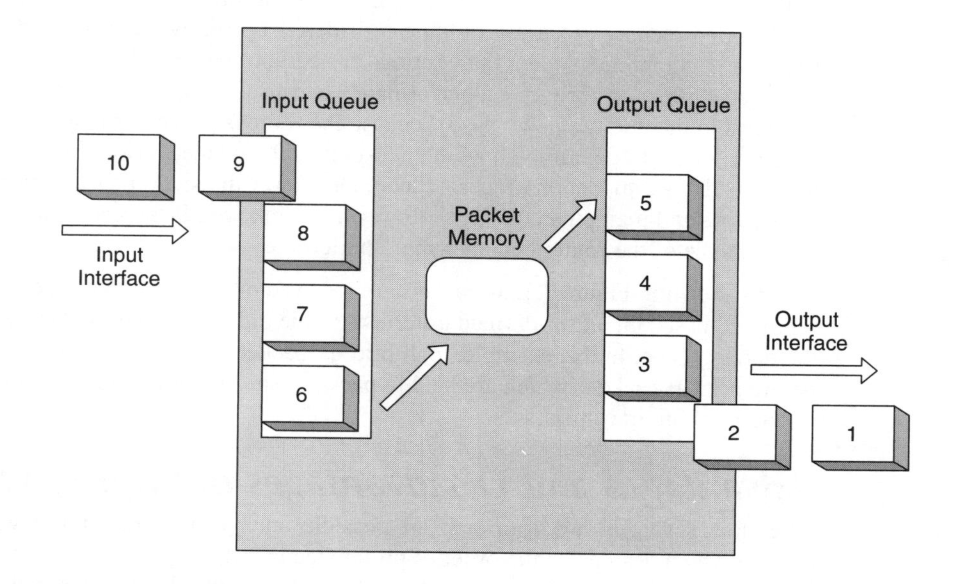

Figure 3.2: *FIFO queuing.*

instead of FIFO, there is a strong possibility that there may very well be some negative impact on forwarding performance and an increase (sometimes dramatically) on the computational overhead of the system. This depends, of course, on the queuing discipline and the quality of the vendor implementation.

Advantages and Disadvantages of FIFO Queuing

When a network operates in a mode with a sufficient level of transmission capacity and adequate levels of switching capability, queuing is necessary only to ensure that short-term, highly transient traffic bursts do not cause packet discard. In such an environment, FIFO queuing is highly efficient because, as long as the queue depth remains sufficiently short, the average packet-queuing delay is an insignificant fraction of the end-to-end packet transmission time.

When the load on the network increases, the transient bursts cause significant queuing delay (significant in terms of the fraction of the total transmission time), and when the queue is fully populated, all subsequent packets are discarded. When the network operates in this mode for extended periods, the offered service level inevitably degenerates. Different queuing strategies can alter this service-level degradation, allowing some services to continue to operate without perceptible degradation while imposing more severe degradation on other services. This is the fundamental principle of using queue management as the mechanism to provide QoS arbitrated differentiated services.

Priority Queuing

One of the first queuing variations to be widely implemented was *priority queuing*. This is based on the concept that certain types of traffic can be identified and shuffled to the front of the output queue so that some traffic always is transmitted ahead of other types of traffic. Priority queuing certainly could be considered a primitive form of traffic differentiation, but this approach is less than optimal for certain reasons. Priority queuing may have an adverse effect on forwarding performance because of packet reordering (non-FIFO queuing) in the output queue. Also, because the router's processor may have to look at each packet in detail to determine how the packet must be queued, priority queuing also may have an adverse impact on processor load. On slower links, a router has more time to closely examine and manipulate packets. However, as link speeds increase, the impact on the performance of the router becomes more noticeable.

As shown in Figure 3.3, as packets are received on the input interface, they are reordered based on a user-defined criteria as to the order in which to place certain packets in the output queue. In this example, high-priority packets are placed in the output queue before normal packets, which are held in packet memory until no further high-priority packets are awaiting transmission.

Advantages and Disadvantages of Priority Queuing

Of course, several levels of priority are possible, such as high, medium, low, and so on—each of which designates the order of preference in output queuing. Also, the granularity in identifying traffic to be classified into each queue is quite flexible. For example, IPX could be queued before IP, IP before SNA, and SNA before AppleTalk. Also, specific services

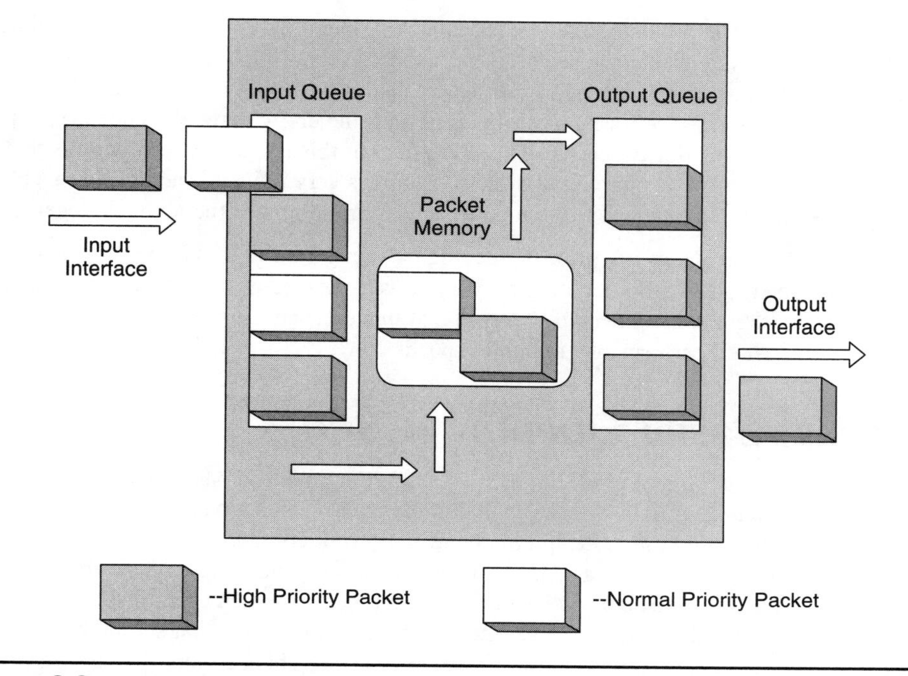

Figure 3.3: *Priorty queuing.*

within a protocol family can be classified in this manner; TCP traffic can be prioritized ahead of UDP traffic, Telnet (TCP port 23) [IETF1983] can be prioritized ahead of FTP [IETF1985] (TCP ports 20 and 21), or IPX type-4 SAPs could be prioritized ahead of type-7 SAPs. Although the level of granularity is fairly robust, the more differentiation attempted, the more impact on computational overhead and packet-forwarding performance.

> **TIP** A listing of Novell Service Advertisement Protocol (SAP) numbers is available at www.novell.com/corp/programs/ncs/toolkit/download/saps.txt.

Another possible vulnerability in this queuing approach is that if the volume of high-priority traffic is unusually high, normal traffic waiting to be queued may be dropped because of *buffer starvation*—a situation that can occur for any number of reasons. Buffer starvation usually occurs because of overflow caused by too many packets waiting to be queued and not enough room in the queue to accommodate them. Another consideration is the adverse impact induced latency may have on applications when traffic sits in a queue for an extended period. It sometimes is hard to calculate how non-FIFO queuing may inject additional latency into the end-to-end round-trip time. In a worst-case scenario, some applications may not function correctly because of added latency or perhaps because some more time-sensitive routing protocols may time-out due to acknowledgments not being received within a predetermined period of time.

The problem here is well known in operating systems design; absolute priority scheduling systems can cause complete resource starvation to all but the highest-priority tasks. Thus, the use of priority queues creates an environment where the degradation of the highest-priority service class is delayed until the entire network is devoted to processing only the highest-priority service. The side-effect of this resource preemption is that the lower levels of service are starved of system resources very quickly, and during this phase of reallocation of critical resources, the total effective throughput of the network degenerates dramatically.

In any event, priority queuing has been used for a number of years as a primitive method of differentiating traffic into various classes of service. Over the course of time, however, it has been discovered that this mechanism simply does not scale to provide the desired performance at higher speeds.

Class-Based Queuing (CBQ)

Another queuing mechanism introduced a couple of years ago is called *Class-Based Queuing* (CBQ) or *custom queuing*. Again, this is a well-known mechanism used within operating system design intended to prevent complete resource denial to any particular class of service. CBQ is a variation of priority queuing, and several output queues can be defined. You also can define the preference by which each of the queues will be serviced and the amount of queued traffic, measured in bytes, that should be drained from each queue on each pass in the servicing rotation. This servicing algorithm is an attempt to provide some semblance of fairness by prioritizing queuing services for certain types of traffic, while not allowing any one class of traffic to monopolize system resources and bandwidth.

The configuration in Figure 3.4, for example, has created three buffers: high, medium, and low. The router could be configured to service 200 bytes from the high-priority queue, 150 bytes from the medium-priority queue, and then 100 bytes from the low-priority queue on each rotation. After traffic in each queue is processed, packets continue to be serviced until the byte count exceeds the configured threshold or the current queue is empty. In this fashion, traffic that has been categorized and classified to be queued into the various queues have a reasonable chance of being transmitted without inducing noticeable amounts of latency and allowing the system to avoid buffer starvation. CBQ also was designed with the concept that certain classes of traffic, or applications, may need minimal queuing latency to function properly; CBQ provides the mechanisms to configure how much traffic can be drained off each queue in a servicing rotation, providing a method to ensure that a specific class does not sit in the outbound queue for too long. Of course, an administrator may have to fumble around with the various queue parameters to gauge whether the desired behavior is achieved. The implementation may be somewhat hit and miss.

Advantages and Disadvantages of Class-Based Queuing

The CBQ approach generally is perceived as a method of allocating dedicated portions of bandwidth to specific types of traffic, but in reality, CBQ provides a more graceful mechanism of preemption, in which the absolute model of service to the priority queue and

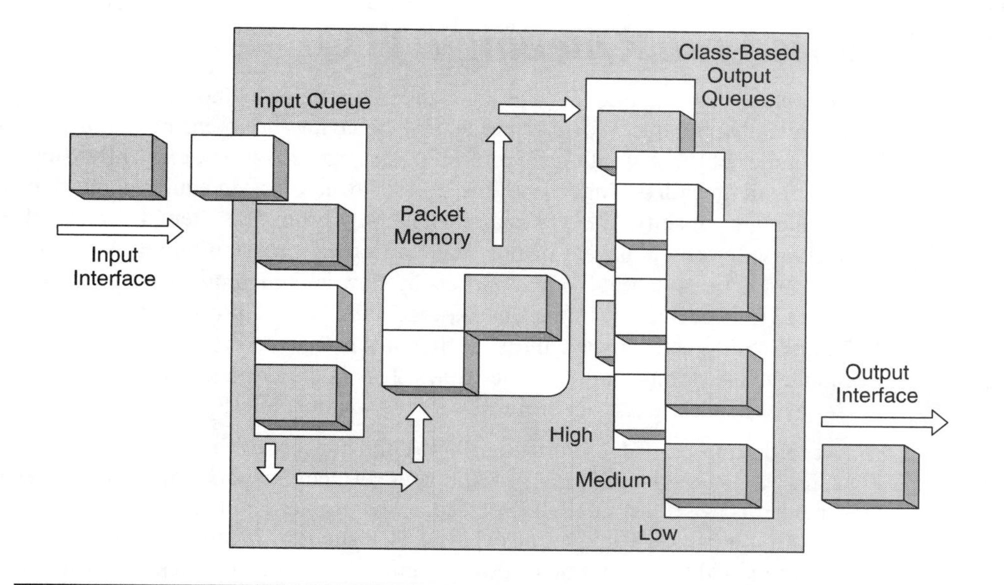

Figure 3.4: *Class-Based Queuing (CBQ).*

resource starvation to other queues in the priority-queuing model are replaced by a more equitable model of an increased level of resource allocation to the higher-precedence queues and a relative decrease to the lower-precedence queues. The fundamental assumption here is that resource denial is far worse than resource reduction. *Resource denial* not only denies data but also denies any form of signaling regarding the denial state. *Resource reduction* is perceived as a more effective form of resource allocation, because the resource level is reduced for low-precedence traffic and the end-systems still can receive the signal of the changing state of the network and adapt their transmission rates accordingly.

CBQ also can be considered a primitive method of differentiating traffic into various classes of service, and for several years, it has been considered a reasonable method of implementing a technology that provides link sharing for *Classes of Service* (CoS) and an efficient method for queue-resource management. However, CBQ simply does not scale to provide the desired performance in some circumstances, primarily because of the computational overhead and forwarding impact packet reordering and intensive queue management imposes in networks with very high-speed links. Therefore, although CBQ does provide the basic mechanisms to provide differentiated CoS, it is appropriate only at lower-speed links, which limits its usefulness.

TIP A great deal of informative research on CBQ has been done by Sally Floyd and Van Jacobson at the Lawrence Berkeley National Laboratory for Network Research. For more detailed information on CBQ, see ftp.ee.lbl.gov/floyd/cbq.html.

Weighted Fair Queuing (WFQ)

Weighted Fair Queuing (WFQ) is yet another popular method of fancy queuing that algorithmically attempts to deliver predictable behavior and to ensure that traffic flows do not encounter buffer starvation (Figure 3.5). WFQ gives low-volume traffic flows preferential treatment and allows higher-volume traffic flows to obtain equity in the remaining amount of queuing capacity. WFQ uses a servicing algorithm that attempts to provide predictable response times and negate inconsistent packet-transmission timing. WFQ does this by sorting and interleaving individual packets by flow and queuing each flow based on the volume of traffic in each flow. Using this approach, the WFQ algorithm prevents larger flows (those with greater byte quantity) from consuming network resources (bandwidth), which could subsequently starve smaller flows. This is the fairness aspect of WFQ—ensuring that larger traffic flows do not arbitrarily starve smaller flows.

According to Partridge [Partridge1994b], the concept of fair queuing was developed by John Nagle [Nagle1985] in 1985 and later refined in a paper by Demers, Keshav, and Shenker [DKS1990] in 1990. In essence, the problem both of these papers address is the situation in which a single, misbehaved TCP session consumes a large percentage of available bandwidth, causing other flows to unfairly experience packet loss and go into TCP slow start. Fair queuing presented an alternative to relegate traffic flows into their own bounded queues so that they could not unfairly consume network resources at the expense of other traffic flows.

The weighted aspect of WFQ is dependent on the way in which the servicing algorithm is affected by other extraneous criteria. This aspect is usually vendor-specific, and at least

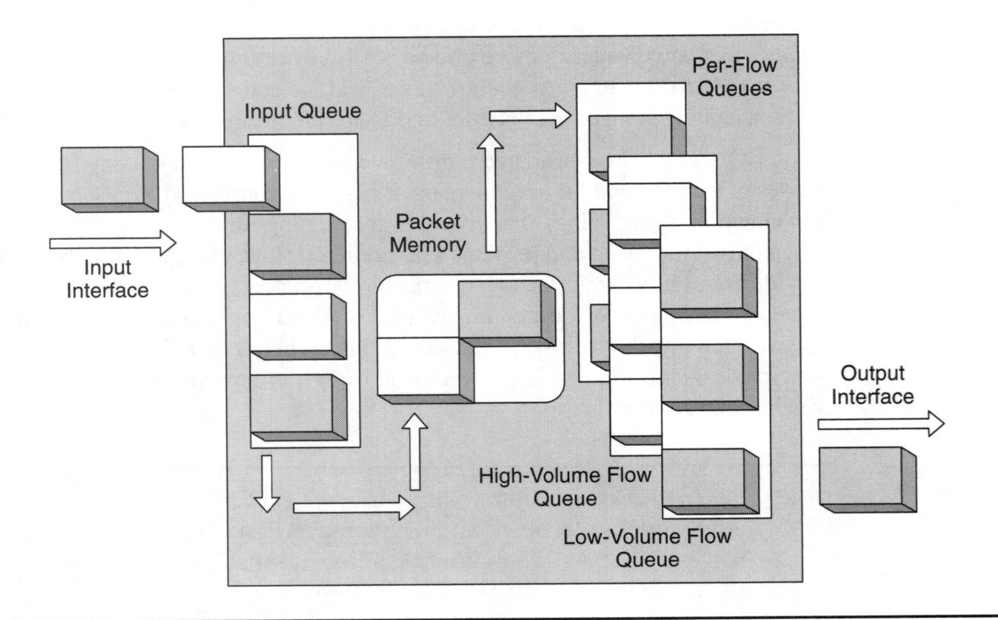

Figure 3.5: *Weighted Fair Queuing (WFQ).*

one implementation uses the IP precedence bits in the TOS (Type of Service) field in the IP packet header to weight the method of handling individual traffic flows. The higher the precedence value, the more access to queue resources a flow is given. You'll look at IP precedence in more depth in Chapter 4, "QoS and TCP/IP: Finding the Common Denominator."

Advantages and Disadvantages of Weighted Fair Queuing

WFQ possesses some of the same characteristics as priority and class-based queuing—it simply does not scale to provide the desired performance in some circumstances, primarily because of the computational overhead and forwarding impact that packet reordering and queue management impose on networks with significantly large volumes of data and very high-speed links. However, if these methods of queuing (priority, CBQ, and WFQ) could be moved completely into silicon instead of being done in software, the impact on forwarding performance could be reduced greatly. The degree of reduction in computational overhead remains to be seen, but if computational processing were not also implemented on the interface cards on a per-interface basis, the computational impact on a central processor probably still would be significant.

Another drawback of WFQ is in the granularity—or lack of granularity—in the control of the mechanisms that WFQ uses to favor some traffic flows over others. By default, WFQ protects low-volume traffic flows from larger ones in an effort to provide equity for all data flows. The weighting aspect is attractive from an unfairness perspective; however, no knobs are available to tune these parameters to alter the behavior of injecting a higher degree of unfairness into the queuing scheme—at least not from the router configuration perspective. Of course, you could assume that if the IP precedence value in each packet were set by the corresponding hosts, for example, they would be treated accordingly. You could assume that higher-precedence packets could be treated with more priority, lesser precedence with lesser priority, and no precedence treated fairly in the scope of the traditional WFQ queue servicing. The method of preferring some flows over others is statically defined in the vendor-specific implementation of the WFQ algorithm, and the degree of control over this mechanism may leave something to be desired.

Traffic Shaping and Admission Control

As an alternative, or perhaps as a complement, to using non-FIFO queuing disciplines in an attempt to control the priority in which certain types of traffic is transmitted on a router interface, there are other methods, such as admission control and traffic shaping, which are used to control what traffic actually is transmitted into the network or the rate at which it is admitted. Traffic shaping and admission control each have very distinctive differences, which you will examine in more detail. There are several schemes for both admission control and traffic shaping, some of which are used as stand-alone technologies and others that are used integrally with other technologies, such as IETF Integrated Services architecture. How each of these schemes is used and the approach each attempts to use in conjunction with other specific technologies defines the purpose each scheme is attempting to serve, as well as the method and mechanics by which each is being used.

It also is important to understand the basic concepts and differences between traffic shaping, admission control, and policing. *Traffic shaping* is the practice of controlling the volume of traffic entering the network, along with controlling the rate at which it is transmitted. *Admission control*, in the most primitive sense, is the simple practice of discriminating which traffic is admitted to the network in the first place. *Policing* is the practice of determining on a hop-by-hop basis within the network beyond the ingress point whether the traffic being presented is compliant with prenegotiated traffic-shaping policies or other distinguishing mechanisms.

As mentioned previously, you should consider the resource constraints placed on each device in the network when determining which of these mechanisms to implement, as well as the performance impact on the overall network system. For this reason, traffic shaping and admission-control schemes need to be implemented at the network edges to control the traffic entering the network. Traffic policing obviously needs one of two things in order to function properly: Each device in the end-to-end path must implement an adaptive shaping mechanism similar to what is implemented at the network edges, or a dynamic signaling protocol must exist that collects path and resource information, maintains state within each transit device concerning the resource status of the network, and dynamically adapts mechanisms within each device to police traffic to conform to shaping parameters.

Traffic Shaping

Traffic shaping provides a mechanism to control the amount and volume of traffic being sent into the network and the rate at which the traffic is being sent. It also may be necessary to identify traffic flows at the *ingress point* (the point at which traffic enters the network) with a granularity that allows the traffic-shaping control mechanism to separate traffic into individual flows and shape them differently.

Two predominate methods for shaping traffic exist: a leaky-bucket implementation and a token-bucket implementation. Both these schemes have distinctly different properties and are used for distinctly different purposes. Discussions of combinations of each scheme follow. These schemes expand the capabilities of the simple traffic-shaping paradigm and, when used in tandem, provide a finer level of granularity than each method alone provides.

Leaky-Bucket Implementation

You use a *leaky bucket* (Figure 3.6) to control the rate at which traffic is sent to the network. A leaky bucket provides a mechanism by which bursty traffic can be shaped to present a steady stream of traffic to the network, as opposed to traffic with erratic bursts of low- and high-volume flows. An appropriate analogy for the leaky bucket is a scenario in which four lanes of automobile traffic converge into a single lane. A regulated admission interval into the single lane of traffic flow helps the traffic move. The benefit of this approach is that traffic flow into the major traffic arteries (the network) is predictable and controlled. The major liability is that when the volume of traffic is vastly greater than the bucket size, in conjunction with the drainage-time interval, traffic backs up in the bucket beyond bucket capacity and is discarded.

The leaky bucket was designed to control the rate at which ATM cell traffic is transmitted within an ATM network, but it has also found uses in the Layer 3 (packet datagram)

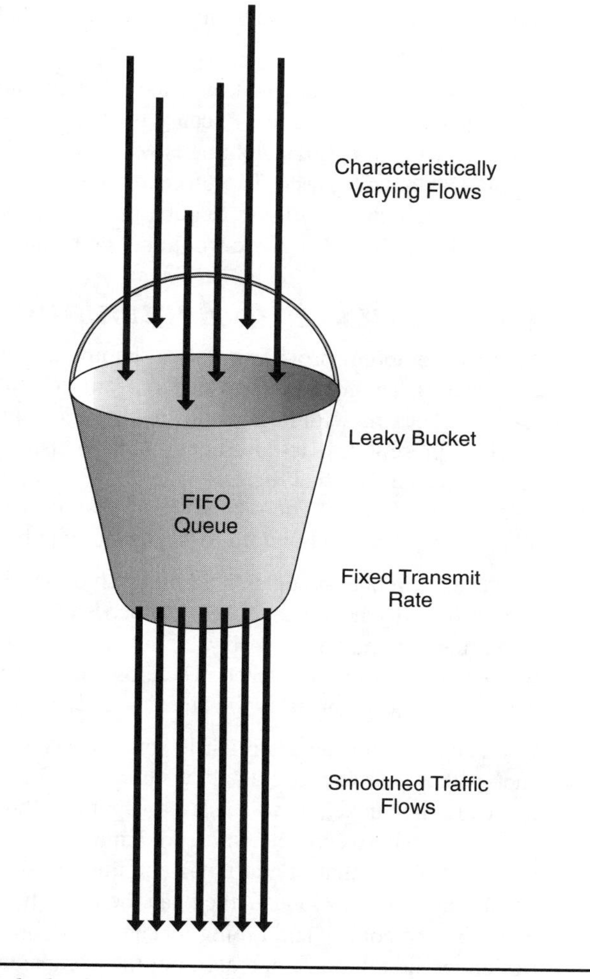

Figure 3.6: *A simple leaky bucket.*

world. The size (depth) of the bucket and the transmit rate generally are user-configurable and measured in bytes. The leaky-bucket control mechanism uses a measure of time to indicate when traffic in a FIFO queue can be transmitted to control the rate at which traffic is leaked into the network. It is possible for the bucket to fill up and subsequent flows to be discarded. This is a very simple method to control and shape the rate at which traffic is transmitted to the network, and it is a fairly straightforward implementation.

The important concept to bear in mind here is that this type of traffic shaping has an important and subtle significance in controlling network resources in the core of the network. Essentially, you could use traffic shaping as a mechanism to conform traffic flows in the network to a use threshold that a network administrator has calculated arbitrarily. This is especially useful if he has oversubscribed the network capacity. However, although this is an effective method of shaping traffic into flows with a fixed rate of admission into the network, it is ineffective in providing a mechanism that provides traffic shaping for variable rates of admission.

It also can be argued that the leaky-bucket implementation does not efficiently use available network resources. Because the leak rate is a fixed parameter, there will be many instances when the traffic volume is very low and large portions of network resources (bandwidth) are not being used. Therefore, no mechanism exists in the leaky-bucket implementation to allow individual flows to burst up to port speed, effectively consuming network resources at times when there would not be resource contention in the network core.

Token-Bucket Implementation

Another method of providing traffic shaping and ingress rate control is the *token bucket* (Figure 3.7). The token bucket differs from the leaky bucket substantially. Whereas the leaky bucket fills with traffic and steadily transmits traffic at a continuous fixed rate when traffic is present, traffic does not actually transit the token bucket. The token bucket is a control mechanism that dictates when traffic can transmitted based on the presence of tokens in the bucket. The token bucket also more efficient uses available network resources by allowing flows to burst up to a configurable burst threshold.

The token bucket contains *tokens*, each of which can represent a unit of bytes. The administrator specifies how many tokens are needed to transmit however many number of bytes; when tokens are present, a flow is allowed to transmit traffic. If there are no tokens in the bucket, a flow cannot transmit its packets. Therefore, a flow can transmit traffic up to its peak burst rate if there are adequate tokens in the bucket and if the burst threshold is configured appropriately.

The token bucket is similar in some respects to the leaky bucket, but the primary difference is that the token bucket allows bursty traffic to continue transmitting while there are tokens in the bucket, up to a user-configurable threshold, thereby accommodating traffic flows with bursty characteristics. An appropriate analogy for the way a token bucket operates is similar to that of a toll plaza on the interstate highway system. Vehicles (packets) are permitted to pass as long as they pay the toll. The packets must be admitted by the toll plaza operator (the control and timing mechanisms), and the money for the toll (tokens) is controlled by the toll plaza operator, not the occupants of the vehicles (packets).

Multiple Token Buckets

A variation of the simple token bucket implementation with a singular bucket and a singular burst rate threshold is one that includes multiple buckets and multiple thresholds (Figure 3.8). In this case, a traffic classifier could interact with separate token buckets, each with a different peak-rate threshold, thus permitting different classes of traffic to be shaped independently. You could use this approach with other mechanisms in the network to provide differentiated CoS; if traffic could be discriminantly tagged after it exceeds a certain threshold, it could be treated differently as it travels through the network. In the next section, you'll look at possible ways of accomplishing this by using IP layer QoS, IP precedence bits, and congestion-control mechanisms.

Another variation on these two schemes is a combination of the leaky bucket and token bucket. The reasoning behind this type of implementation is that, while a token bucket allows individual flows to burst to their peak rate, it also allows individual flows to dominate

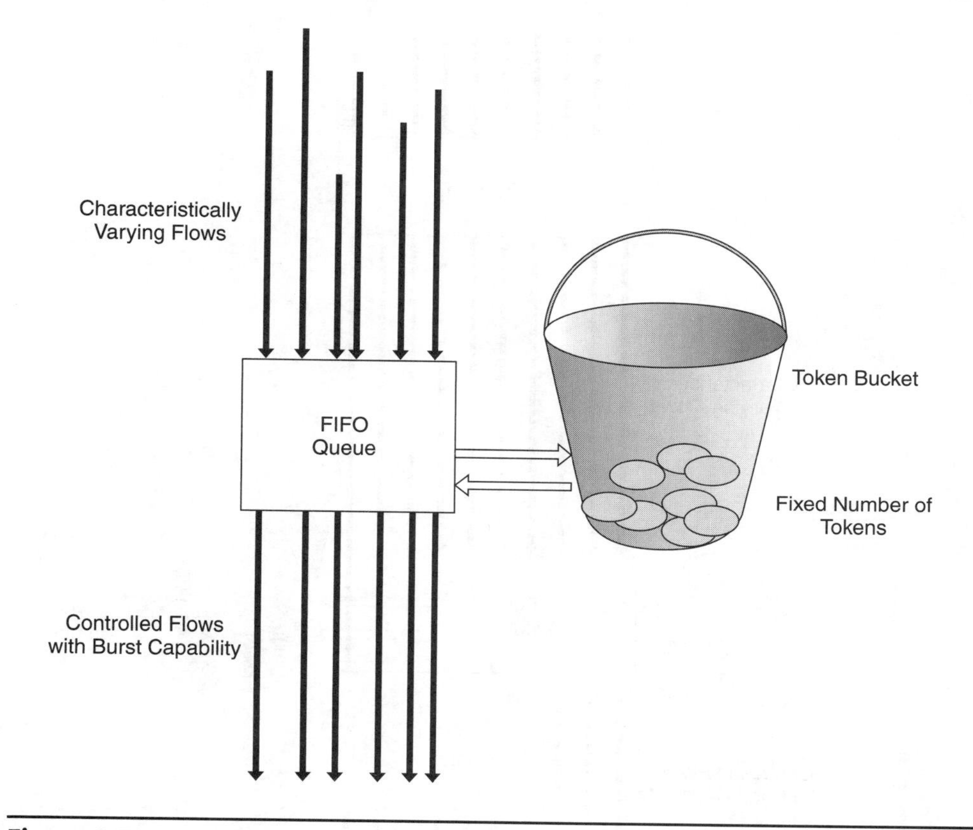

Figure 3.7: *A simple token bucket.*

network resources for the time during which tokens are present in the token bucket or up to the defined burst threshold. This situation, if not configured correctly, allows some traffic to consume more bandwidth than other flows and may interfere with the capability of other traffic flows to transmit adequately due to resource contention in the network. Therefore, a leaky bucket could be used to smooth the traffic rates at a user-configurable threshold after the traffic has been subject to the token bucket. In this fashion, both tools can be used together to create an altogether different traffic shaping mechanism. This is described in more detail in the following paragraph.

Combining Traffic-Shaping Mechanisms

A twist on both these schemes is to combine the two approaches so that traffic is shaped first with a token-bucket mechanism and then placed into a leaky bucket, thus limiting all traffic to a predefined admission rate. If multiple leaky buckets are used for multiple corresponding token buckets, a finer granularity is possible for controlling the types of traffic entering the network. It stands to reason, then, that the use of a token bucket and a leaky bucket makes even more efficient use of available network resources by allowing flows to

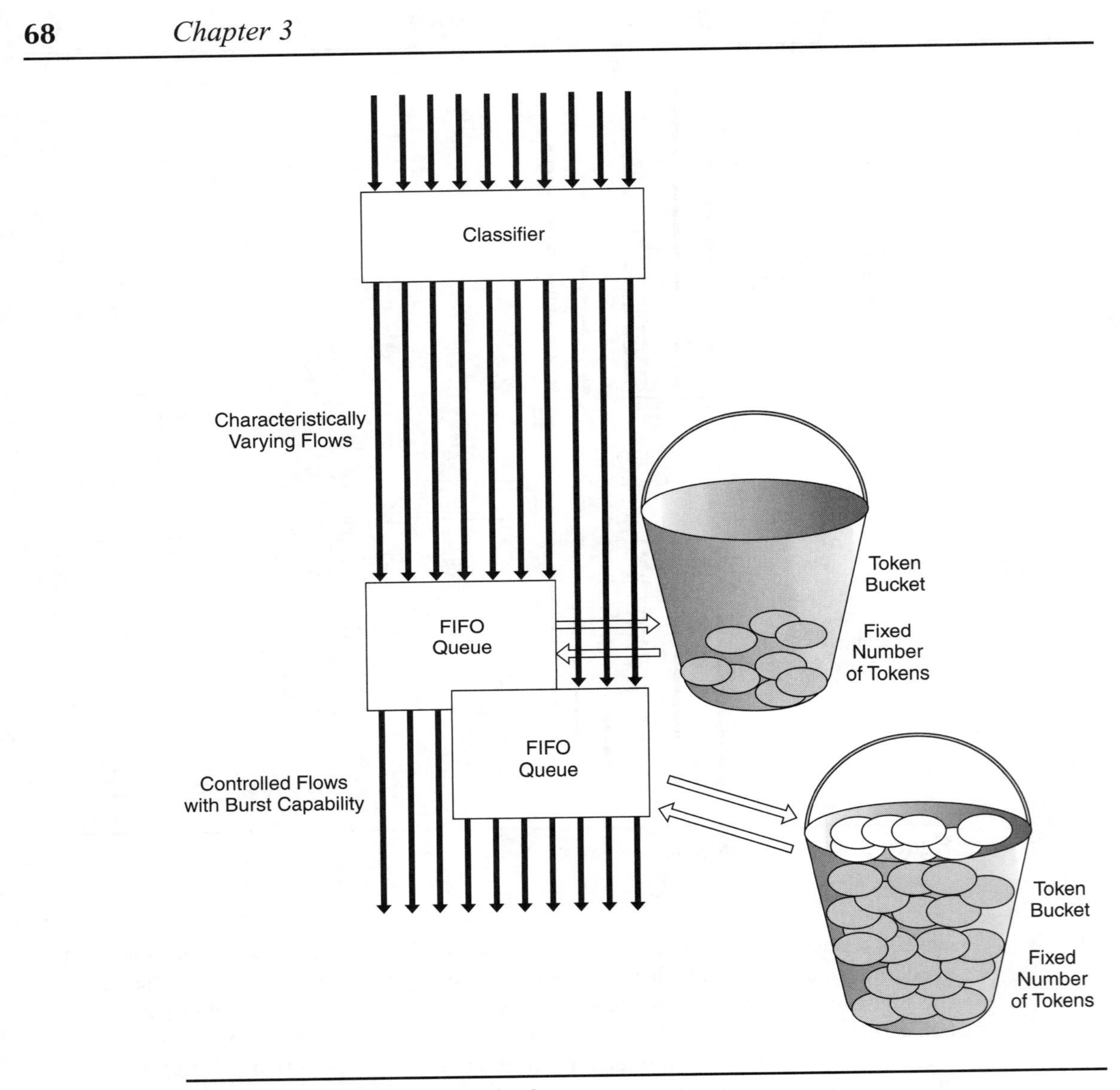

Figure 3.8: *Multiple token buckets.*

burst up to a configurable burst threshold, and an administrator can subsequently define the upper bound for the rate at which different classes of traffic enter the network. One leaky bucket admission rate could be set arbitrarily low for lower-priority traffic, for example, while another could be set arbitrarily high for traffic that an administrator knows has bursty characteristics, allowing for more consumption of available network resources.

The bottom line is that traffic-shaping mechanisms are yet another tool in controlling how traffic flows into (and possibly out of) the network. By themselves, these traffic-shaping tools are inadequate to provide QoS, but coupled with other tools, they can provide useful mechanisms for a network administrator to provide differentiated CoS.

Network-Admission Policy

Network-admission policy is an especially troublesome topic and one for which there is no clear agreed-on approach throughout the networking industry. In fact, *network-admission policy* has various meanings, depending on the implementation and context. By and large, the most basic definition is one in which admission to the network, or basic access to the network itself, is controlled by the imposition of a policy constraint. Some traffic (traffic that conforms to the policy) may be admitted to the network, and the remainder may not. Some traffic may be admitted under specific conditions, and when the conditions change, the traffic may be disallowed. By contrast, admission may be based on identity through the use of an authentication scheme.

Also, you should not confuse the difference between *access control* and *admission policy*—subtle differences, as well as similarities, exist in the scope of each approach.

Access Control

Access control can be defined as allowing someone or some thing (generally a remote daemon or service) to gain access to a particular machine, device, or virtual service. By contrast, *admission policy* can be thought of as controlling what type of traffic is allowed to enter or transit the network. One example of primitive access control dates back to the earliest days of computer networking: the password. If someone cannot provide the appropriate password to a computer or other networking device, he is denied access. Network dial-up services have used the same mechanism; however, several years ago, authentication control mechanisms began to appear that provided more granular and manageable remote access control. With these types of mechanisms, access to network devices is not reliant on the password stored on the network devices but instead on a central authentication server operated under the supervision of the network or security administrator.

Admission Policy

Admission policy can be a very important area of networking, especially in the realm of providing QoS. Revisiting earlier comments on architectural principles and service qualifications, if you cannot identify services (e.g., specific protocols, sources, and destinations), the admission-control aspect is less important. Although the capability to limit the sheer amount of traffic entering the network and the volume at which it enters is indeed important (to some degree), it is much less important if the capability to distinguish services is untenable, because no real capability exists to treat one type of traffic differently than another.

An example of a primitive method of providing admission policy is through the use of traffic filtering. Traffic filters can be configured based on device port (e.g., a serial or Ethernet interface), IP host address, TCP or UDP port, and so on. Traffic that conforms to the filtering is permitted or denied access to the network. Of course, these traffic filters could be used with a leaky- or token-bucket implementation to provide a more granular method of controlling traffic; however, the level of granularity still may leave something to be desired.

Admission policy is a crucial part of any QoS implementation. If you cannot control the traffic entering the network, you have no control over the introduction of congestion into the network system and must rely on congestion-control and avoidance mechanisms to maintain stability. This is an undesirable situation, because if the traffic originators have the capability to induce severe congestion situations into the network, the network may be ill-conceived and improperly designed, or admission-control mechanisms may be inherently required to be implemented with prejudice. The prejudice factor can be determined by economics. Those who pay for a higher level of service, for example, get a shot at the available bandwidth first, and those who do not get throttled, or dropped, when utilization rates reach a congestion state. In this fashion, admission control could feasibly be coupled with traffic shaping.

Admission policy, within the context of the IETF Integrated Services architecture, determines whether a network node has sufficiently available resources to supply the requested QoS. If the originator of the traffic flow requests QoS levels from the network using parameters within the RSVP specification, the admission-control module in the RSVP process checks the requester's TSpec and RSpec (described in more detail in Chapter 7) to determine whether the resources are available along each node in the transit path to provide the requested resources. In this vein, the QoS parameters are defined by the IETF Integrated Services working group, which is discussed in Chapter 7, "The Integrated Services Architecture."

chapter 4

QoS and TCP/IP: Finding the Common Denominator

As mentioned several times in this book, wildly varying opinions exist on how to provide differentiation of *Classes of Service* (CoS) and where it can be provided most effectively in the network topology.

After many heated exchanges, it is safe to assume that there is sufficient agreement on at least one premise: The most appropriate place to provide differentiation is within the most common denominator, where *common* is defined in terms of the level of end-to-end deployment in today's networks. In this vein, it becomes an issue of which component has the most prevalence in the end-to-end traffic path. In other words, what is the common bearer service? Is it ATM? Is it IP? Is it the end-to-end application protocol?

The TCP/IP Protocol Suite

In the global Internet, it is undeniable that the common bearer service is the TCP/IP protocol suite—therefore, IP is indeed the common denominator. (The TCP/IP protocol suite usually is referred to simply as *IP*; this has become the networking vernacular used to describe IP, as well as ICMP, TCP, and UDP.) This thought process has several supporting lines of reason. The common denominator is chosen in the hope of using the most pervasive and ubiquitous protocol in the network, whether it be Layer 2 or Layer 3. Using the most pervasive protocol makes implementation, management, and troubleshooting much easier and yields a greater possibility of successfully providing a QoS implementation that actually works.

It is also the case that this particular technology operates in an end-to-end fashion, using a signaling mechanism that spans the entire traversal of the network in a consistent fashion. IP is the end-to-end transportation service in most cases, so that although it is possible to create QoS services in substrate layers of the protocol stack, such services cover only part of the end-to-end data path. Such partial measures often have their effects masked by the effects of the traffic distortion created from the remainder of the end-to-end path in which they are not present, and hence the overall outcome of a partial QoS structure often is ineffectual.

Asynchronous Transfer Mode (ATM)

The conclusion of IP as the common denominator is certainly one that appears to be at odds with the one consistently voiced as a major benefit of ATM technologies. ATM is provided with many features described as QoS *enablers*, allowing for constant or variable bit rate virtual circuits, burst capabilities, and traffic shaping. However, although ATM indeed provides a rich set of QoS knobs, it is not commonly used as an end-to-end application transport protocol. In accordance with the *commonly deployed as an end-to-end technology* rule of thumb, ATM is not the ideal candidate for QoS deployment. Although situations certainly exist in which the end-to-end path may consist of pervasive ATM, end-stations themselves rarely are ATM-attached.

When the end-to-end path does not consist of a single pervasive data-link layer, any effort to provide differentiation within a particular link-layer technology most likely will not provide the desired result. This is the case for several reasons. In the Internet, for example, an IP packet may traverse any number of heterogeneous link-layer paths, each of which may or may not possess characteristics that inherently provide methods to provide traffic differentiation. However, the packet also inevitably traverses links that cannot provide any type of differentiated services at the data-link layer. With data-link layer technologies such as Ethernet and token ring, for example, only a framing encapsulation is provided, and in which a MAC (Media Access Control) address is rewritten in the frame and then forwarded on its way toward its destination. Of course, solutions have been proposed that may provide enhancements to specific link-layer technologies (e.g., Ethernet) in an effort to manage bandwidth and provide flow-control. (This is discussed in Chapter 9, "QoS and Future Possibilities.") However, for the sake of this discussion, you will examine how you can use IP for differentiated CoS because, in the Internet, an IP packet is an IP packet is an IP packet. The only difference is that a few bytes are added and removed by link-layer encapsulations (and subsequently are unencapsulated) as the packet travels along its path to its destination.

Of course, the same can be said of other routable Layer 3 and Layer 4 protocols, such as IPX/SPX, AppleTalk, DECnet, and so on. However, you should recognize that the TCP/IP suite is the fundamental, de facto protocol base used in the global Internet. Evidence also exists that private networks are moving toward TCP/IP as the ubiquitous end-to-end protocol, but migrating legacy network protocols and host applications sometimes is a slow and tedious process. Also, some older legacy network protocols may not have inherent *hooks* in the protocol specification to provide a method to differentiate packets as to how they are treated as they travel hop by hop to their destination.

Tip A very good reference on the way in which packet data is encapsulated by the various devices and technologies as it travels through heterogeneous networks is provided in *Interconnections: Bridges and Routers,* by Radia Perlman, published by Addison-Wesley Publishing.

Differentiation by Class

Several interesting proposals exist on providing differentiation of CoS within IP; two particular approaches merit mention.

Per-Flow Differentiation

One approach proposes per-flow differentiation. You can think of a *flow* in this context as a sequence of packets that share some unique information. This information consists of a 4-tuple (source IP address, source UDP or TCP port, destination IP address, destination UDP or TCP port) that can uniquely identify a particular end-to-end application-defined flow or conversation. The objective is to provide a method of extracting information from the IP packet header and have some capability to associate it with previous packets. The intended result is to identify the end-to-and application stream of which the packet is a member. Once a packet can be assigned to a flow, the packet can be forwarded with an associated class of service that may be defined on a per-flow basis.

This is very similar in concept to assigning CoS characteristics to Virtual Circuits (VCs) within a frame relay or ATM network. The general purpose of per-flow differentiation is to be able to define similar CoS characteristics to particular IP end-to-end sessions, allowing real-time flows, for example, to be forwarded with CoS parameters different from other non-real-time flows. However, given the number of flows that may be active in the core of the Internet at any given time (hint: in many cases, there are in excess of 256,000 active flows), this approach is widely considered to be impractical as far as scaleability is concerned. Maintaining state and manipulating flow information for this large a number of flows would require more computational overhead than is practical or desired. This is primarily the approach that RSVP takes, and it has a lot of people gravely concerned that it will not be able to scale in a sufficiently large network, let alone the Internet. Thus, a simpler and more scaleable approach may be necessary for larger networks.

Using IP Precedence for Classes of Service

Several proposals and, in fact, a couple of actual implementations use the *IP precedence* bits in the IP TOS (Type of Service) [IETF1992b] field in the IP header as a method of marking packets and distinguishing them, based on these bit settings, as they travel through the network to their destinations. Figure 4.1 shows these fields. This is the second approach to CoS within IP.

The TOS field has been a part of the IP specification since the beginning and has been little used in the past. The semantics of this field are documented in RFC1349, "Type of

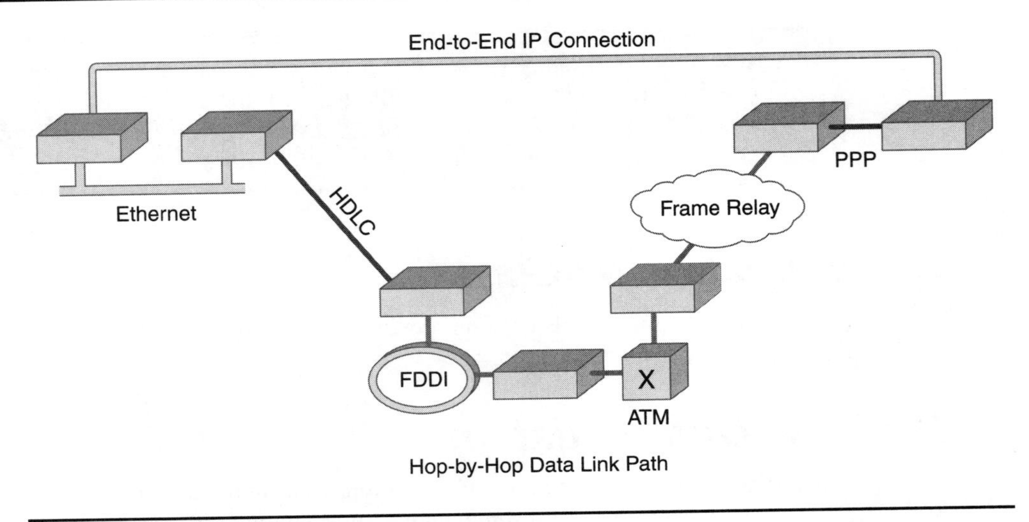

Figure 4.1: *A heterogeneous end-to-end path.*

Service in the Internet Protocol Suite" [IETF1992b], which suggests that the values specified in the TOS field could be used to determine how packets are treated with monetary considerations. As an aside, at least two routing protocols, including OSPF (Open Shortest Path First) [IETF1994a] and Integrated IS-IS [IETF1990b], can be configured to compute paths separately for each TOS value specified.

In contrast to the use of this 4-bit TOS field as described in the original 1981 DARPA Internet Protocol Specification RFC791 [DARPA19982b], where each bit has its own meaning, RFC1349 attempts to redefine this field as a set of bits that should be considered collectively rather than individually. RFC1349 redefined the semantics of these field values as depicted here:

Binary Value	*Characterization*
1000	Minimize delay
0100	Maximize throughput
0010	Maximize reliability
0001	Minimize monetary cost
0000	Normal service

In any event, the concept of using the 4-bit TOS values in this fashion did not gain in popularity, and no substantial protocol implementation (aside from the two routing protocols mentioned earlier) has made any tangible use of this field. It is important to remember this, because you will revisit the topic of using the IP TOS field during the discussion of QoS-based routing schemes in Chapter 9, "QoS and Future Possibilities."

On the other hand, at least one I-D (Internet Draft) in the IETF has proposed using the IP precedence in a protocol negotiation, of sorts, for transmitting traffic across administrative boundaries [ID1996e]. Also, you can be assured that components in the IP TOS field are supported by all IP-speaking devices, because use of the precedence bits in the TOS field is required to be supported in the IETF "Requirements for IP Version 4 Routers" document, RFC1812 [IETF1995b]. Also, RFC1812 discusses

- Precedence-ordered queue service (section 5.3.3.1 of RFC1812), which (among other things) provides a mechanism for a router to specifically queue traffic for the forwarding process based on highest precedence.
- Precedence-based congestion control (section 5.3.6 of RFC1812), which causes a router to drop packets based on precedence during periods of congestion.
- Data-link-layer priority features (section 5.3.3.2 of RFC1812), which cause a router to select service levels of the lower layers in an effort to provide preferential treatment.

Although RFC1812 does not explicitly describe how to perform these IP precedence-related functions in detail, it does furnish references on previous related works and an overview of some possible implementations where administration, management, and inspection of the IP precedence field may be quite useful.

The issue here with the proposed semantics of the IP precedence fields is that, instead of the router attempting to perform an initial classification of a packet into one of potentially many thousands of active flows and then apply a CoS rule that applies to that form of flow, you can use the IP precedence bits to reduce the scope of the task considerably. There is not a wide spectrum of CoS actions the interior switch may take, and the alternative IP-precedence approach is to mark the IP packet with the desired CoS when the packet enters the network. On all subsequent interior routers, the required action is to look up the IP precedence bits and apply the associated CoS action to the packet. This approach can scale quickly and easily, given that the range of CoS actions is a finite number and does not grow with traffic volume, whereas flow identification is a computational task related to traffic volume.

The format and positioning of the TOS and IP precedence fields in the IP packet header are shown in Figure 4.2.

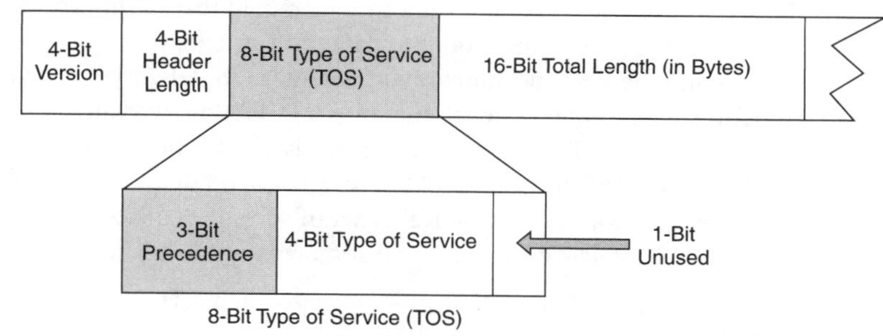

Figure 4.2: *The IP TOS field and IP precedence.*

Reviewing Topological Significance

When examining real-world implementations of differentiated CoS using the IP precedence field, it is appropriate to review the *topological significance* of where specific functions take place—where IP precedence is set, where it is policed, and where it is used to administer traffic transiting the network core. As mentioned previously, architectural principles exist regarding where certain technologies should be deployed in a network. Some technologies, or implementations of a particular technology, have more of an adverse impact on network performance than other technologies.

Therefore, it is important to push the implementation of these technologies toward the edges of the network to avoid imposing a performance penalty on a larger percentage of the overall traffic in the network. Technologies that have a nominal impact on performance can be deployed in the network core. It is important to make the distinction between setting or policing the IP precedence, which could be quite resource intensive, and simply administering traffic based on the IP precedence as it travels through the network. Setting the IP precedence requires more overhead, because it requires going into the packet, examining the precedence contents, and possibly rewriting the precedence values. By contrast, simply looking at this field requires virtually no additional overhead, because every router must look at the contents of the IP header anyway to determine where the packet is destined, and it is within the IP header that the IP precedence field is located.

As discussed in the section on implementing policy in Chapter 3, and again within the section on admission control and traffic shaping, this is another example of a technology implementation that needs to be performed at the edge of the network. In the network core, you can simply examine the precedence field and make some form of forwarding, queuing, and scheduling decision. Once you understand why the architectural principles are important, it becomes an issue of understanding how using IP precedence can actually deliver a way of differentiating traffic in a network.

Understanding CoS at the IP Layer

Differentiated classes of service at the IP layer are not as complex as you might imagine. At the heart of the confusion is the misperception that you can provide some technical guarantee that traffic will reach its destination and that it will do so better than traditional best-effort traffic. It is important to understand that in the world of TCP/IP networking, you are operating in a connectionless, datagram-oriented environment—the TCP/IP protocol suite—and there are no functional guarantees in delivery. Instead of providing a specific functional performance-level guarantee, TCP/IP networking adopts a somewhat different philosophy. A TCP/IP session dynamically probes the current state of resource availability of the sender, the network, and the receiver and attempts to maximize the rate of reliable data transmission to the extent this set of resources allows. It also is important to understand that TCP was designed to be somewhat graceful in the face of packet loss.

TCP is considered to be *self-clocking*; when a TCP sender determines that a packet has been lost (because of a loss or time-out of a packet-receipt acknowledgment), it backs off, ceases transmitting, shrinks its transmission window size, and retransmits the lost packet at what effectively can be considered a slower rate in an attempt to complete the original data transfer. After a reliable transmission rate is established, TCP commences to probe whether more resources have become available and, to the extent permitted by the sender and receiver, attempts to increase the transmission rate until the limit of available network resources is reached. This process commonly is referred to as a TCP *ramp up* and comes in two flavors: the initial aggressive compound doubling of the rate in *slow-start mode* and the more conservative linear increase in the rate in *congestion-avoidance mode*. The critical signal that this maximum level has been exceeded is the occurrence of packet loss, and after this occurs, the transmitter again backs off the data rate and reprobes. This is the nature of TCP self-clocking.

TCP Congestion Avoidance

One major drawback in any high-volume IP network is that when there are congestion hot spots, uncontrolled congestion can wreak havoc on the overall performance of the network to the point of congestion collapse. When thousands of flows are active at the same time, and a congestion situation occurs within the network at a particular bottleneck, each flow conceivably could experience loss at approximately the same time, creating what is known as *global synchronization*. *Global*, in this case, has nothing to do with an all-encompassing planetary phenomenon; instead, it refers to all TCP flows in a given network that traverse a common path. Global synchronization occurs when hundreds or thousands of flows back off and go into TCP slow start at roughly the same time. Each TCP sender detects loss and reacts accordingly, going into slow-start, shrinking its window size, pausing for a moment, and then attempting to retransmit the data once again. If the congestion situation still exists, each TCP sender detects loss once again, and the process repeats itself over and over again, resulting in network gridlock [Zhang1990].

Uncontrolled congestion is detrimental to the network systems—behavior becomes unpredictable, system buffers fill up, packets ultimately are dropped, and the result is a large number of retransmits.

Random Early Detection (RED)

Van Jacobson discussed the basic methods of implementing congestion avoidance in TCP in 1988 [Jacobson1988]. However, Jacobson's approach was more suited for a small number of TCP flows, which is much less complex to manage. In 1993, Sally Floyd and Van Jacobson documented the concept of RED (Random Early Detection), which provides a mechanism to avoid congestion collapse by randomly dropping packets from arbitrary flows in an effort to avoid the problem of global synchronization and, ultimately, congestion collapse [Floyd1993]. The principal goal of RED is to avoid a situation in which all TCP flows experience congestion at the same time, and subsequent packet loss, thus avoiding global synchronization. RED monitors the queue depth, and as the queue begins to fill, it begins to randomly select individual TCP flows from which to drop packets, in order to signal the receiver to slow down (Figure 4.3). The threshold at which RED begins to drop packets generally is configurable by the network administrator, as well as the rate at which drops occur in relation to how quickly the queue fills. The more it fills, the greater the number of flows selected, and the greater the number of packets dropped (Figure 4.4). This results in signaling a greater number of senders to slow down, thus resulting in congestion avoidance.

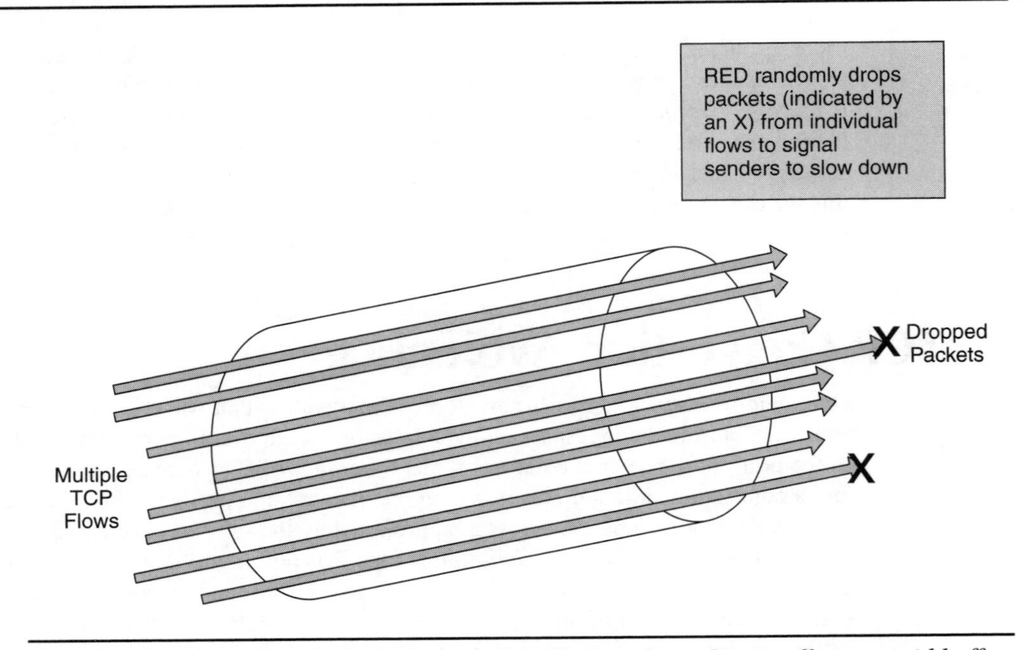

Figure 4.3: *RED selects traffic from random flows to discard in an effort to avoid buffer overflow.*

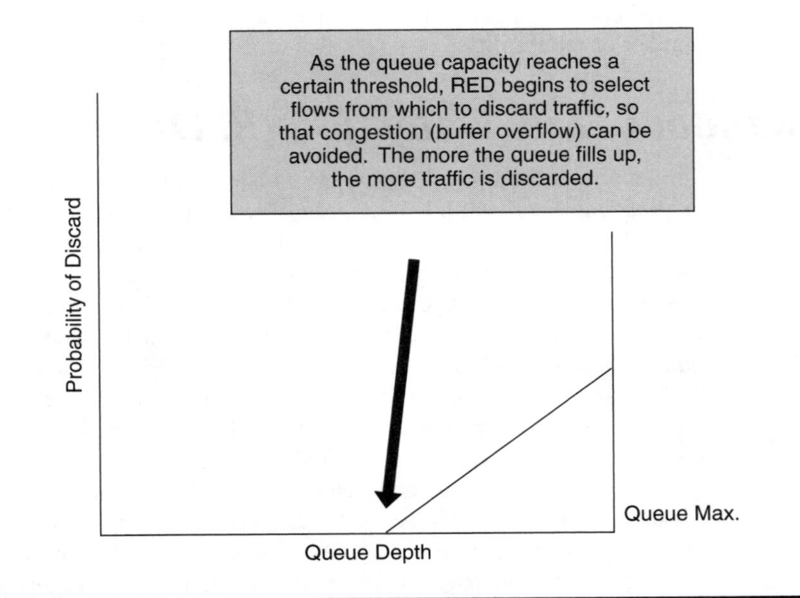

Figure 4.4: *RED: The more the queue fills, the more traffic is discarded.*

The RED approach does not possess the same undesirable overhead characteristics as some of the non-FIFO queuing techniques discussed earlier. With RED, it is simply a matter of who gets into the queue in the first place—no packet reordering or queue management takes place. When packets are placed into the outbound queue, they are transmitted in the order in which they are queued. Priority, class-based, and weighted-fair queuing, however, require a significant amount of computational overhead because of packet reordering and queue management. RED requires much less overhead than fancy queuing mechanisms, but then again, RED performs a completely different function.

TIP You can find more detailed information on RED at www.nrg.ee.lbl.gov/floyd/red.html.

IP rate-adaptive signaling happens in units of end-to-end Round Trip Time (RTT). For this reason, when network congestion occurs, it can take some time to clear, because transmission rates will not immediately back off in response to packet loss. The signaling involved is that packet loss will not be detected until the receiver's timer expires, and the transmitter will not see the signal until the receiver's NAK (negative acknowledgment) arrives back at the sender. Hence, when congestion occurs, it is not cleared quickly. The objective of RED is to start the congestion-signaling process at a slightly earlier time than queue saturation. The use of random selection of flows to drop packets could be argued to favor dropping packets from flows in which the rate has opened up and flows are at their longest duration; these flows generally are considered to be the greatest contributor to the longevity of normal congestion situations.

Introducing Unfairness

RED can be said to be fair: It chooses random flows from which to discard traffic in an effort to avoid global synchronization and congestion collapse, as well as to maintain equity in which traffic actually is discarded. Fairness is all well and good, but what is really needed here is a tool that can induce unfairness—a tool that can allow the network administrator to predetermine what traffic is dropped first or last when RED starts to select flows from which to discard packets. You can't differentiate services with fairness.

An implementation that already has been deployed in several networks uses a combination of the technologies already discussed to include multiple token buckets for traffic shaping. However, exceeded token-bucket thresholds also provide a mechanism for marking IP precedence. As precedence is set or policed when traffic enters the network (at ingress), a weighted congestion-avoidance mechanism implemented in the core routers determines which traffic should be dropped first when congestion is anticipated due to queue-depth capacity. The higher the precedence indicated in a packet, the lower the probability of drop. The lower the precedence, the higher the probability of drop. When the congestion avoidance is not actively discarding packets, all traffic is forwarded with equity.

Weighted Random Early Detection (WRED)

Of course, for this type of operation to work properly, an intelligent congestion-control mechanism must be implemented on each router in the transit path. A least one mechanism is available that provides an unfair or weighted behavior for RED. This deviation of RED yields the desired result for differentiated traffic discard in times of congestion and is called *Weighted Random Early Detection* (WRED). A similar scheme, called *enhanced RED*, is documented in a paper authored by Feng, Kandlur, Saha, and Shin [FKSS1997].

Active and Passive Admission Control

As depicted in Figure 4.5, the network simply could rely on the end-stations setting the IP precedence themselves. Alternatively, the ingress router (router A) could check and force the traffic to conform to administrative policies. The former is called passive admission

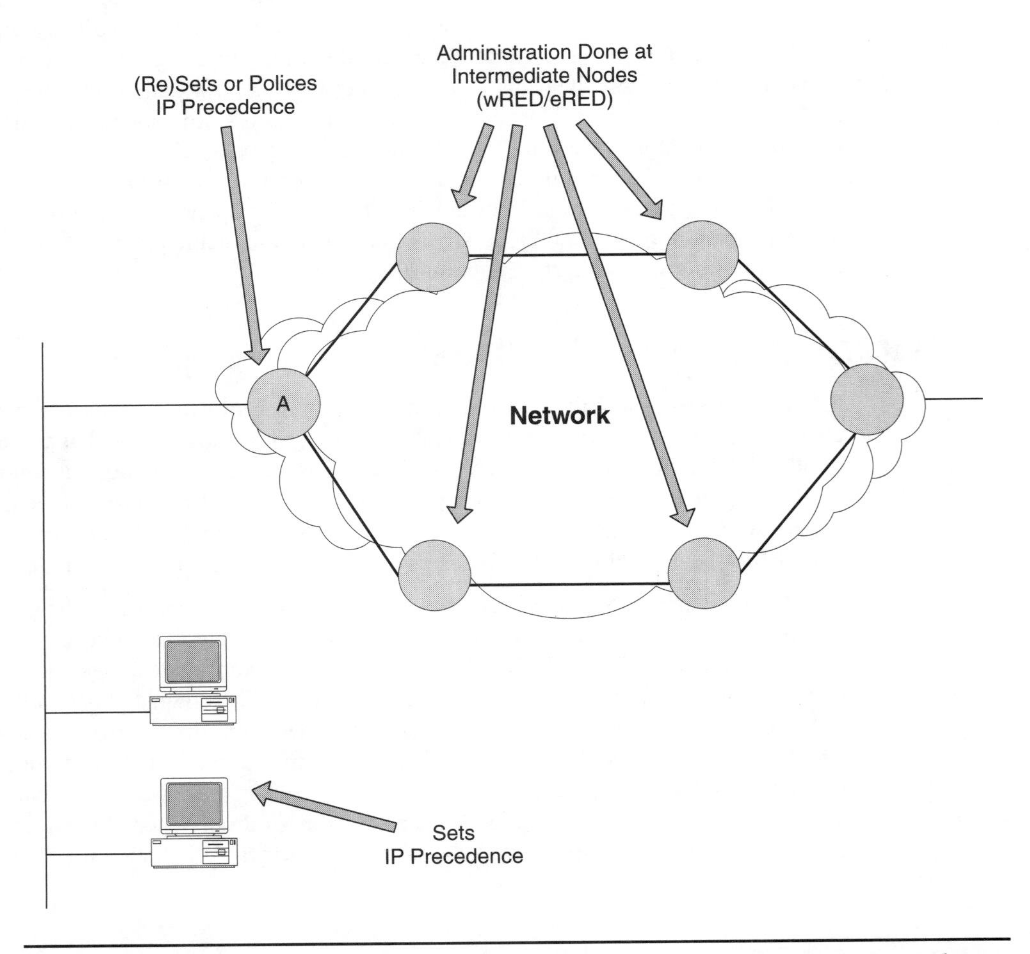

Figure 4.5: *Using IP precedence to indicate drop preference with congestion avoidance.*

control. The latter, of course, is called policing through an active admission control mechanism. Policing is a good idea for several reasons. First, it is usually not a good idea to explicitly trust all downstream users to know how and when their traffic should be marked at a particular precedence. The danger always exists that users will try to attain a higher service classification than entitled.

This is a political issue, not a technical one, and network administrators need to determine whether they will passively admit traffic with precedence already set and simply bill accordingly for the service or actively police traffic as it enters their network. The mention of billing for any service assumes that the necessary tools are in place to adequately measure traffic volumes generated by each downstream subscriber. Policing assumes that the network operator determines the mechanisms for allocation of network resources to competing clients, and presumably the policy enforced by the policing function yields the most beneficial economic outcome to the operator.

The case can be made, however, that the decision as to which flow is of greatest economic value to the customer is a decision that can be made only by the customer and that IP precedence should be a setting that can be passed only into the network by the customer and not set (or reset) by the network. Obviously, within this scenario, the transportation of a packet in which precedence is set attracts a higher transport fee than if the precedence is not set, regardless of whether the network is in a state of congestion. It is not immediately obvious which approach yields the greatest financial return for the network operator. On the one hand, the policing function creates a network that exhibits congestion loading in a relatively predictable manner, as determined by the policies of the network administrator. The other approach allows the customer to identify the traffic flows of greatest need for priority handling, and the network manages congestion in a way that attempts to impose degradation on precedence traffic to the smallest extent possible.

Threshold Triggering

The interesting part of the threshold-triggering approach is in the traffic-shaping mechanisms and the associated thresholds, which can be used to provide a method to mark a packet's IP precedence. As mentioned previously, precedence can be set in two places: by the originating host or by the ingress router that polices incoming traffic. In the latter case, you can use token-bucket thresholds to set precedence. You can implement a token bucket to define a particular bit-rate threshold, for example, and when this threshold is exceeded, mark packets with a lower IP precedence. Traffic transmitted within the threshold can be marked with a higher precedence. This allows traffic that conforms to the specified bit rate to be marked as a *lower probability of discard* in times of congestion. This also allows traffic in excess of the configured bit-rate threshold to burst up to the port speed, yet with a higher probability of discard than traffic that conforms to the threshold.

You can realize additional flexibility by adding multiple token buckets, each with similar or dissimilar thresholds, for various types of traffic flows. Suppose that you have an interface connected to a 45-Mbps circuit. You could configure three token buckets: one for FTP, one for HTTP, and one for all other types of traffic. If, as a network administrator, you want to provide a better level of service for HTTP, a lesser quality of service for FTP, and a yet lesser quality for all other types of traffic, you could configure each token bucket independently. You also could select precedence values based on what you believe to be reasonable service levels or based on agreements between yourself and your customers.

For example, you could strike a service-level agreement that states that all FTP traffic up to 10 Mbps is reasonably important, but that all traffic (with the exception of HTTP) in excess of 10 Mbps simply should be marked as best effort (Figure 4.6). In times of congestion, this traffic is discarded first.

This gives you, as the network administrator, a great deal of flexibility in determining the value of the traffic as well as deterministic properties in times of congestion. Again, this approach does not require a great deal of computational overhead as do the fancy queuing mechanisms discussed earlier, because RED is still the underlying congestion-avoidance mechanism, and it does not have to perform packet reordering or complicated queue-management functions.

Varying Levels of Best Effort

The approaches discussed above use existing mechanisms in the TCP/IP protocol suite theto provide varying degrees of best-effort delivery. In effect, there is arguably no way to guarantee traffic delivery in a Layer 813 IP environment. Delivering differentiated classes of service becomes as simple as ensuring that best-effort traffic has a higher probability of being dropped than does premium traffic during times of congestion (the differentiation is made by setting the IP precedence bits). If there is no congestion, everyone's traffic is delivered (hopefully, in a reasonable amount of time and negligible latency), and everyone is happy. This is a very straightforward and simple approach to delivering differentiated CoS. You can apply these same principles to IP unicast or multicast traffic.

	Classification	Threshold (Mbps)	Under Threshold Precedence	Over Threshold Precedence
Token Bucket 0	HTTP	30	6	5
Token Bucket 1	FTP	10	4	0
Token Bucket 2	All Other Traffic	10	3	0

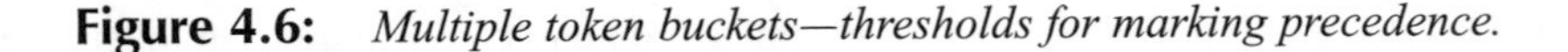

Figure 4.6: *Multiple token buckets—thresholds for marking precedence.*

Of course, a large contingent of people will want guaranteed delivery, and this is what RSVP attempts to provide. You'll look at this in more detail in Chapter 7, "The Integrated Services Architecture," as well as some opinions on how these two approaches stack up.

chapter 5

QoS and Frame Relay

What is *Frame Relay*? It has been described as an "industry-standard, switched data link-layer protocol that handles multiple virtual circuits using HDLC encapsulation between connected devices. Frame Relay is more efficient than X.25, the protocol for which it is generally considered a replacement. See also X.25" [Cisco1995].

Perhaps you should examine "see also X.25" and then look at Frame Relay as a transport technology. This will provide the groundwork for looking at QoS mechanisms that are supportable across a Frame Relay network. So to start, here's a thumbnail sketch of the X.25 protocol.

The X.25 Protocol

X.25 is a transport protocol technology developed by the telephony carriers in the 1970s. The technology model used for the protocol was a close analogy to telephony, using many of the well-defined constructs within telephony networks. The network model is therefore not an internetwork model but one that sees each individual connected computer as a call-initiation or termination point. The networking protocol supports switched virtual circuits, where one connected computer can establish a point-to-point dedicated connection with another computer (equivalent to a *call* or a *virtual circuit*). Accordingly, the X.25 model essentially defines a telephone network for computers.

The X.25 protocol specifications do not specify the internals of the packet-switched network but do specify the interface between the network and the client. The network boundary point is the Data Communications Equipment (DCE), or the network switch, and the Customer Premise Equipment (CPE) is the Data Termination Equipment (DTE), where the appropriate equipment is located at the customer premise.

About Precision

Many computer and network protocol descriptions suffer from "acronym density." The excuse given is that to use more generic words would be too imprecise in the context of the protocol definition, and therefore each protocol proudly develops its own terminology and related acronym set. X.25 and Frame Relay are no exception to this general rule. This book stays with the protocol-defined terminology and acronyms for the same reason of precision and brevity of description. We ask for your patience as we work through the various protocol descriptions.

X.25 specifies the DCE/DTE interaction in terms of framing and signaling. A complementary specification, X.75, specifies the switch-to-switch interaction for X.25 switches within the interior of the X.25 network.

The major control operations in X.25 are *call setup*, *data transfer*, and *call clear*. Call setup establishes a virtual circuit between two computers. The call setup operation consists of a handshake: One computer initiates the call, and the call is answered from the remote end as it returns a signal confirming receipt of the call. Each DCE /DTE interface then has a locally defined Local Channel Identifier (LCI) associated with the call. All further references to the call use the LCI as the identification of the call instead of referring to the called remote DCE.

The LCI is not a constant value within the network—each data-link circuit within the end-to-end path uses a unique LCI so that the initial DTE/DCE LCI is mapped into successive channel identifiers in each X.25 switch along the end-to-end call path. Thus, when the client computer refers to an LCI as the prefix for a data transfer, it is only a locally significant reference. Then, when a frame is transmitted out on a local LCI, the DTE simply assumes that it is correctly being transmitted to the appropriate switch because of the configured association between the LCI and the virtual circuit. The X.25 switch receives the HDLC (High-Level Data Link Control) frame and then passes it on to the next hop in the path, using local lookup tables to determine how to similarly switch the frame out on one of various locally defined channels. Because each DCE uses only the LCI as the identification for this particular virtual circuit, there is no need for the DCE to be aware of the remote LCI.

X.25 and QoS

X.25 was designed to be a reliable network transport protocol, so all data frames are sequenced and checked for errors. X.25 switch implementations are flow controlled, so each interior switch within the network does not consider that a frame has been transmitted successfully to the next switch in the path until it explicitly acknowledges the transfer. The transfer also is verified by each intermediate switch for dropped and duplicated packets by a sequence number check. Because all packets within an X.25 virtual circuit must follow the same internal path through the network, out-of-sequence packets are readily detected within the network's packet switches. Therefore, the X.25 network switches implement switch-to-switch flow control and switch-to-switch error detection and retransmission, as well as preserve packet sequencing and integrity.

This level of network functionality allows relatively simple end-systems to make a number of assumptions about the transference of data. In the normal course of operation, all data passed to the network will be delivered to the call destination in order and without error, with the original framing preserved.

"Smartness"

In the same way that telephony uses simple peripherial devices (*dumb* handsets) and a complex interior switching system (*smart* network), X.25 attempts to place the call- and flow-management complexity into the interior of the network and create a simple interface with minimal functional demands on the connected peripheral devices. TCP (Transmission Control Protocol) implements the opposite technology model, with a simple best-effort datagram delivery network (dumb network) and flow control and error detection and recovery in the end-system (smart perhiperals).

No special mechanisms really exist that provide Quality of Service (QoS) within the X.25 protocol. Therefore, any differentiation of services needs to be at the network layer (e.g., IP) or by preferential queuing (e.g., priority, CBQ, or WFQ). This is a compelling reason to consider Frame Relay instead of X.25 as a wide-area technology for implementing QoS-based services.

Frame Relay

So how does Frame Relay differ from X.25? Frame Relay has been described as being faster, more streamlined, and more efficient as a transport protocol than X.25. The reality is that Frame Relay removes the switch-to-switch flow control, sequence checking, and error detection and correction from X.25, while preserving the connection orientation of data calls as defined in X.25. This allows for higher-speed data transfers with a lighter-weight transport protocol.

> **TIP** You can find approved and pending Frame Relay technical specifications at the Frame Relay Forum Web site, located at www.frforum.com.

Frame Relay's origins lie in the development of ISDN (Integrated Services Digital Network) technology, where Frame Relay originally was seen as a packet-service technology for ISDN networks. The Frame Relay rationale proposed was the perceived need for the efficient relaying of HDLC framed data across ISDN networks. With the removal of data link-layer error detection, retransmission, and flow control, Frame Relay opted for end-to-end signaling at the transport layer of the protocol stack model to undertake these functions. This allows the network switches to consider data-link frames as being forwarded without waiting for positive acknowledgment from the next switch. This in turn allows the switches to operate with less memory and to drive faster circuits with the reduced switch functionality required by Frame Relay.

However, like X.25, Frame Relay has a definition of the interface between the client and the Frame Relay network called the UNI (User-to-Network Interface). Switches within the confines of the Frame Relay network may use varying technologies, such as cell relay or HDLC frame passing. However, whereas interior Frame Relay switches have no requirement to undertake error detection and frame retransmission, the Frame Relay specification does specify that frames must be delivered in their original order, which is most commonly implemented using a connection-oriented interior switching structure.

Current Frame Relay standards address only permanent virtual circuits that are administratively configured and managed in the Frame Relay network; however, Frame Relay Form standards-based work currently is underway to support Switched Virtual Circuits (SVCs). Additionally, work recently was completed within the Frame Relay Forum to define Frame Relay high-speed interfaces at HSSI (52 Mbps), T3 (45 Mbps) and E3 (34 Mbps) speeds, augmenting the original T1/E1 specifications.

Figure 5.1 illustrates a simple Frame Relay network.

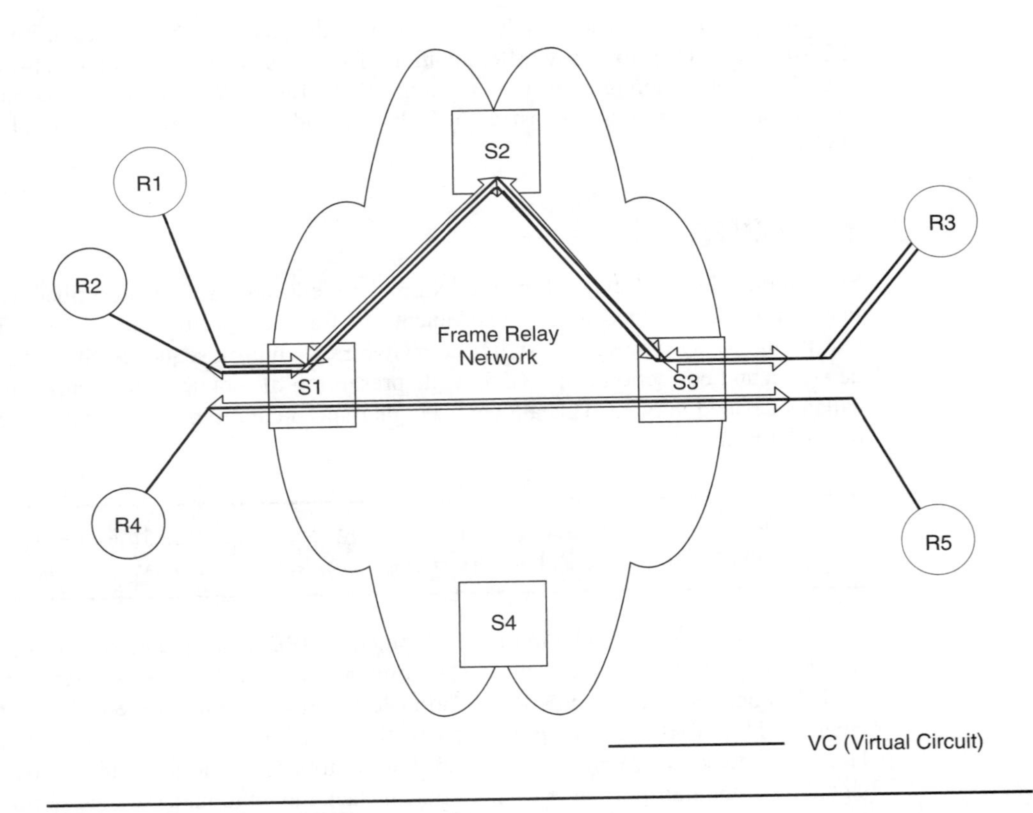

Figure 5.1: *A Frame Relay network.*

The original framing format of Frame Relay is defined in format by CCITT Recommendation Q.921/I.441 [ANSI T1S1 "DSSI Core Aspects of Frame Relay," March 1990]. Figure 5.2 shows this format. The minimum, and the default Frame Address field, is 16 bits. In the Frame Address field, the Data Link Connection Identifier (DLCI) is addressed using 10 bits, the extended address field is 2 bits, the Forward Explicit Congestion Notification (FECN) is 1 bit, the Backward Explicit Congestion Notification (BECN) is 1 bit, and the Discard Eligible (DE) field is 1 bit.

These final 3 bits within the Frame Relay header are perhaps the most significant components of Frame Relay when examining QoS possibilities.

Committed Information Rate

The Discard Eligible (DE) bit is used to support the engineering notion of a bursty connection. Frame Relay defines this type of connection using the concepts of Committed Information Rate (CIR) and traffic bursts and applies these concepts to each Virtual Circuit (VC) at the interface between the client (DTE) and the network (DCE). (Frame Relay could never be called acronym-light!) Each VC is configured with an administratively assigned

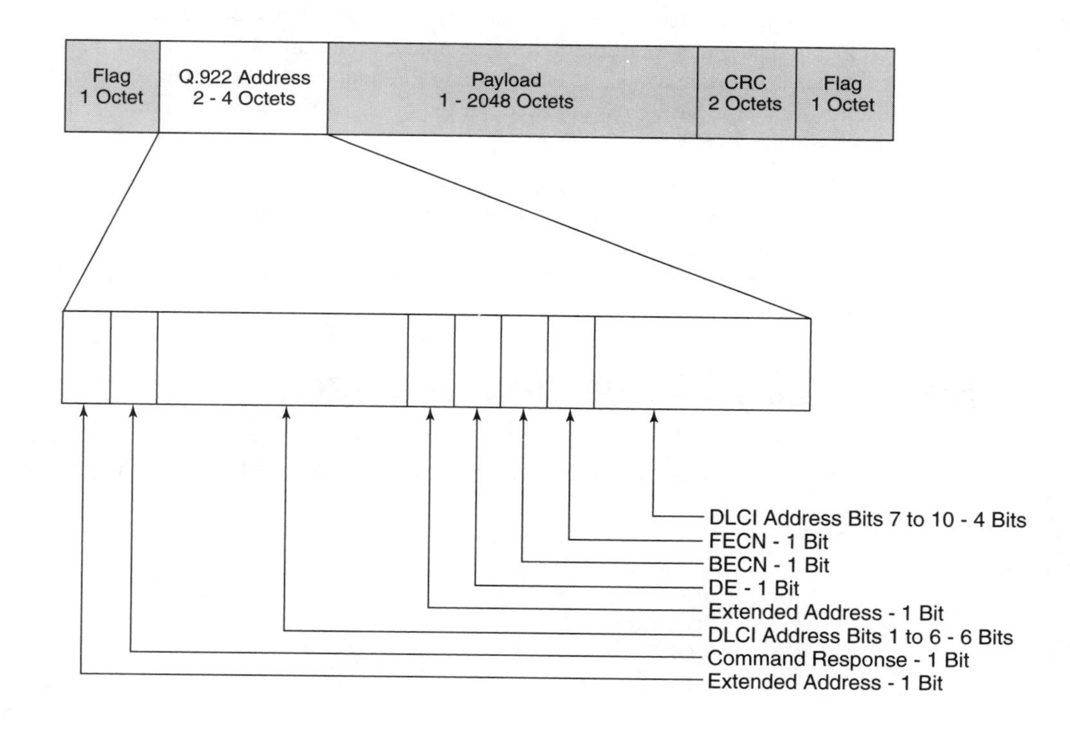

Figure 5.2: *The Frame Relay defined frame format.*

information transfer rate, or *committed rate*, which is referred to as the CIR of the virtual circuit. All traffic that is transmitted in excess of the CIR is marked as DE.

The first-hop Frame Relay switch (DCE) has the responsibility of enforcing the CIR at the ingress point of the Frame Relay network. When the information rate is exceeded, frames are marked as exceeding the CIR. This allows the network to subsequently enforce the committed rate at some point internal to the network. This is implemented using a rate filter on incoming frames. When the frame arrival rate at the DCE exceeds the CIR, the DCE marks the excess frames with the Discard Eligible bit set to 1 (DE = 1). The DE bit instructs the interior switches of the Frame Relay network to select those frames with the DE bit set as discard eligible in the event of switch congestion and discard these frames in preference of frames with their DE field set to 0 (DE = 0).

As long as the overall capacity design of the Frame Relay network is sufficiently robust to allow the network to meet the sustained requirements for all PVCs operating at their respective CIRs, bandwidth in the network may be consumed above the CIR rate up to the port speed of each network-attached DTE device. This mechanism of using the DE bit to discard frames as congestion is introduced into the Frame Felay network provides a method to accommodate traffic bursts while providing capacity protection for the Frame Relay network.

No signaling mechanism (to speak of) is available between the network DCE and the DTE to indicate that a DE marked frame has been discarded (Figure 5.3). This is an extremely important aspect of Frame Relay to understand. The job of recognizing that frames somehow have been discarded in the Frame Relay network is left to higher-layer protocols, such as TCP.

The architecture of this ingress rate *tagging* is a useful mechanism. The problem with Frame Relay, however, is that the marking of the DE bit is not well integrated with the higher-level protocols. Frames normally are selected for DE tagging by the DCE switch without any signaling from the higher-level application or protocol engine that resides in the DTE device.

Frame Relay Congestion Management

Frame Relay congestion control is handled in two ways: congestion avoidance and congestion recovery. *Congestion avoidance* consists of a Backward Explicit Congestion

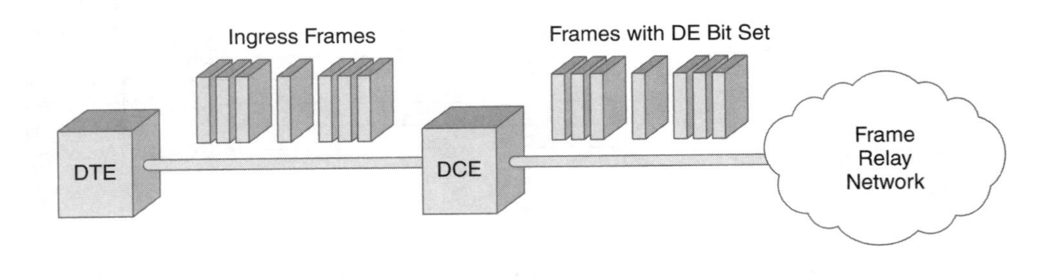

Figure 5.3: *Frames with the DE bit set.*

Notification (BECN) bit and a Forward Explicit Congestion Notification (FECN) bit. The BECN bit provides a mechanism for any switch in the Frame Relay network to notify the *originating* node (sender) of potential congestion when there is a build-up of queued traffic in the switch's queues. This informs the sender that the transmission of additional traffic (frames) should be restricted. The FECN bit notifies the *receiving* node of potential future delays, informing the receiver to use possible mechanisms available in a higher-layer protocol to alert the transmitting node to restrict the flow of frames.

These mechanisms traditionally are implemented within a Frame Relay switch so that it typically uses three queue thresholds for frames held in the switch queues, awaiting access to the transmission-scheduling resources.

When the frame queue exceeds the first threshold, the switch sets the FECN and BECN bits of all frames. Both bits are not set simultaneously—the precise action of whether the notification is *forward* or *backward* is admittedly somewhat arbitrary and appears to depend on whether the notification is generated at the egress from the network (FECN) or at the ingress (BECN). The intended result is to signal the sender or receiver on the UNI interface that there is congestion in the interior of the network. No specific action is defined for the sending or receiving node on receipt of this signal, although the objective is that the node recognizes that congestion may be introduced if the present traffic level is sustained and that some avoidance action may be necessary to reduce the level of transmitted traffic.

If the queue length continues to grow past the second threshold, the switch then discards all frames that have the Discard Eligible (DE) bit set. At this point, the switch is functionally enforcing the CIR levels on all VCs that pass through the switch in an effort to reduce queue depth. The intended effect is that the sending or receiving nodes recognize that traffic has been discarded and subsequently throttle traffic rates to operate within the specified CIR level, at least for some period before probing for the availability of burst capacity. The higher-level protocol is responsible for detecting lost frames and retransmitting them and also is responsible for using this discard information as a signal to reduce transmission rates to help the network back off from the congestion point.

The third threshold is the queue size itself, and when the frame queue reaches this threshold, all further frames are discarded (Figure 5.4).

Frame Relay and QoS

Frame Relay has gained much popularity in the data networking market because of the concept of allowing a maximum burst rate, which provides subscriber traffic to exceed the sustained (and presumably tariffed) committed rate when the network had excess capacity, up to the port speed of the local circuit loop. The concept of a free lunch always has been a powerful marketing incentive.

However, can Frame Relay congestion management mechanisms be used so that the end user can set IP QoS policies, which can in turn provide some direction to the congestion management behavior of the underlying Frame Relay network? Also, can the Frame Relay congestion signals (FECN and BECN) be used to trigger IP layer congestion management behavior?

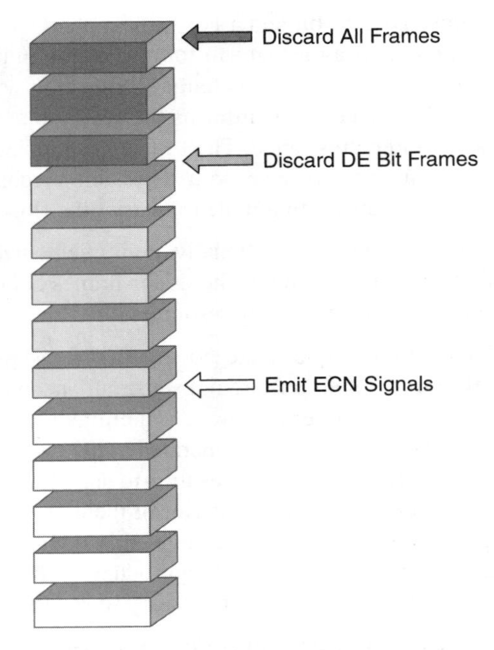

Figure 5.4: *Frame Relay switch queue thresholds.*

VC Congestion Signals

When considering such questions, the first observation to be made is that Frame Relay uses connection-based Victual Circuits (VCs). Frame Relay per VC congestion signaling flows along these fixed paths, whereas IP flows do not use fixed paths through the network. Given that Frame Relay signals take some time to propagate back through the network, there is always the risk that the end-to-end IP path may be dynamically altered before the signal reaches its destination, and that the consequent action may be completely inappropriate for handling the current situation. Countering this is the more pragmatic observation that the larger the network, the greater the pressure to dampen the dynamic nature of routing-induced, logical topology changes. The resulting probability of a topology change occurring within any single TCP end-to-end session then becomes very low indeed. Accordingly, it is relevant to consider whether any translation between IP QoS signaling and Frame Relay QoS signaling is feasible.

FECN and BECN Signals

The FECN and BECN signaling mechanisms are intended to notify the end-to-end transport protocol mechanisms about the likely onset of congestion and potential data loss. The use of selectively discarding frames using a DE flag is intended to allow an additional mechanism to reduce load *fairly,* as well as to provide a secondary signaling mechanism to indicate the possibility of more serious congestion. Certainly, it first appears that there are effective places where signals can be generated into the IP protocol stack; however, this is not the case. BECN and FECN signaling is not explicitly recognized by the protocols in the TCP/IP protocol suite, nor is the discarding of DE frames explicitly signaled into a TCP/IP protocol as a congestion indicator.

The BECN and FECN signals are analogous to the ICMP (Internet Control Message Protocol) source quench signal. They are intended to inform the transmitter's protocol stack that congestion is being experienced in the network and that some reduction of the transmission rate is advisable. However, this signal is not used in the implementation of IP over Frame Relay for good reason. As indicated in the IETF (Internet Engineering Task Force) document, "Requirements for IP Version 4 Routers," RFC1812 [IETF1995b], "Research seems to suggest that Source Quench consumes network bandwidth but is an ineffective (and unfair) antidote to congestion." In the case of BECN and FECN, no additional bandwidth is being consumed (the signal is a bit set in the Frame Relay header so that there is no additional traffic overhead), but the issue of effectiveness and fairness is relevant. Although these notifications can indeed be signaled back (or forward, as the case may be) to the CPE, where the transmission rate corresponding to the DLCI can be reduced, such an action must be done at the Frame Relay layer. To translate this back up the protocol stack to the IP layer, a subsequent reaction would be necessary for the Frame Relay interface equipment, upon receipt of a BECN or FECN signal, to set a condition that generates an ICMP source quench for all IP packets that correspond to such signaled frames. In this situation, the cautionary advice of RFC1812 is particularly relevant.

Discarding of Nonconformant Traffic

The discard of DE packets as the initial step in traffic-load reduction by the frame switches allows a relatively rapid feedback to the end-user TCP stack to reduce transmission rates. The discarded packet causes a TCP NAK to be delivered from the destination back to the sender upon reception of the subsequent packet, allowing the sender to receive a congestion-experienced signal within one Round Trip Time (RTT). Like RED schemes, the behavior of such random marking of packets leads to a higher probability of having DE marked packets, and the enforcement of the DE discarding leads to a likely trimming of the highest-rate TCP packets as the first measure in congestion management.

However, the most interesting observation about the interaction of Frame Relay and IP is one that indicates what is missing rather than what is provided. The DE bit is a powerful mechanism that allows the interior of the Frame Relay network to take rapid and predictable actions to reduce traffic load when under duress. The challenge is to relate this action to the environment of the end-user of the IP network.

Within the UNI specification, the Frame Relay specification allows the DTE (router) to set the DE bit in the frame header before passing it to the DCE (switch), and in fact, this is possible in a number of router vender implementations. This allows a network administrator to specify a simple binary priority. However, this rarely is done in a heterogeneous network, simply because it is somewhat self-defeating if no other subscriber undertakes the same action.

Effectiveness of Frame Relay Discard Eligibility

In the course of pursuing this line of thought, you can make the observation that in a heterogeneous network that uses a number of data-link technologies to support end-to-end data paths, the Frame Relay DE bit is not a panacea; it does not provide for end-to-end signaling, and the router is not necessarily the system that provides the end-to-end protocol stack. The router is more commonly performing IP packet into Frame-Relay encapsulation. With this in mind, a more functional approach to user selection of discard eligible traffic is possible—one that uses a field in the IP header to indicate a defined quality level, and allows this designation to be carried end-to-end across the entire network path. With this facility, it then is logical to allow the DTE IP router (which performs the encapsulation of an IP datagram into a Frame Relay frame) to set the DE bit according to the bit setting indicated in the IP header field, and then pass the frame to the DCE, which then can confirm or clear the DE bit in accordance with the procedures outlined earlier.

Without coherence between the data-link transport-signaling structures and the higher-level protocol stack, the result is somewhat more complex. Currently, the Frame Relay network works within a locally defined context of using selective frame discard as a means of enforcing rate limits on traffic as it enters the network. This is done as the primary response to congestion. The basis of this selection is undertaken without respect to any hints provided by the higher-layer protocols. The end-to-end TCP protocol uses packet loss as the primary signaling mechanism to indicate network congestion, but it is recognized only by the TCP session originator. The result is that when the network starts to reach a congestion state, the method in which end-system applications are degraded matches no particular imposed policy, and in this current environment, Frame Relay offers no great advantage over any other data-link transport technology in addressing this.

However, if the TOS field in the IP header were used to allow a change to the DE bit semantics in the UNI interface, it is apparent that Frame Relay would allow a more graceful response to network congestion by attempting to reduce load in accordance with upper-layer protocol policy directives. In other words, you can construct an IP over Frame Relay network that adheres to QoS policies if you can modify the standard frame relay mode of operation.

chapter 6

QoS and ATM

Asynchronous Transfer Mode (ATM) is undeniably the only technology that provides data-transport speeds in excess of OC-3 (155 Mbps) today. This is perhaps the predominant reason why ATM has enjoyed success in the Internet backbone environment. In addition to providing a high-speed bit-rate clock, ATM also provides a complex subset of traffic-management mechanisms, Virtual Circuit (VC) establishment controls, and Quality of Service (QoS) parameters. It is important to understand why these underlying mechanisms are not being exploited by a vast number of organizations that are using ATM as a data-transport mechanism for Internet networks in the wide area. The predominate use of ATM in today's Internet networks is simply because of the high data-clocking rate and multiplexing flexibility available with ATM implementations.

This chapter is not intended to be a detailed description of the inner workings of ATM networking, but a glimpse at the underlying mechanics is necessary to understand why ATM and certain types of QoS are inextricably related. The majority of information in this chapter is condensed from the ATM Forum Traffic Management Specification Version 4.0 [AF1996a] and the ATM Forum Private Network-Network Interface Specification Version 1.0 [AF1996c].

ATM Background

Historically, organizations have used TDM (Time Division Multiplexing) equipment to combine, or *mux*, different data and voice streams into a single physical circuit (Figure 6.1), and subsequently *de-mux* the streams on the receiving end, effectively breaking them out into their respective connections on the remote customer premise. Placing a mux on both ends of the physical circuit in this manner provided a means to an end; it was considered economically more attractive to mux multiple data streams together into a single physical circuit than it was to purchase different individual circuits for each application. This economic principle still holds true today.

There are, of course, some drawbacks to using the TDM approach, but the predominant liability is that once multiplexed, it is impossible to manage each individual data stream. This is sometimes an unacceptable paradigm in a service-provider environment, especially when management of data services is paramount to the service provider's economic livelihood. By the same token, it may not be possible to place a mux on both ends of a circuit because of the path a circuit may traverse. In the United States, for example, it is common that one end of a circuit may terminate in one RBOC's (Regional Bell Operating Company) network, whereas the other end may terminate in another RBOC's network on the other side of the country. Since the breakup of AT&T in the mid-1980s, the resulting RBOC's (Pacific Bell, Bell Atlantic, U.S. West, et al.) commonly have completely different policies, espouse different network philosophies and architectures, provide various services, and often deploy noninteroperable and diverse hardware platforms. Other countries have differing regulatory

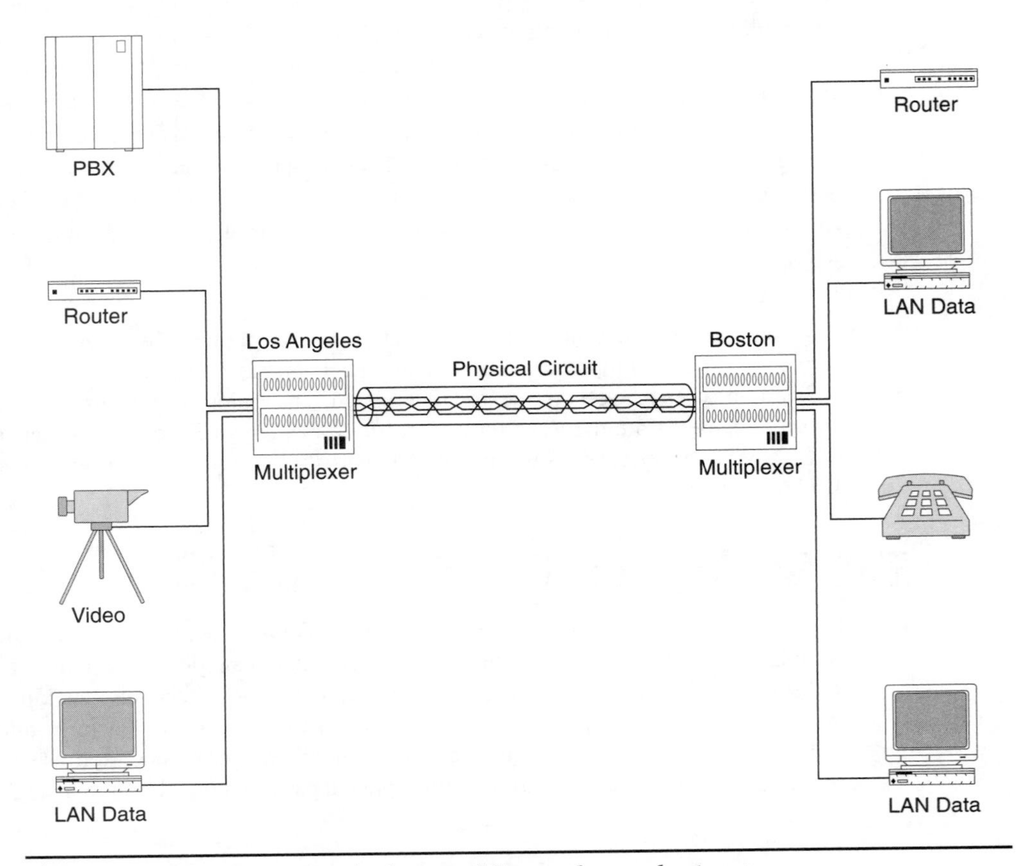

Figure 6.1: *Traditional multiplexing of diverse data and voice streams.*

restrictions on different classes of traffic, particularly where voice is concerned, and the multiplexing of certain classes of traffic may conflict with such regulations. Another drawback to this approach is that the multiplexors represent a single point of failure.

ATM Switching

The introduction of ATM in the early 1990s provided an alternative method to traditional multiplexing, in which the basic concept (similar to its predecessor, frame relay) is that multiple Virtual Channels (VCs) or Virtual Paths (VPs) now could be used for multiple data streams. Many VCs can be delivered on a single VP, and many VPs can be delivered on a single physical circuit (Figure 6.2). This is very attractive from an economic, as well as a management, perspective. The economic allure is obvious. Multiple discrete traffic paths can be configured and directed through a wide-area ATM switched network, while

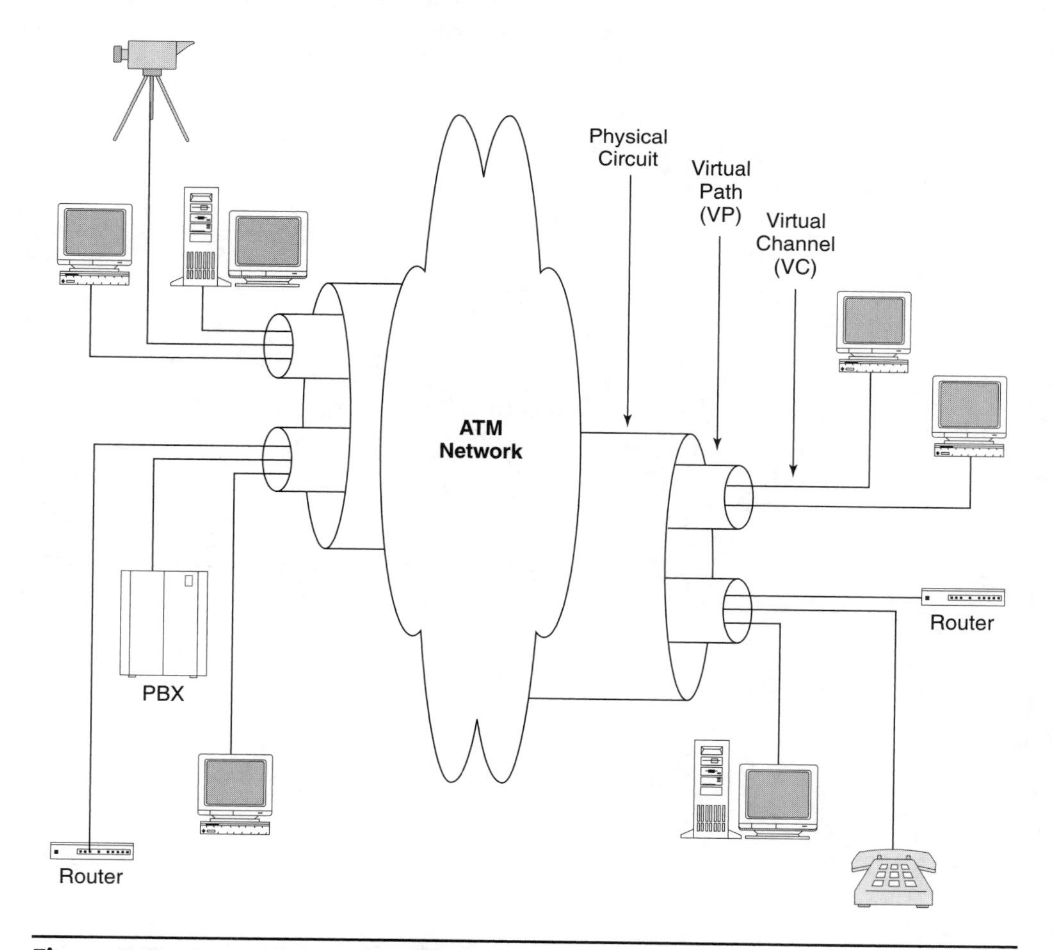

Figure 6.2: *ATM assuming the role of multiplexor.*

avoiding the monthly costs of several individual physical circuits. Traffic is switched end-to-end across the network—a network consisting of several ATM switches. Each VC or VP can be mapped to a specific path through the network, either statically by a network administrator or dynamically via a switch-to-switch routing protocol used to determine the best path from one end of the ATM network to the other, as simplified in Figure 6.3.

The primary purpose of ATM is to provide a high-speed, low-delay, low-jitter multiplexing and switching environment that can support virtually any type of traffic, such as voice, data, or video applications. ATM segments and multiplexes user data into 53-byte cells. Each cell is identified with a VC and a VP Identifier (VCI and VPI, respectively) which indicate how the cell is to be switched from its origin to its destination in the ATM switched network.

The ATM switching function is fairly straightforward. Each device in the ATM end-to-end path rewrites the VPI/VCI value, because it is only locally significant. That is, the VPI/VCI value is used only on a switch to indicate which local interface and/or VPI/VCI a cell is to be forwarded on. An ATM switch, or router, receives a cell on an incoming inter-

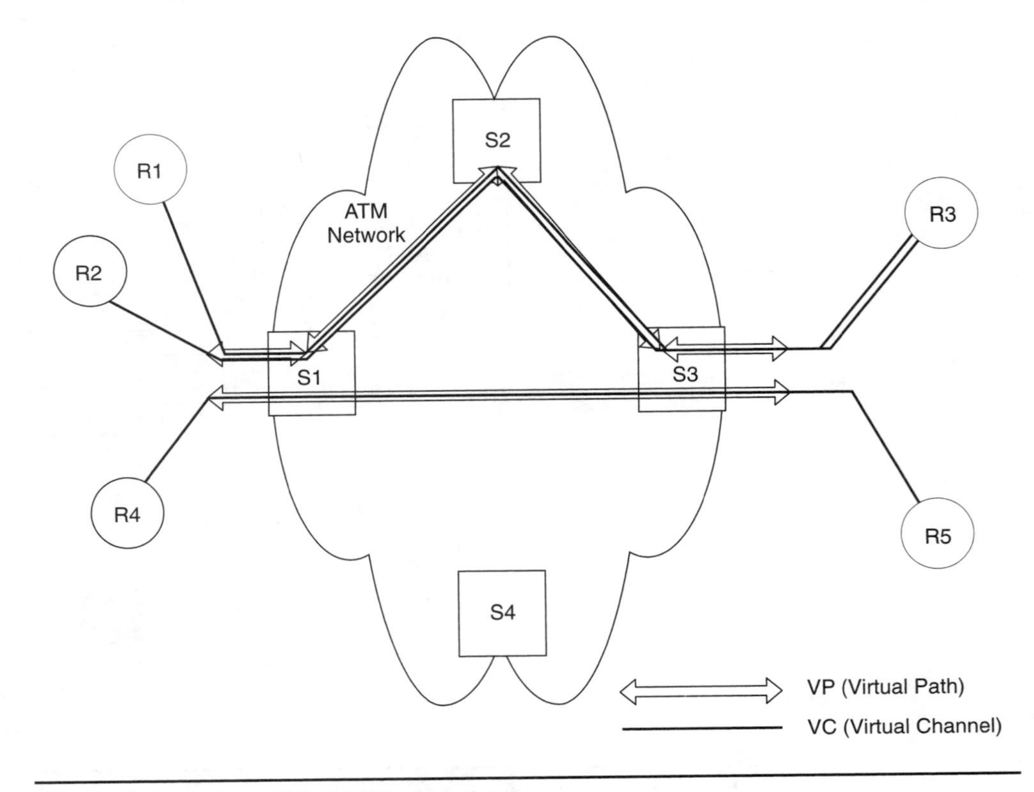

Figure 6.3: *A simplified ATM network.*

Foiled by Committee

Why 53 bytes per ATM cell? This is a good example of why technology choices should not be made by committee vote. The 53 bytes are composed of a 5-byte cell header and a 48-byte payload. The original committee work refining the ATM model resulted in two outcomes, with one group proposing 128 bytes of payload per cell and the other 16 bytes of payload per cell. Further negotiation within the committee brought these two camps closer, until the two proposals were for 64 and 32 bytes of payload per cell. The proponents of the smaller cell size argued that the smaller size reduced the level of nework-induced jitter and the level of signal loss associated with the drop of a single cell. The proponents of the large cell size argued that the large cell size permitted a higher data payload in relation to the cell header. Interestingly enough, both sides were proposing data payload sizes that were powers of 2, allowing for relatively straightforward memory mapping of data structures into cell payloads with associated efficiency of payload descriptor fields. The committee resolved this apparent impasse simply by taking the median value of 48 for the determined payload per cell. The committee compromise of 48 really suited neither side. It is considered too large for voice use and too small for data use; current measurements indicate that there is roughly a 20 percent efficiency overhead in using ATM as the transport substrate for an IP network. Sometimes, technology committees can concentrate too heavily on reaching a consensus and lose sight of their major responsibility to define rational technology.

face with a known VPI/VCI value, looks up the value in a local translation table to determine the outbound interface and the corresponding VPI/VCI value, rewrites the VPI/VCI value, and then switches the cell onto the outbound interface for retransmission with the appropriate connection identifiers.

Frame-based traffic, such as Ethernet-framed IP datagrams, is segmented into 53-byte cells by the ingress (entrance) router and transported along the ATM network until it reaches the egress (exit) router, where the frames are reassembled and forwarded to their destination. The Segmentation and Reassembly (SAR) process requires substantial computational resources, and in modern router implementations, this process is done primarily in silicon on specially designed Application Specific Integrated Circuit (ASIC) firmware.

ATM Connections

ATM networks essentially are connection oriented; a virtual circuit must be set up and established across the ATM network before any data can be transferred across it. There are two types of ATM connections: *Permanent Virtual Connections* (PVCs) and *Switched Virtual Connections* (SVCs). PVCs generally are configured statically by some external mechanism—usually, a network-management platform of some sort. PVCs are configured by a network administrator. Each incoming and outgoing VPI/VCI, on a switch-by-switch basis, must be configured for each end-to-end connection.

Obviously, when a large number of VCs must be configured, PVCs require quite a bit of administrative overhead. SVCs are set up automatically by a signaling protocol, or rather, the interaction of different signaling protocols. There are also *soft PVCs*; the end-points of the soft PVC—that is, the segment of the VC between the ingress or egress switch to the end-system or router—remain static. However, if a VC segment in the ATM network (between switches) becomes unavailable, experiences abnormal levels of cell loss, or becomes overly congested, an interswitch-routing protocol reroutes the VC within the confines of the ATM network. Thus, to the end-user, there is no noticeable change or availability on the status of the local PVC.

ATM Traffic-Management Functions

As mentioned several times earlier, certain architectural choices in any network design may impact the success of a network. The same principles ring just as true with regard to ATM as they would with other networking technologies. Being able to control traffic in the ATM network is crucial to ensuring the success of delivering differentiated QoS to the various applications that request and rely on the controls themselves. The primary responsibility of traffic-management mechanisms in the ATM network is to promote network efficiency and avoid congestion situations so that the overall performance of the network does not degenerate. It is also a critical design objective of ATM that the network utilization imposed by transporting one form of application data does not adversely impact the capability to efficiently transport other traffic in the network. It may be critically important, for example, that the transport of bursty traffic does not introduce an excessive amount of jitter into the transportation of constant bit rate, real-time traffic for video, or audio applications.

To deliver this stability, the ATM Forum has defined the following set of functions to be used independently or in conjunction with one another to provide for traffic management and control of network resources:

Connection Admission Control (CAC). Actions taken by the network during call setup to determine whether a connection request can be accepted or rejected.

Usage Parameter Control (UPC). Actions taken by the network to monitor and control traffic and to determine the validity of ATM connections and the associated traffic transmitted into the network. The primary purpose of UPC is to protect the network from traffic misbehavior that can adversely impact the QoS of already established connections. UPC detects violations of negotiated traffic parameters and takes appropriate actions—either tagging cells as CLP = 1 or discarding cells altogether.

Cell Loss Priority (CLP) control. If the network is configured to distinguish the indication of the CLP bit, the network may selectively discard cells with their CLP bit set to 1 in an effort to protect traffic with cells marked as a higher priority (CLP = 0). Different strategies for network resource allocation may be applied, depending on whether CLP = 0 or CLP = 1 for each traffic flow.

Traffic shaping. As discussed in Chapter 3, "QoS and Queuing Disciplines, Traffic Shaping, and Admission Control," ATM devices may control traffic load by implementing leaky-bucket traffic shaping to control the rate at which traffic is transmitted into the network. A standardized algorithm called GCRA (Generic Cell Rate Algorithm) is used to provide this function.

Network-resource management. Allows the logical separation of connections by Virtual Path (VP) according to their service criteria.

Frame discard. A congested network may discard traffic at the AAL (ATM Adaptation Layer) frame level, rather than at the cell level, in an effort to maximize discard efficiency.

ABR flow control. You can use the Available Bit Rate (ABR) flow-control protocol to adapt subscriber traffic rates in an effort to maximize the efficiency of available network resource utilization. You can find ABR flow-control details in the ATM Forum Traffic Management Specification 4.0 [AF1996a]. ABR flow control also provides a *crankback* mechanism to reroute traffic around a particular node when loss or congestion is introduced, or when the traffic contract is in danger of being violated as a result of a local CAC (connection admission control) determination. With the crankback mechanism, an intervening node signals back to the originating node that it no longer is viable for a particular connection and no longer can deliver the committed QoS.

Simplicity and Consistency

The major strengths of any networking technology are simplicity and consistency. Simplicity yeilds scaleable implementations that can readily interoperate. Consistency results in a set of capabilites that are complementary. The preceding list of ATM functions may look like a grab bag of fashionable tools for traffic management without much regard for simplicity or consistency across the set of functions. This is no accident. Again, as an outcome of the committee process, the ATM technology model is inclusive, without the evidence of operation of a filter of consistency. It is left as an exercise for the network operator to take a subset of these capabilities and create a stable set of network services.

ATM Admission Control and Policing

Each ingress ATM switch provides the functions of admission control and traffic policing. The admission-control function is called Connection Admission Control (CAC) and is the decision process an ingress switch undertakes when determining whether an SVC or PVC establishment request should be honored, negotiated, or rejected. Based on this CAC decision process, a connection request is entertained only when sufficient resources are available at each point within the end-to-end network path. The CAC decision is based on various parameters, including service category, traffic contract, and requested QoS parameters.

The ATM policing function is called Usage Parameter Control (UPC) and also is performed at the ingress ATM switch. Although connection monitoring at the public or private UNI is referred to as UPC and connection monitoring at a NNI (Network-to-Network Interface) can be called NPC (Network Parameter Control), UPC is the generic reference commonly used to describe either one. UPC is the activity of monitoring and controlling traffic in the network at the point of entry.

The primary objective of UPC is to protect the network from malicious, as well as unintentional, misbehavior that can adversely affect the QoS of other, already established connections in the network. The UPC function checks the validity of the VPI and/or VCI values and monitors the traffic entering the network to ensure that it conforms to its negotiated traffic contract. The UPC actions consist of allowing the cells to pass unmolested, tagging the cell with CLP = 1 (marking the cell as discard eligible), or discarding the cells altogether. No priority scheme to speak of is associated with ATM connection services. However, an explicit bit in the cell header indicates when a cell may be dropped—usually, in the face of switch congestion. This bit is called the CLP (Cell Loss Priority) bit. Setting the CLP bit to 1 indicates that the cell may be dropped in preference to cells with the CLP bit set to 0. Although this bit may be set by end-systems, it is set predominantly by the network in specific circumstances. This bit is advisory and not mandatory. Cells with the CLP set to 1 are not dropped when switch congestion is not present. Cells with CLP set to 0 may be dropped if there is switch congestion. The function of the CLP bit is a two-level prioritization of cells used to determine which cells to discard first in the event of switch congestion.

ATM Signaling and Routing

There are two basic types of ATM signaling: the User-to-Network Interface (UNI) and the Network-to-Network Interface (NNI), which sometimes is referred to as the Network-to-Node Interface. UNI signaling is used between ATM-connected end-systems, such as routers and ATM-attached workstations, as well as between separate, interconnected private ATM networks. A public UNI signaling is used between an end-system and a public ATM network or between different private ATM networks; a private UNI signaling is used between an end-system and a private ATM network. NNI signaling is used between ATM switches within the same administrative ATM switch network. A public NNI signaling protocol called B-ICI (BISDN or Broadband ISDN) Inter Carrier Interface, which is depicted in Figure 6.4 and described in [AF1995a], is used to communicate between public ATM networks.

The UNI signaling request is mapped by the ingress ATM switch into NNI signaling, and then is mapped from NNI signaling back to UNI signaling at the egress switch. An end-system UNI request, for example, may interact with an interswitch NNI signaling protocol, such as PNNI (Private Network-to-Network Interface).

PNNI is a dynamic signaling and routing protocol that is run within the ATM network between switches and sets up SVCs through the network. PNNI uses a complex algorithm to determine the best path through the ATM switch network and to provide rerouting services when a VC failure occurs. The generic PNNI reference is specified in [AF1996c].

The original specification of PNNI, Phase 0, also is called IISP (Interim Interswitch

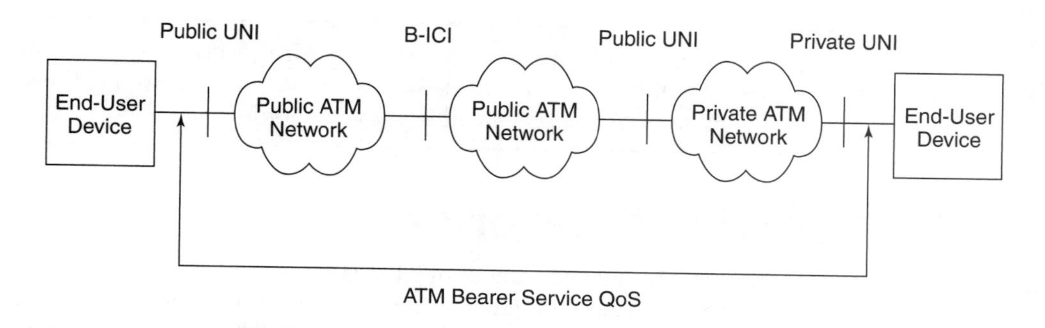

Figure 6.4: *ATM signaling reference [AF1996a].*

Signaling Protocol) . The name change is intended to avoid confusion between PNNI Phase 0 and PNNI Phase 1.

PNNI Phase 1 introduces support for QoS-based VC establishment (routing) and crankback mechanisms. This refinement to PNNI does not imply that IISP is restrictive in nature, because the dynamics of PNNI Phase 1 QoS-based routing actually are required only to support ATM VBR (Variable Bit Rate) services.

PNNI provides for highly complex VC path-selection services that calculate paths through the network based on the cost associated with each interswitch link. The costing can be configured by the network administrator to indicate preferred links in the switch topology. PNNI is similar to OSPF (Open Shortest Path First) in many regards—both are fast convergence link-state protocols. However, whereas PNNI is used only to route signaling requests across the ATM network and ultimately provide for VC establishment, OSPF is used at the network layer in the OSI (Open Systems Interconnection) reference model to calculate the best path for packet forwarding. PNNI does not forward packets, and PNNI does not forward cells; it simply provides routing and path information for VC establishment.

PNNI does provide an aspect of QoS within ATM, however. Unlike other link-state routing protocols, PNNI not only advertises the link metrics of the ATM network, it also advertises information about each node in the ATM network, including the internal state of each switch and the transit behavior of traffic between switches in the network. PNNI also performs *source routing* (also known as *explicit routing*), in which the ingress switch determines the entire path to the destination, as opposed to path calculation being done on a hop-by-hop basis. This behavior is one of the most attractive features of ATM dynamic VC establishment and path calculation—the capability to determine the state of the network, to determine the end-to-end path characteristics (such as congestion, latency, and jitter), and to build connections according to this state. With the various ATM service categories (listed in the following section), as well as the requested QoS parameters (e.g., cell delay, delay variance, and loss ratio), PNNI also provides an admission-control function (CAC). When a connection is requested, the ingress switch determines whether it can honor the request based on the traffic parameters included in the request. If it cannot, the connection request is rejected.

TIP You can find approved and pending ATM technical specifications, including the Traffic Management 4.0 and PNNI 1.0 Specifications, at the ATM Forum Web site, located at www.atmforum.com.

ATM Service Categories

One of the more unexploited features of ATM is the capability to request different service levels during SVC connection setup. Of course, a VC also can be provisioned as a PVC with the same traffic class. However, the dynamics of SVCs appear to be more appealing than the static configuration and provisioning of PVCs. Although PVCs must be provisioned manually, it is not uncommon to discover that SVCs also are manually provisioned, to a certain extent, in many instances. Although it certainly is possible to allow PNNI to determine the best end-to-end paths for VCs in the ATM network, it is common for network administrators to manually define administrative link parameters called *administrative weights* to enable PNNI to favor a particular link over another.

Currently, five ATM Forum-defined service categories exist (Table 6.1):

Constant Bit Rate (CBR)

Real-Time Variable Bit Rate (rt-VBR)

Non-Real-Time Variable Bit Rate (nrt-VBR)

Available Bit Rate (ABR)

Unspecified Bit Rate (UBR)

Table 6.1: *ATM Forum Traffic Services [AF1997]*

ATM Forum Traffic Management 4.0 ATM Service Category	*ITU-T I.371 ATM Transfer Capability*	*Typical Use*
Constant Bit Rate (CBR)	Deterministic Bit Rate (DBR)	Real-time, QoS guarantees
Real-Time Variable Bit Rate (rt-VBR)	(For further study)	Statistical mux, real time
Non-Real-Time Variable Bit Rate (nrt-VBR)	Statistical Bit Rate (SBR)	Statistical mux
Available Bit Rate (ABR)	Available Bit Rate (ABR)	Resource exploitation, feedback control
Unspecified Bit Rate (UBR)	(No equivalent)	Best effort, no guarantees
(No equivalent)	ATM Block Transfer (ABT)	Burst level feedback control

The basic differences among these service categories are described in the following sections.

Constant Bit Rate (CBR)

The CBR service category is used for connections that transport traffic at a consistent bit rate, where there is an inherent reliance on time synchronization between the traffic source and destination. CBR is tailored for any type of data for which the end-systems require predictable response time and a static amount of bandwidth continuously available for the lifetime of the connection. The amount of bandwidth is characterized by a Peak Cell Rate (PCR). These applications include services such as video conferencing; telephony (voice services); or any type of on-demand service, such as interactive voice and audio. For telephony and native voice applications, AAL1 (ATM Adaptation Layer 1) and CBR service is best suited to provide low-latency traffic with predictable delivery characteristics. In the same vein, the CBR service category typically is used for circuit emulation. For multimedia applications, such as video, you might want to choose the CBR service category for a compressed, frame-based, streaming video format over AAL5 for the same reasons.

Real-Time Variable Bit Rate (rt-VBR)

The rt-VBR service category is used for connections that transport traffic at variable rates —traffic that relies on accurate timing between the traffic source and destination. An example of traffic that requires this type of service category are variable rate, compressed video streams. Sources that use rt-VBR connections are expected to transmit at a rate that varies with time (e.g., traffic that can be considered bursty). Real-time VBR connections can be characterized by a Peak Cell Rate (PCR), Sustained Cell Rate (SCR), and Maximum Burst Size (MBS). Cells delayed beyond the value specified by the maximum CTD (Cell Transfer Delay) are assumed to be of significantly reduced value to the application.

Non-Real-Time Variable Bit Rate (nrt-VBR)

The nrt-VBR service category is used for connections that transport variable bit rate traffic for which there is no inherent reliance on time synchronization between the traffic source and destination, but there is a need for an attempt at a guaranteed bandwidth or latency. An application that might require an nrt-VBR service category is Frame Relay interworking, where the Frame Relay CIR (Committed Information Rate) is mapped to a bandwidth guarantee in the ATM network. No delay bounds are associated with nrt-VBR service.

You can use the VBR service categories for any class of applications that might benefit from sending data at variable rates to most efficiently use network resources. You could use Real-Time VBR (rt-VBR), for example, for multimedia applications with *lossy* properties—applications that can tolerate a small amount of cell loss without noticeably degrading the quality of the presentation. Some multimedia protocol formats may use a lossy compression scheme that provides these properties. You could use Non-Real-Time VBR (nrt-VBR), on the other hand, for transaction-oriented applications, such as interactive reservation systems, where traffic is sporadic and bursty.

Available Bit Rate (ABR)

The ABR service category is similar to nrt-VBR, because it also is used for connections that transport variable bit rate traffic for which there is no reliance on time synchronization between the traffic source and destination, and for which no required guarantees of bandwidth or latency exist. ABR provides a best-effort transport service, in which flow-control mechanisms are used to adjust the amount of bandwidth available to the traffic originator. The ABR service category is designed primarily for any type of traffic that is not time sensitive and expects no guarantees of service. ABR service generally is considered preferable for TCP/IP traffic, as well as other LAN-based protocols, that can modify its transmission behavior in response to the ABR's rate-control mechanics.

ABR uses Resource Management (RM) cells to provide feedback that controls the traffic source in response to fluctuations in available resources within the interior ATM network. The specification for ABR flow control uses these RM cells to control the flow of cell traffic on ABR connections. The ABR service expects the end-system to adapt its traffic rate in accordance with the feedback so that it may obtain its fair share of available network resources. The goal of ABR service is to provide fast access to available network resources at up to the specified Peak Cell Rate (PCR).

Unspecified Bit Rate (UBR)

The UBR service category also is similar to nrt-VBR, because it is used for connections that transport variable bit rate traffic for which there is no reliance on time synchronization between the traffic source and destination. However, unlike ABR, there are no flow-control mechanisms to dynamically adjust the amount of bandwidth available to the user. UBR generally is used for applications that are very tolerant of delay and cell loss. UBR has enjoyed success in the Internet LAN and WAN environments for store-and-forward traffic, such as file transfers and e-mail. Similar to the way in which upper-layer protocols react to ABR's traffic-control mechanisms, TCP/IP and other LAN-based traffic protocols can modify their transmission behavior in response to latency or cell loss in the ATM network.

These service categories provide a method to relate traffic characteristics and QoS requirements to network behavior. ATM network functions such as VC/VP path establishment, CAC, and bandwidth allocation are structured differently for each category. The service categories are characterized as being real-time or non-real-time. There are two real-time service categories: CBR and rt-VBR, each of which is distinguished by whether the traffic descriptor contains only the Peak Cell Rate (PCR) or both the PCR and the Sustained Cell Rate (SCR) parameters. The remaining three service categories are considered non-real-time services: nrt-VBR, UBR, and ABR. Each service class differs in its method of obtaining service guarantees provided by the network and relies on different mechanisms implemented in the end-systems and the higher-layer protocols to realize them. Selection of an appropriate service category is application specific.

ATM Traffic Parameters

Each ATM connection contains a set of parameters that describes the traffic characteristics of the source. These parameters are called *source traffic parameters*. Source traffic parameters, coupled with another parameter called the CDVT (Cell Delay Variation Tolerance) and a conformance-definition parameter, characterize the traffic properties of an ATM connection. Not all these traffic parameters are valid for each service category. When an end-system requests an ATM SVC (Switched Virtual Connection) to be set up, it indicates to the ingress ATM switch the type of service required, the traffic parameters of each data flow (in both directions), and the QoS parameters requested in each direction. These parameters form the *traffic descriptor* for the connection. You just examined the service categories; the traffic parameters consist of the following:

Peak Cell Rate (PCR). The maximum allowable rate at which cells can be transported along a connection in the ATM network. The PCR is the determining factor in how often cells are sent in relation to time in an effort to minimize jitter. PCR generally is coupled with the CDVT, which indicates how much jitter is allowable.

Sustainable Cell Rate (SCR). A calculation of the average allowable, long-term cell transfer rate on a specific connection.

Maximum Burst Size (MBS). The maximum allowable burst size of cells that can be transmitted contiguously on a particular connection.

Minimum Cell Rate (MCR). The minimum allowable rate at which cells can be transported along an ATM connection.

QoS parameters. Discussed in the following section.

ATM Topology Information and QoS Parameters

As mentioned earlier, ATM service parameters are negotiated between an end-system and the ingress ATM switch prior to connection establishment. This negotiation is accomplished through UNI signaling and, if negotiated successfully, is called a *traffic contract*. The traffic contract contains the traffic descriptor, which also includes a set of QoS parameters for each direction of the ATM connection. This signaling is done in one of two ways. In the case of SVCs, these parameters are signaled through UNI signaling and are acted on by the ingress switch. In the case of PVCs (Permanent Virtual Connections), the parameters are configured statically through a Network Management System (NMS) when the connections are established. SVCs are set up dynamically and torn down in response to signaling requests. PVCs are permanent or semipermanent, because after they are configured and set up, they are not torn down until manual intervention. (Of course, there also are *soft PVCs*, but for the sake of this overview, they are not discussed here.)

Two other important aspects of topology information carried around in PNNI routing updates are topology attributes and topology metrics. A *topology metric* is the cumulative information about each link in the end-to-end path of a connection. A *topology attribute* is the information about a single link. The PNNI path-selection process determines whether a link is acceptable or desirable for use in setting up a particular connection based on the topology attributes of a particular link or node. These parameters are where you begin to get into the mechanics of ATM QoS. The topology metrics follow:

Cell Delay Variation (CDV). An algorithmic determination for the variance in the cell delay, primarily intended to determine the amount of jitter. The CDV is a required metric for CBR and rt-VBR service categories. It is not applicable to nrt-VBR, ABR, and UBR service categories.

Maximum Cell Transfer Delay (maxCTD). A cumulative summary of the cell delay on a switch-by-switch basis along the transit path of a particular connection, measured in microseconds. The maxCTD is a required topology metric for CBR, rt-VBR, and nrt-VBR service categories. It is not applicable to UBR and ABR service.

Cell Loss Ratio (CLR). CLR is the ratio of the number of cells unsuccessfully transported across a link, or to a particular node, compared to the number of cells successfully transmitted. CLR is a required topology attribute for CBR, rt-VBR, and nrt-VBR service categories and is not applicable to ABR and UBR service categories. CLR is defined for a connection as

$$\text{CLR} = \frac{\text{Lost Cells}}{\text{Total Transmitted Cells}}$$

Administrative Weight (AW). The AW is a value set by the network administrator to indicate the relative preference of a link or node. The AW is a required topology metric for all service categories, and when one is not specified, a default value is assumed. A higher AW value assigned to a particular link or node is less preferable than one with a lower value.

The topology attributes consist of the following:

Maximum Cell Rate (maxCR). maxCR is the maximum capacity available to connections belonging to the specified service category. maxCR is a required topology attribute for ABR and UBR service categories and an optional attribute for CBR, rt-VBR, and nrt-VBR service categories. maxCR is measured in units of cells per second.

Available Cell Rate (AvCR). AvCR is a measure of effective available capacity for CBR, rt-VBR, and nrt-VBR service categories. For ABR service, AvCR is a measure of capacity available for Minimum Cell Rate (MCR) reservation.

Cell Rate Margin (CRM). CRM is the difference between effective bandwidth allocation and the allocation for Sustained Cell Rate (SCR) measured in units of cells per second. CRM is an indication of the safety margin allocated above the aggregate sustained cell rate. CRM is an optional topology attribute for rt-VBR and nrt-VBR service categories and is not applicable to CBR, ABR, and UBR service categories.

Variance Factor (VF). VF is a variance measurement calculated by obtaining the square of the cell rate normalized by the variance of the sum of the cell rates of all existing connections. VF is an optional topology attribute for rt-VBR and nrt-VBR service categories and is not applicable to CBR, ABR, and UBR service categories.

Figure 6.5 provides a chart illustrating the PNNI topology state parameters. Figure 6.6 shows a matrix of the various ATM service categories and how they correspond to their respective traffic and QoS parameters.

The ATM Forum's Traffic Management Specification 4.0 specifies six QoS service parameters that correspond to network-performance objectives. Three of these parameters may be negotiated between the end-system and the network, and one or more of these parameters may be offered on a per-connection basis.

PNNI Topology State Information		
Topology Metrics	Topology Attributes	
	Performance/Resource Related	Policy Related
Cell Delay Variation (CDV)	Cell Loss Ratio for CLP=0 (CLR_0)	Restricted Transit Flag
Maximum Cell Transfer Delay (maxCTD)	Cell Loss Ratio for CLP=0+1 (CLR_{0+1})	
Administrative Weight (AW)	Maximum Cell Rate (maxCR)	
	Available Cell Rate (AvCR)	
	Variance Factor (VF)	
	Restricted Branching Flag	

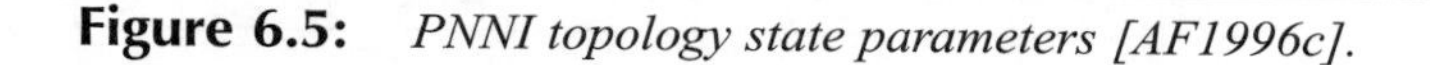

Figure 6.5: *PNNI topology state parameters [AF1996c].*

The following three negotiated QoS parameters were described earlier; they are repeated here because they also are topology metrics carried in the PNNI Topology State Packets (PTSPs). Two of these negotiated QoS parameters are considered *delay parameters* (CDV and maxCTD), and one is considered a *dependability parameter* (CLR):

- Peak-to-Peak Cell Delay Variation (Peak-to-Peak CDV)
- Maximum Cell Transfer Delay (maxCTD)
- Cell Loss Ratio (CLR)

The following three QoS parameters are not negotiated:

Cell Error Ratio (CER). Successfully transferred cells and errored cells contained in cell blocks counted as SECBR (Severely Errored Cell Block Rate) cells should be excluded in this calculation. The CER is defined for a connection as

$$\text{CER} = \frac{\text{Errored Cells}}{\text{Successfully Transferred Cells}} + \text{errored cells}$$

Severely Errored Cell Block Ratio (SECBR). A cell block is a sequence of consecutive cells transmitted on a particular connection. A severely errored cell block determination occurs when a specific threshold of errored, lost, or misinserted cells are observed. The SECBR is defined as

$$\text{SECBR} = \frac{\text{Severely Errored Cell Blocks}}{\text{Total Transmitted Cell Blocks}}$$

Attribute	ATM Layer Service Categories				
	CBR	rt-VBR	nrt-VBR	UBR	ABR
Traffic Parameters:	Performance/Resource Related			Policy Related	
PCR and CDVT [4,5]	specified			specified [2]	specified [3]
SCR, MBS, CDVT [4,5]	n/a	specified		n/a	
MCR [4]	n/a				specified
QoS Parameters:					
peak-to-peak CDV	specified		unspecified		
maxCTD	specified		unspecified		
CLR [4]	specified			unspecified	See Note 1
Other Attributes:					
Feedback	unspecified				specified [6]

Notes:

1. CLR is low for sources that adjust cell flow in response to control information. Whether a quantitative value for CLR is specified is network specific.
2. May not be subject to CAC and UPC procedures.
3. Represents the maximum rate at which the ABR source may send. The actual rate is subject to the control information.
4. These parameters are either explicitly or implicitly specified for PVCs or SVCs.
5. CDVT refers to the cell delay variation tolerance. CDVT is not signaled. In general, CDVT need not have a unique value for a connection.
   Different values may apply at each interface along the path of a connection.
6. See: [AF1996b]

Figure 6.6: *ATM Forum service category attributes [AF1996a].*

Cell Misinsertion Rate (CMR). The CMR most often is caused by an undetected error in the header of a cell being transmitted. This performance parameter is defined as a rate rather than a ratio, because the mechanism that produces misinserted cells is independent of the number of transmitted cells received. The SECBR should be excluded when calculating the CMR. The CMR can be defined as

$$\text{CMR} = \frac{\text{Misinserted Cells}}{\text{Time Interval}}$$

Table 6.2 lists the cell-transfer performance parameters and their corresponding QoS characterizations.

Table 6.2: *Performance Parameter QoS Characterizations [AF1996d]*

Cell-Transfer Performance Parameter	*QoS Characterization*
Cell Error Ratio (CER)	Accuracy
Severely Errored Cell Block Rate (SECBR)	Accuracy
Cell Loss Ratio (CLR)	Dependability
Cell Misinsertion Rate (CMR)	Accuracy
Cell Transfer Delay (CTD)	Speed
Cell Delay Variation (CDV)	Speed

ATM QoS Classes

There are two types of ATM QoS classes: one that explicitly specifies performance parameters (*specified QoS class*) and one for which no performance parameters are specified (*unspecified QoS class*). QoS classes are associated with a particular connection and specify a set of performance parameters and objective values for each performance parameter specified. Examples of performance parameters that could be specified in a given QoS class are CTD, CDV, and CLR.

An ATM network may support several QoS classes. At most, however, only one unspecified QoS class can be supported by the network. It also stands to reason that the performance provided by the network overall should meet or exceed the performance parameters requested by the ATM end-system. The ATM connection indicates the requested QoS by a particular class specification. For PVCs, the Network Management System (NMS) is used to indicate the QoS class across the UNI signaling. For SVCs, a signaling protocol's information elements are used to communicate the QoS class across the UNI to the network.

A correlation for QoS classes and ATM service categories results in a general set of service classes:

Service class A. Circuit emulation, constant bit rate video

Service class B. Variable bit rate audio and video

Service class C. Connection-oriented data transfer

Service class D. Connectionless data transfer

Currently, the following QoS classes are defined:

QoS class 1. Supports a QoS that meets service class A performance requirements. This should provide performance comparable to digital private lines.

QoS class 2. Supports a QoS that meets service class B performance requirements. Should provide performance acceptable for packetized video and audio in teleconferencing and multimedia applications.

QoS class 3. Supports a QoS that meets service class C performance requirements. Should provide acceptable performance for interoperability connection-oriented protocols, such as Frame Relay.

QoS class 4. Supports a QoS that meets service class D performance requirements. Should provide for interoperability of connectionless protocols, such as IP.

The primary difference between specified and unspecified QoS classes is that with an unspecified QoS class, no objective is specified for the performance parameters. However, the network may determine a set of internal QoS objectives for the performance parameters, resulting in an implicit QoS class being introduced. For example, a UBR connection may select best-effort capability, an unspecified QoS class, and only a traffic parameter for the PCR with a CLP = 1. This criteria then can be used to support data capable of adapting the traffic flow into the network based on time-variable resource fluctuation.

ATM and IP Multicast

Although it does not have a direct bearing on QoS issues, it is nonetheless important to touch on the basic elements that provide for the interaction of IP multicast and ATM. These concepts will surface again later and be more relevant when discussing the IETF Integrated Services architecture, the RSVP (Resource ReSerVation Protocol), and ATM.

Of course, there are no outstanding issues when IP multicast is run with ATM PVCs, because all ATM end-systems are static and generally available at all times. Multicast receivers are added to a particular multicast group as they normally would in any point-to-point or shared media environment.

The case of ATM SVCs is a bit more complex. There are basically two methods for using ATM SVCs for IP multicast traffic. The first is the establishment of an end-to-end VC for each sender-receiver pair in a the multicast group. This is fairly straightforward; however, depending on the number of nodes participating in a particular multicast group, this approach has obvious scaling issues associated with it. The second method uses ATM SVCs to provide an ingenious mechanism to handle IP multicast traffic by *point-to-multipoint* VCs. As multicast receivers are added to the multicast tree, new branches are added to the point-to-multipoint VC tree (Figure 6.7).

When PIM (Protocol Independent Multicast) [ID1996b, ID1996c, ID1996d] is used, certain vendor-specific implementations may provide for dynamic signaling between multicast end-systems and ATM UNI signaling to build point-to-multipoint VCs.

ATM-based IP hosts and routers may alternatively use a Multicast Address Resolution Server (MARS) [IETF1996b] to support RFC 1112 [IETF1989a] style Level 2 IP multicast over the ATM Forum's UNI 3.0/3.1 point-to-multipoint connection service. The MARS server is an extension of the ATM ARP (Address Resolution Protocol) server described in RFC1577 [IETF1994c], and for matters of practicality, the MARS functionality can be incorporated into the router to facilitate multicast-to-ATM host address resolution services. MARS messages support the distribution of multicast group membership information between the MARS server and multicast end-systems. End-systems query the MARS server when an IP address needs to be resolved to a set of ATM endpoints making up the multicast group, and end-systems inform MARS when they need to join or leave a multicast group.

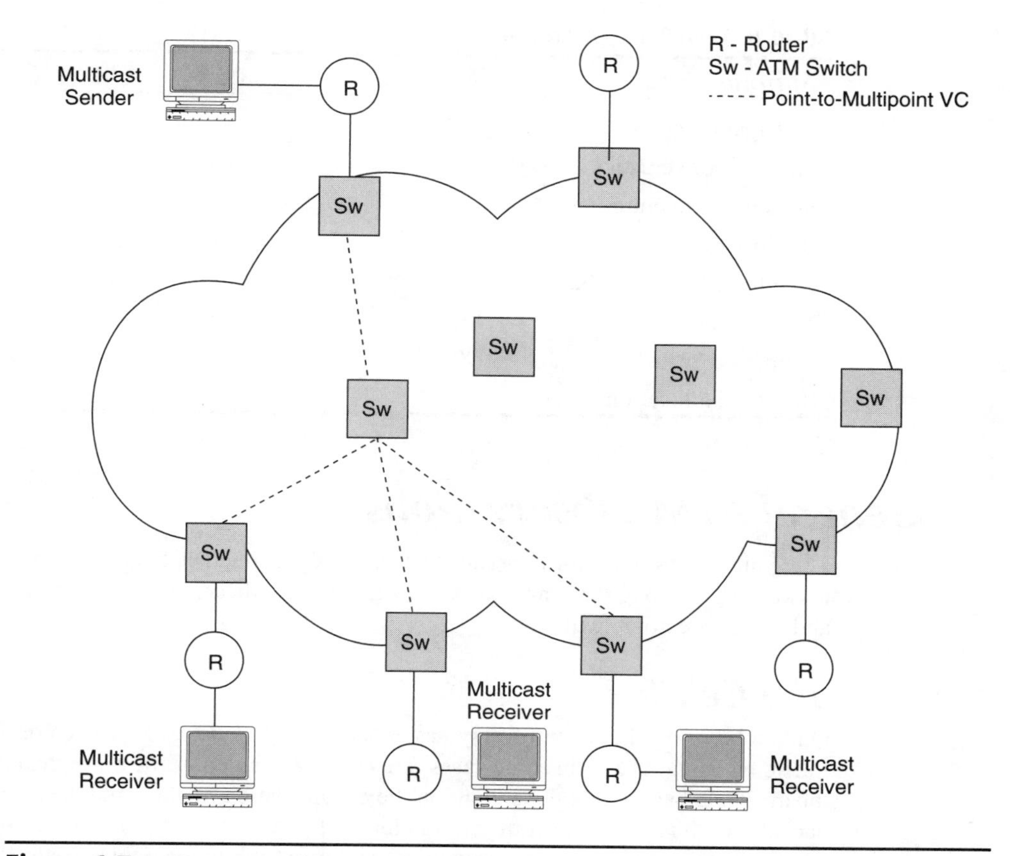

Figure 6.7: *Point-to-multipoint VCs.*

Factors That May Affect ATM QoS Parameters

It is important to consider factors that may have an impact on QoS parameters—factors that may be the result of undesirable characteristics of a public or private ATM network. As outlined in [AF1996d], there are several reasons why QoS might become degraded, and certain network events may adversely impact the network's capability to provide qualitative QoS. One of the principal reasons why QoS might become degraded is because of the ATM switch architecture itself. The ATM switching matrix design may be suboptimal, or the buffering strategy may be shared across multiple ports, as opposed to providing per-port or per-VC buffering. Buffering capacity therefore may be less than satisfactory, and as a result, congestion situations may be introduced into the network. Other sources of QoS degradation include media errors; excessive traffic load; excessive capacity reserved for a particular set of connections; and failures introduced by port, link, or switch loss. Table 6.3 lists the QoS parameters associated with particular degradation scenarios.

Table 6.3: *QoS Degradation*

Attribute	CER	SECBR	CLR	CMR	CTD	CDV
Progagation delay					X	
Media error statistics	X	X	X	X		
Switch architecture			X		X	X
Buffer capacity		X	X		X	X
Number of tandem nodes	X	X	X	X	X	X
Traffic load			X	X	X	X
Failures	X	X	X			
Resource allocation			X		X	X

General ATM Observations

There are several noteworthy issues that make ATM somewhat controversial as a method of networking. Among these are excessive signaling overhead, encapsulation inefficiencies, and inordinate complexity.

The Cell Tax

Quite a bit of needless controversy has arisen over the overhead imposed on frame-based traffic by using ATM—in some cases, the overhead can consume more than 20 percent of the available bandwidth, depending on the encapsulation method and the size of the packets. With regard to IP traffic in the Internet, the controversy has centered around the overhead imposed because of segmenting and reassembling variable-length IP packets in fixed-length ATM cells.

The last cell of an AAL5 frame, for example, will contain anywhere between 0 to 39 bytes of padding, which can be considered wasted bandwidth. Assuming that a broad range of packet sizes exists in the Internet, you could conclude that the average waste is about 20 bytes per packet. Based on an average packet size of 200 bytes, for example, the waste caused by cell padding is about 10 percent. However, because of the broad distribution of packet sizes, the actual overhead may vary substantially. Note that this 10 percent is in addition to the 10 percent overhead imposed by the 5-byte ATM cell headers (5 bytes subtracted from the cell size of 53 bytes is approximately a 10 percent overhead) and various other overhead (some of which also is present in frame-over-SONET, Synchronous Optical Network, schemes).

Suppose that you want to estimate the ATM bandwidth available on an OC3 circuit. With OC-3 SONET, 155.520 Mbps is reduced to 149.760 Mbps due to section, line, and SONET path overhead. Next, you reduce this figure by 10 percent, because an average of 20 bytes per 200-byte packet is lost (due to ATM cell padding), which results in 134.784 Mbps. Next, you can subtract 9.43 percent due to ATM headers of 5 bytes in 53-byte packets. Thus, you end up with a 122.069-Mbps available bandwidth figure, which is about 78.5 percent of the nominal OC-3 capacity. Of course, this figure may vary depending on the size of the packet data and the amount of padding that must be done to segment the packet into a 48-byte cell payload and fully populate the last cell with padding. Additional overhead is added for AAL5 (the most common ATM adaptation layer used to transmit data across ATM networks), framing (4 bytes length and 4 bytes CRC), and LLC (Link Layer Control) SNAP (SubNetwork Access Protocol) (8 bytes) encapsulation of frame-based traffic.

When you compare this scenario to the 7 bytes of overhead for traditional PPP (Point-to-Point Protocol) encapsulation, which traditionally is run on point-to-point circuits, you can see how this produces a philosophical schism between IP engineering purists and ATM proponents. Although conflicts in philosophies regarding engineering efficiency clearly exist, once you can get beyond the *cell tax*, as the ATM overhead is called, ATM does provide interesting traffic-management capabilities. This is why you can consider the philosophical arguments over the cell tax as fruitless: After you accept the fact that ATM does indeed consume a significant amount of overhead, you still can see that ATM provides more significant benefits than liabilities.

More Rope, Please

Developing network technologies sometimes is jokingly referred to as being in the business of selling rope: The more complex the technology, the more rope someone has to hang himself with. This also holds true in some cases in which the technology provides an avenue of convenience that, when taken, may yield further problems (more rope).

In this vein, you can see that virtual multiplexing technologies such as ATM and frame relay also provide the necessary tools that enable people to build sloppy networks—poor designs that attempt to create a flat network, in which all end-points are virtually one hop away from one another, regardless of how many physical devices are in the transit path. This design approach is not a reason for concern in ATM networks, in which an insignificantly small number of possible ATM end-points exists. However, in networks in which a substantially large number of end-points exists, this design approach presents a reason for serious concern over scaling the Layer 3 routing system. Many Layer 3 routing protocols require that routers maintain adjacencies or peering relationships with other routers to exchange routing and topology information. The more peers or adjacencies, the greater the computational resources consumed by each device. Therefore, in a flat network topology, which has no hierarchy, a much larger number of peers or adjacencies exists. Failure to introduce a hierarchy into a large network in an effort to promote scaling sometimes can be suicidal.

ATM QoS Observations

Several observations must be made to realize the value of ATM QoS and its associated complexity. This section attempts to provide an objective overview of the problems associated with relying solely on ATM to provide quality of service on a network. However, it sometimes is difficult to quantify the significance of some issues because of the complexity involved in the ATM QoS delivery mechanisms and their interactions with higher-layer protocols and applications. In fact, the inherent complexity of ATM and its associated QoS mechanisms may be a big reason why people are reluctant to implement those QoS mechanisms.

Many people feel that ATM is excessively complex and that when tested against the principle of Occam's Razor, ATM by itself would not be the choice for QoS services, simply because of the complexity involved compared with other technologies that provide similar results. However, the application of Occam's Razor does not provide assurances that the desired result will be delivered; instead, it simply expresses a preference for simplicity.

Occam's Razor

William of Occam (or Ockham) was a philosopher (presumed dates 1285–1349) who coined Occam's Razor, which states that "Entities are not to be multiplied beyond necessity." The familiar modern English version is "Make things as simple as possible—but no simpler." A popular translation frequently used in the engineering community for years is "All things being equal, choose the solution that is simpler." Or, it's vain to do with more what can be done with less.

ATM enthusiasts correctly point out that ATM is complex for good reason; to provide predictive, proactive, and real-time services, such as dynamic network resource allocation, resource guarantees, virtual circuit rerouting, and virtual circuit path establishment to accommodate subscriber QoS requests, ATM's complexity is unavoidable.

It also has been observed that higher-layer protocols, such as TCP/IP, provide the end-to-end transportation service in most cases, so that although it is possible to create QoS services in a lower layer of the protocol stack, namely ATM in this case, such services may cover only part of the end-to-end data path. This gets to the heart of the problem in delivering QoS with ATM, when the true end-to-end bearer service is not pervasive ATM. Such partial QoS measures often have their effects masked by the effects of the traffic distortion created from the remainder of the end-to-end path in which they do not reside, and hence the overall outcome of a partial QoS structure often is ineffectual.

In other words, if ATM is not pervasively deployed end-to-end in the data path, efforts to deliver QoS using ATM can be ineffectual. The traffic distortion is introduced into the ATM landscape by traffic-forwarding devices that service the ATM network and upper-layer protocols such as IP, TCP, and UDP, as well as other upper-layer network protocols. Queuing and buffering introduced into the network by routers and non-ATM-attached hosts skew the accuracy with which the lower-layer ATM services calculate delay and delay variation. Routers also may introduce needless congestion states, dependent on the quality of the hardware platform or the network design.

Differing Lines of Reasoning

An opposing line of reasoning suggests that end-stations simply could be ATM-attached. However, a realization of this suggestion introduces several new problems, such as the inability to aggregate downstream traffic flows and provide adequate bandwidth capacity in the ATM network. Efficient utilization of bandwidth resources in the ATM network continues to be a primary concern for network administrators, nonetheless. Yet another line of reasoning suggests that upper-layer protocols are unnecessary, because they tend to render

ATM QoS mechanisms ineffectual by introducing congestion bottlenecks and unwanted latency into the equation. The flaw in this line of reasoning is that native ATM applications do not exist for the majority of popular, commodity, off-the-shelf software applications, and even if they did, the capability to build a hierarchy and the separation of administrative domains into the network system is diminished severely. Scaleability such as exists in the global Internet is impossible with native ATM.

On a related note, some have suggested that most traffic on ATM networks would be primarily UBR or ABR connections, because higher-layer protocols and applications cannot request specific ATM QoS service classes and therefore cannot fully exploit the QoS capabilities of the VBR service categories. A cursory examination of deployed ATM networks and their associated traffic profiles reveals that this is indeed the case, except in the rare instance when an academic or research organization has developed its own native ATM-aware applications that can fully exploit the QoS parameters available to the rt-VBR and nrt-VBR service categories. Although this certainly is possible and has been done on many occasions, real-world experience reveals that this is the proverbial exception and not the rule.

It is interesting to note the observations published by Jagannath and Yin [ID1997a6], which suggest that "it is not sufficient to have a lossless ATM subnetwork from the end-to-end performance point of view." This observation is due to the fact that two distinct control loops exist: ABR and TCP (Figure 6.8). Although it generally is agreed that ABR can effectively control the congestion in the ATM network, ABR flow control simply pushes the congestion to the edges of the network (i.e., the router), where performance degradation or packet loss may occur as a result. Jagannath and Yin also point out that "one may argue that the reduction in buffer requirements in the ATM switch by using ABR flow control may be at the expense of an increase in buffer requirements at the edge device (e.g., ATM router interface, legacy LAN to ATM switches)." Because most applications use the flow control provided by TCP, you might question the benefit of using ABR flow control at the subnetwork layer, because UBR (albeit with Early Packet Discard) is equally effective and much

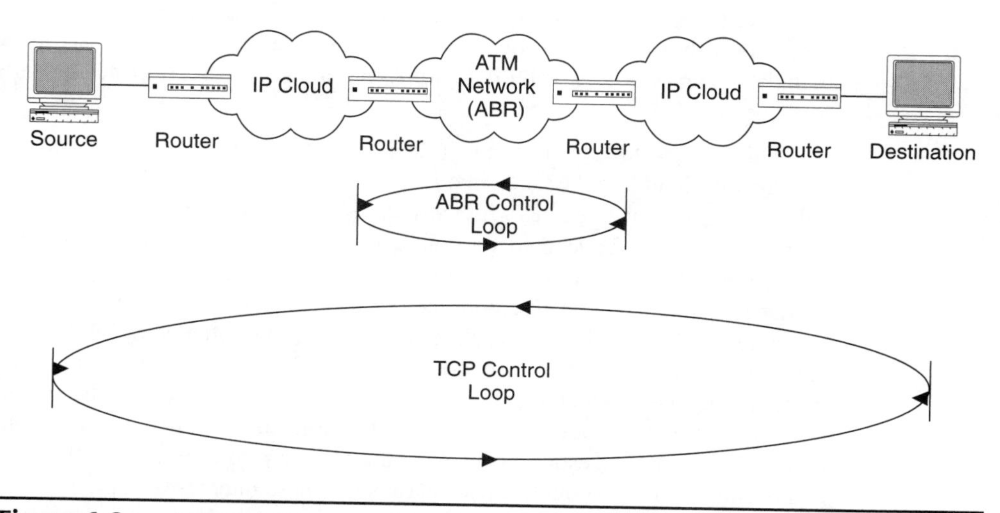

Figure 6.8: *ABR and TCP control loops.*

less complex. ABR flow control also may result in longer feedback delay for TCP control mechanisms, and this ultimately exacerbates the overall congestion problem in the network.

Aside from traditional data services that may use UBR, ABR, or VBR services, it is clear that circuit-emulation services that may be provisioned using the CBR service category clearly can provide the QoS necessary for telephony communications. However, this becomes an exercise in comparing apples and oranges. Delivering voice services on virtual digital circuits using circuit emulation is quite different from delivering packet-based data found in local area and wide-area networks.

By the same token, providing QoS in these two environments is substantially different; it is substantially more difficult to deliver QoS for data, because the higher-layer applications and protocols do not provide the necessary hooks to exploit the QoS mechanisms in the ATM network. As a result, an intervening router must make the QoS request on behalf of the application, and thus the ATM network really has no way of discerning what type of QoS the application may truly require. This particular deficiency has been the topic of recent research and development efforts to address this shortcoming and investigate methods of allowing the end-systems to request network resources using RSVP [IETF1997f], and then map these requests to native ATM QoS service classes as appropriate. You will revisit this issue in Chapter 7, "The Integrated Services Architecture."

Clarification and Understanding

Finally, you should understand that ATM QoS commitments are only probable estimates and are intended only to provide a first-order approximation of the performance the network expects to offer over the duration of the ATM connection. Because there is no limit to the duration of connections, and the ATM network can make decisions based only on the information available to it at the time the connection is established, the actual QoS may vary over the duration of the connection's lifetime. In particular, transient events (including uncontrollable failures in the transmission systems) can cause short-term performance to be worse than the negotiated QoS commitment. Therefore, the QoS commitments can be evaluated only over the long term and with other connections that have similar QoS commitments. The precision with which the various QoS values can be specified may be significantly greater than the accuracy with which the network can predict, measure, or deliver for a given performance level.

This leads to the conclusion that the gratuitous use of the term *guarantee* is quite misleading and should not be taken in a literal sense. Although ATM certainly is capable of delivering QoS when dealing with native cell-based traffic, the introduction of packet-based traffic (i.e., IP) and Layer 3 forwarding devices (routers) into this environment may have an adverse impact on the ATM network's capability to properly deliver QoS and certainly may produce unpredictable results. With the upper-layer protocols, there is no equivalent of a guarantee. In fact, packet loss is expected to occur to implicitly signal the traffic source that errors are present or that the network or the specified destination is not capable of accepting traffic at the rate at which it is being transmitted. When this occurs, these discrete mechanisms that operate at various substrates of the protocol's stack (e.g., ATM traffic parameter monitoring, TCP congestion avoidance, random early detection, ABR flow control) may well demonstrate self-defeating behavior because of these internetworking discrepancies—the inability for these different mechanisms to explicitly communicate with one another. ATM provides desirable properties with regard to increased speed of data-transfer rates, but in most cases, the underlying signaling and QoS mechanisms are viewed as excess baggage when the end-to-end bearer service is not ATM.

ATM and IP Design

Given these observations about QoS and ATM, the next question is, "What sort of QoS can be provided specifically by ATM in a heterogeneous network such as the Internet, which uses (in part) an ATM transport level?"

When considering this question, the basic differences in the design of ATM and IP become apparent. The prevailing fundamental design philosophy for the Internet is to offer coherent end-to-end data delivery services that are not reliant on any particular transport technology and indeed can function across a path that uses a diverse collection of transport technologies. To achieve this functionality, the basic TCP/IP signaling mechanism uses two very basic parameters for end-to-end characterization: a dynamic estimate of end-to-end Round Trip Time (RTT) and packet loss. If the network exhibits a behavior in which congestion occurs within a window of the RTT, the end-to-end signaling can accurately detect and adjust to the dynamic behavior of the network.

ATM, like many other data-link layer transport technologies, uses a far richer set of signaling mechanisms. The intention here is to support a wider set of data-transport applications, including a wide variety of real-time applications and traditional non-real-time applications. This richer signaling capability is available simply because of the homogenous nature of the ATM network, and the signaling capability can be used to support a wide variety of traffic-shaping profiles that are available in ATM switches. However, this richer signaling environment, together with the use of a profile adapted toward real-time traffic with very low jitter tolerance, can create a somewhat different congestion paradigm. For real-time traffic, the response to congestion is immediate load reduction, on the basis that queuing data can dramatically increase the jitter and lengthen the congestion event duration. The design objective in a real-time environment is the immediate and rapid discarding of cells to clear the congestion event. Given the assumption that integrity of real-time traffic is of critical economic value, data that requires integrity will use end-to-end signaling to detect and retransmit the lost data; hence, the longer recovery time for data transfer is not a significant economic factor to the service provider.

The result of this design objective is that congestion events in an ATM environment occur and are cleared (or at the very least, are attempted to be cleared) within time intervals that generally are well within a single end-to-end IP round-trip time. Therefore, when the ATM switch discards cells to clear local queue overflow, the resultant signaling of IP packet loss to the destination system (and the return signal of a NAK for missing a packet) takes a time interval of up to one RTT. By the time the TCP session reduces the transmit window in response to this signaling, the ATM congestion event is cleared. The resulting observation indicates that it is a design challenge to define the ATM traffic-shaping characteristics for IP-over-ATM traffic paths in order for end-to-end TCP sessions to sustain maximal data-transfer rates. This, in turn, impacts the overall expectation that ATM provide a promise of increased cost efficiency through multiplexing different traffic streams over a single switching environment; it is countered by the risks of poor payload delivery efficiency.

❖ ❖ ❖

The QoS objective for networks similar in nature to the Internet lies principally in directing the network to alter the switching behavior at the IP layer so that certain IP packets are delayed or discarded at the onset of congestion to delay (or completely avoid if at all possible) the impact of congestion on other classes of IP traffic. When looking at IP-over-ATM, the issue (as with IP-over-frame relay) is that there is no mechanism for mapping such IP-level directives to the ATM level, nor is it desirable, given the small size of ATM cells and the consequent requirement for rapid processing or discard. Attempting to increase the complexity of the ATM cell discard mechanics to the extent necessary to preserve the original IP QoS directives by mapping them into the ATM cell is counterproductive.

Thus, it appears that the default IP QoS approach is best suited to IP-over-ATM. It also stands to reason that if the ATM network is adequately dimensioned to handle burst loads without the requirement to undertake large-scale congestion avoidance at the ATM layer, there is no need for the IP layer to invoke congestion-management mechanisms. Thus, the discussion comes full circle to an issue of capacity engineering, and not necessarily one of QoS within ATM.

chapter 7

The Integrated Services Architecture

Anyone faced with the task of reviewing, understanding, and comparing approaches in providing Quality of Service might at first be overwhelmed by the complexities involved in the IETF (Internet Engineering Task Force) Integrated Services architecture, described in detail in [IETF1994b]. The Integrated Services architecture was designed to provide a set of extensions to the best-effort traffic delivery model currently in place in the Internet. The framework was designed to provide special handling for certain types of traffic and to provide a mechanism for applications to choose between multiple levels of delivery services for its traffic.

A Note about Internet Drafts

The IETF I-Ds (Internet Drafts) referenced in this book should be considered works in progress because of the ongoing work by their authors (and respective working group participants) to refine the specifications and semantics contained in them. The I-D document versions may change over the course of time, and some of these drafts may be advanced and subsequently published as Requests for Comments (RFCs) as they are finalized. Links to updated and current versions of these documents, proposals, and technical specifications mentioned in this chapter generally can be found on the IETF Web site, located at www.ietf.org, within the Integrated Services (IntServ), Integrated Services over Specific Link Layers (ISSLL), or Resource ReSerVation Setup Protocol (RSVP) working groups sections.

The Integrated Services architecture, in and of itself, is amazingly similar in concept to the technical mechanics espoused by ATM—namely, in an effort to provision "guaranteed" services, as well as differing levels of best effort via a "controlled-load" mechanism. In fact, the Integrated Services architecture and ATM are somewhat analogous; IntServ provides signaling for QoS parameters at Layer 3 in the OSI (Open Systems Interconnection) reference model, and ATM provides signaling for QoS parameters at Layer 2.

A Background on Integrated Services Framework

The concept of the Integrated Services framework begins with the suggestion that the basic underlying Internet architecture does not need to be modified to provide customized support for different applications. Instead, it suggests that a set of extensions can be developed that provide services beyond the traditional best-effort service. The IETF IntServ working group charter [INTSERVa] articulates that efforts within the working group are focused on three primary goals:

Clearly defining the services to be provided. The first task faced by this working group was to define and document this "new and improved" enhanced Internet service model.

Defining the application service, router scheduling, and link-layer interfaces or "subnets." The working group also must define at least three high-level interfaces: one that expresses the application's end-to-end requirements, one that defines what information is made available to individual routers within the network, and one that handles the additional expectations (if any) the enhanced service model has for specific link-layer technologies. The working group will define these abstract interfaces and coordinate with and advise other appropriate IP-over-subnet working groups efforts (such as IP-over-ATM).

Developing router validation requirements to ensure that the proper service is provided. The Internet will continue to contain a heterogeneous set of routers, run different routing protocols, and use different forwarding algorithms. The working group must seek to define a minimal set of additional router requirements that ensure that the Internet can support the new service model. Instead of presenting specific scheduling- and admission-control algorithms that must be supported, these requirements will likely take the form of behavioral tests that measure the capabilities of routers in the integrated services environment. This approach is used because no single algorithm seems likely to be appropriate in all circumstances at this time.

Contextual QoS Definitions

Before discussing the underlying mechanisms in the Integrated Services model, a review of a couple of definitions is appropriate.

Quality of Service, in the context of the Integrated Services framework, refers to the nature of the packet delivery service provided by the network, as characterized by parameters such as achieved bandwidth, packet delay, and packet loss rates [ITEF1997e]. A *network node* is any component of the network that handles data packets and is capable of imposing QoS control over data flowing through it. Nodes include routers, subnets (the underlying link-layer transport technologies), and end-systems. A *QoS-capable* or *IS-capable* node can be described as a network node that can provide one or more of the services defined in the Integrated Services model. A *QoS-aware* or *IS-aware* node is a network node that supports the specific interfaces required by the Integrated Services service definitions but cannot provide the requested service. Although a QoS-aware node may not be able to provide any of the QoS services themselves, it can simply understand the service request parameters and deny QoS service requests accordingly.

Service or *QoS control service* refers to a coordinated set of QoS control capabilities provided by a single node. The definition of a service includes a specification of the functions to be performed by the node, the information required by the node to perform these functions, and the information made available by a specific node to other nodes in the network

Components of the Integrated Services Model

The IETF Integrated Services model assumes that, first and foremost, resources (i.e., bandwidth) in the network must be controlled in order to deliver QoS. It is a fundamental building block of the Integrated Services architecture that traffic managed under this model must be subject to admission-control mechanisms. In addition to admission control, the IntServ model makes provisions for a resource-reservation mechanism, which is equally important in providing differentiated services. Integrated Services proponents argue that real-time applications, such as real-time video services, cannot be accommodated without resource guarantees, and that resource guarantees cannot be realized without resource reservations.

As stated in [IETF1994b], however, the term *guarantee* must be afforded loose interpretation in this context—guarantees may be *approximated* and *imprecise*. This is a disturbing mixing of words and is somewhat contradictory. The term *guarantee* really should never be used as an approximation. It implies an absolute state, and as such, should not be used to describe anything less. However, the Integrated Services model continues to define the guaranteed service level as a predictable service that a user can request from the network for the duration of a particular session.

The Integrated Services architecture consists of five key components: QoS requirements, resource-sharing requirements, allowances for packet dropping, provisions for usage feedback, and a resource reservation protocol (in this case, RSVP).

Each of these components is discussed in more detail in the following sections.

QoS Requirements

The Integrated Services model is concerned primarily with the *time-of-delivery* of traffic; therefore, per-packet delay is the central theme in determining QoS commitments.

Understanding the characterization of *real-time* and *non-real-time*, or elastic, applications and how they behave in the network is an important aspect of the Integrated Services model.

Real-Time and Non-Real-Time (Elastic) Traffic

The Integrated Services model predominantly focuses on real-time classes of application traffic. Real-time applications generally can be defined as those with *playback* characteristics—in other words, a data stream that is packetized at the source and transported through the network to its destination, where it is de-packetized and *played back* by the receiving application. As the data is transported along its way through the network, of course, latency is inevitably introduced at each point in the transit path.

The amount of latency introduced is variable, because latency is the cumulative sum of the transmission times and queuing hold times (where queuing hold-times can be highly variable). This variation in latency or jitter in the real-time signal is what must be *smoothed* by the playback. The receiver compensates for this jitter by buffering the received data for a period of time (an offset delay) before playing back the data stream, in an attempt to negate the effects of the jitter introduced by the network. The trick is in calculating the offset delay, because having an offset delay that is too short for the current level of jitter effectively renders the original real-time signal pretty much worthless.

The ideal scenario is to have a mechanism that can dynamically calculate and adjust the offset delay in response to fluctuations in the average jitter induced. An application that can adjust its offset delay is called an *adaptive* playback application. The predominate trait of a real-time application is that it does not wait for the late arrival of packets when playing back the data signal at the receiver; it simply imposes an offset delay prior to processing.

Tolerant and Intolerant Real-Time Applications

The classification of real-time applications is further broken down into two subcategories: those that are tolerant and those that are intolerant of induced jitter. Tolerant applications can be characterized as those that can function in the face of nominal induced jitter and still produce a reasonable signal quality when played back. Examples of such tolerant real-time applications are various packetized audio- or video-streaming applications. Intolerant applications can be characterized as those in which induced jitter and packet delay result in enough introduced distortion to effectively render the playback signal quality unacceptable or to render the application nonfunctional. Examples of intolerant applications are two-way telephony applications (interactive voice, as opposed to noninteractive audio playback) and circuit-emulation services.

For tolerant applications, the Integrated Services model recommends the use of a *predictive* service, otherwise known as a *controlled-load* service. For intolerant applications, IntServ recommends a *guaranteed* service model. The fundamental difference in these two models is that one provides a reliable upper bound on delay (guaranteed) and the other (controlled-load) provides a less-than-reliable delay bound.

Non-real-time, or elastic, applications differ from real-time applications in that elastic applications always wait for packets to arrive before the application actually processes the data. Several types of elastic applications exist: interactive burst (e.g., Telnet), interactive bulk transfer (e.g., File Transfer Protocol, or FTP), and asynchronous bulk transfer (e.g., Simple Mail Transport Protocol, or SMTP). The delay sensitivity varies dramatically with each of the types, so they can be referred to as belonging to a *best-effort* service class.

Real-Time versus Elastic Applications

The fundamental difference in the two data-streaming models lies in flow-control and error-handling mechanisms. In terms of flow control, real-time traffic is intended to be self-paced, in that the sending data rate is determined by the characteristics of the sending applications. The task of the network is to attempt to impose minimal distortion on this sender-pacing of the data so that flow control is nonadaptive and is based on some condition imposed by the sender. Elastic applications typically use an adaptive flow-control algorithm in which the data flow rate is based on the sender's view of available network capacity and available space at the reciever's end, instead of on some external condition. With respect to error handling, real-time traffic has a timeliness factor associated with it, because if the packet cannot be delivered within a certain elapsed time frame and within the sender-sequencing format, it is irrelevant and can be discarded. Packets that are in error, are delivered out of sequence, or arrive outside of the delivery time frame, must be discarded. Elastic applications do not have a timeliness factor, and accordingly, there is no time by which the packet is irrelevant. The relevance and importance of delivery of the packet depend on other factors within such applications, and in general, elastic applications use error-detection and retransmission-recovery mechanisms at the application layer to recover cleanly from data-transmission errors (such as the TCP protocol).

The effect of real-time traffic flows is to lock the sender and receiver into a common clocking regime, where the only major difference is the offset of the receiver's clock, as determined by the network-propagtion delay. This state must be imposed on the network as distinct from the sender adapting the rate clocking to the current state of the network.

Integrated Services Control and Characterization

The Integrated Services architecture uses a set of general and service-specific control parameters to characterize the QoS service requested by the end-system's application. The *general* parameters are ones that appear in all QoS service-control services. *Service-specific* parameters are ones used only with the controlled-load or guaranteed service classes—for example, a node that may need to occasionally export a service-specific value that differs from the default. In the case of the guaranteed service class, if a node is restricted by some sort of platform- or processing-related overhead, it may have to export a service-specific value (such as smaller maximum packet size) that is different from the default value, so that applications using the guaranteed service will function properly. These parameters are used to characterize the QoS capabilities of nodes in the path of a packet flow and are described in detail in [IETF1997g]. A brief overview of these control parameters is presented here so that you can begin to understand how the Integrated Services model fits together.

NON_IS_HOP

The NON_IS_HOP parameter provides information about the presence of nodes that do not implement QoS control services along the data path. In this vein, the *IS* portion of this and other parameters also means *Integrated Services–aware*, in that an IS-aware element is one that conforms to the requirements specified in the Integrated Services architecture. A flag is set in this object if a node does not implement the relevant QoS control service or knows that there is a break in the traffic path of nodes that implement the service. This also is called a *break bit*, because it represents a break in the chain of network elements required to provide an end-to-end traffic path for the specified QoS service class.

NUMBER_OF_IS_HOPS

The NUMBER_OF_IS_HOPS parameter is represented by a counter that is a cumulative total incremented by 1 at each IS-aware hop. This parameter is used to inform the flow end-points of the number of IS-aware nodes which lie in the data path. Valid values for this parameter range from 1 to 255, and in practice, is limited by the bound on the IP hop count.

AVAILABLE_PATH_BANDWIDTH

The AVAILABLE_PATH_BANDWIDTH parameter provides information about the available bandwidth along the path followed by a data flow. This is a local parameter and provides an estimate of the bandwidth nodes available for traffic following the path. Values for this parameter are measured in bytes per second and range in value from 1 byte per second to 40 terabytes per second (which is believed to be the theoretical maximum bandwidth of a single strand of fiber).

MINIMUM_PATH_LATENCY

The MINIMUM_PATH_LATENCY local parameter is a representation of the latency in the forwarding process associated with the node, where the latency is defined to be the smallest possible packet delay added by the node itself. This delay results from speed-of-light propagation delay, packet-processing limitations, or both. It does not include any variable queuing delay that may be introduced. The purpose of this parameter is to provide a baseline minimum path latency figure to be used with services that provide estimates or bounds on additional path delay, such as the guaranteed service class. Together with the queuing delay bound offered by the guaranteed service class, this parameter gives the application a priori knowledge of both the minimum and maximum packet-delivery delay. Knowing both minimum and maximum latencies experienced by traffic allows the receiving application to attempt to accurately compute buffer requirements to remove network-induced jitter.

PATH_MTU

The PATH_MTU parameter is a representation of the Maximum Transmission Unit (MTU) for packets traversing the data path, measured in bytes. This parameter informs the end-point of the packet MTU size that can traverse the data path without being fragmented. A correct and valid value for this parameter must be specified by all IS-aware nodes. This value is required to invoke QoS control services that require the IP packet size to be strictly limited to a specific MTU. Existing MTU discovery mechanisms cannot be used, because they provide information only to the sender, and they do not directly allow for QoS control services to specify MTUs smaller than the physical MTU. The local parameter is the IP MTU, where the MTU of the node is defined as the maximum size the node can transmit without fragmentation, including upper-layer and IP headers but excluding link-layer headers.

TOKEN_BUCKET_TSPEC

The TOKEN_BUCKET_TSPEC parameter describes traffic parameters using a simple token-bucket filter and is used by data senders to characterize the traffic it expects to generate. This parameter also is used by QoS control services to describe the parameters of traffic for which the subsequent reservation should apply. This parameter takes the form of a token-bucket specification plus a peak rate, a minimum policed unit, and a maximum packet size. The token-bucket specification itself includes an average token rate and a bucket depth.

The token rate (*r*) is measured in bytes of IP datagrams per second and may range in value from 1 byte per second to 40 terabytes per second. The token-bucket depth (*b*) is measured in bytes and values range from 1 byte to 250 gigabytes. The peak traffic rate (*p*) is measured in bytes of IP datagrams per second and may range in value from 1 byte per second to 40 terabytes per second.

The minimum policed unit (*m*) is an integer measured in bytes. The purpose of this parameter is to allow a reasonable estimate of the per-packet resources needed to process a flow's packets; the maximum packet rate can be computed from the values expressed in *b* and *m*. The size includes the application data and all associated protocol headers at or above the IP layer. It does not include the link-layer headers, because these may change in size as a packet traverses different portions of a network. All datagrams less than size *m* are treated as being of size *m* for the purposes of resource allocation and policing.

The maximum packet size (*M*) is the largest packet that will conform to the traffic specification, also measured in bytes. Packets transmitted that are larger than *M* may not receive QoS-controlled service, because they are considered to be nonconformant with the traffic specification.

The range of values that can be specified in these parameters is intentionally designed large enough to allow for future network technologies—a node is not expected to support the full range of values.

The Controlled Load Service Class

The Integrated Services definition for *controlled-load service* [IETF1997h] attempts to provide end-to-end traffic behavior that closely approximates traditional best-effort services within the environmental parameters of unloaded or lightly utilized network conditions. In other words, *better than best-effort* delivery. That is, it attempts to provide traffic delivery within the same bounds as an unloaded network in the same situation. Assuming that the network is functioning correctly (i.e., routing and forwarding), applications may assume that a very high percentage of transmitted packets will be delivered successfully and that any latency introduced into the network will not greatly exceed the minimum delay experienced by any successfully transmitted packet.

To ensure that this set of conditions is met, the application requesting the controlled-load service provides the network with an estimation of the traffic it will generate—the *TSpec* or *traffic specification*. The controlled-load service uses the TOKEN_BUCKET_TSPEC to describe a data flow's traffic parameters and therefore is synonymous with the term *TSpec* referenced hereafter. In turn, each node handling the controlled-load service request ensures that sufficient resources are available to accommodate the request. The amount of accuracy with which the *TSpec* matches available resources in the network does not have to be precise. If the requested resources fall outside the bounds of what is available, the traffic originator may experience a negligible amount of induced delay or possibly dropped packets because of congestion situations. However, the degree at which traffic may be dropped or delayed should be slight enough for the adaptive real-time applications to function without noticeable degradation.

The controlled-load service does not accept or use specific values for control parameters that include information about delay or loss. Acceptance of a controlled-load request implies a commitment to provide a better-than-best-effort service that approximates network behavior under nominal network-utilization conditions.

The method a node uses to determine whether adequate resources are available to accommodate a service request is purely a local matter and may be implementation dependent; only the control parameters and message formats are required to be interoperable.

Links on which the controlled-load service is run are not allowed to fragment packets. Packets larger that the MTU of the link must be treated as nonconformant with the *TSpec*.

The controlled-load service is provided to a flow when traffic conforms to the *TSpec* given at the time of flow setup. When nonconformant packets are presented with a controlled-load flow, the node must ensure that three things happen. First, the node must ensure that it continues to provide the contracted QoS to those controlled-load flows that are conformant. Second, the node should prevent nonconformant traffic in a controlled-load flow from unfairly impacting other conformant controlled-load flows. Third, the node must attempt to forward nonconformant traffic on a best-effort basis if sufficient resources are available. Nodes should not assume that nonconformant traffic is indicative of an error, because large numbers of packets may be nonconformant as a matter of course. This nonconformancy occurs because some downstream nodes may not police extended bursts of traffic to conform with the specified *TSpec* and in fact will borrow available bandwidth resources to clear traffic bursts that have queued up. If a flow obtains its exact fixed-token rate in the presence of an extended burst, for example, there is a danger that the queue will fill up to the point of packet discard. To prevent this situation, the controlled-load node may allow the flow to exceed its token rate in an effort to reduce the queue buildup. Thus, nodes should be prepared to accommodate bursts larger than the advertised *TSpec*.

The Guaranteed QoS Service Class

The guaranteed service class [IETF1997i] provides a framework for delivering traffic for applications with a bandwidth guarantee and delay bound—applications that possess intolerant real-time properties. The guaranteed service only computes the queuing delay in the end-to-end traffic path; the fixed delay in the traffic path is introduced by factors other than queuing, such as speed-of-light propagation and setup mechanisms used to negotiate end-to-end traffic parameters. The guaranteed service framework mathematically asserts that the queuing delay is a function of two factors—primarily, the token-bucket depth (b) and the data rate (r) the application requests. Because the application controls these values, it has an *a priori* knowledge of the queuing delay provided by the guaranteed service.

The guaranteed service guarantees that packets will arrive within a certain delivery time and will not be discarded because of queue overflows, provided that the flow's traffic stays within the bounds of its specified traffic parameters. The guaranteed service does not control the minimal or average delay of traffic, and it doesn't control or minimize *jitter* (the variance between the minimal and maximal delay)—it only controls the maximum queuing delay.

The guarantee service is invoked by a sender specifying the flow's traffic parameters (the *TSpec*) and the receiver subsequently requesting a desired service level (the *RSpec*). The guaranteed service also uses the TOKEN_BUCKET_TSPEC parameter as the *TSpec*. The *RSpec* (reservation specification) consists of a data rate (R) and a *slack term* (S), where R must be greater than or equal to the token-bucket data rate (r). The rate (R) is measured in bytes of IP datagrams per second and has a value range of between 1 byte per second to 40 terabytes per second. The slack term (S) is measured in microseconds. The *RSpec* rate can be larger than the *TSpec* rate, because higher rates are assumed to reduce queuing delay. The slack term represents the difference between the desired delay and the delay obtained by using a reservation level of R. The slack term also can be used by the network to reduce its resource reservation for the flow.

Because of the end-to-end and hop-by-hop calculation of two error terms (C and D), every node in the data path must implement the guaranteed service for this service class to function.. The first error term (C) provides a cumulative representation of the delay a packet might experience because of rate parameters of a flow, also referred to as *packet serialization*. The error term (C) is measured in bytes. The second error term (D) is a rate-independent, per-element representation of delay imposed by time spent waiting for transmission through a node. The error term (D) is measured in units of 1 microsecond. The cumulative end-to-end calculation of these error terms (C_{tot} and D_{tot}) represent a flow's deviation from the *fluid model*. The *fluid model* states that service flows within the available total service model can operate independently of each other.

As with the controlled-load service, links on which the guaranteed service is run are not allowed to fragment packets. Packets larger than the MTU of the link must be treated as nonconformant with the *TSpec*.

Two types of traffic policing are associated with the guaranteed service: simple policing and reshaping. Policing is done at the edges of the network, and reshaping is done at intermediate nodes within the network. *Simple policing* is comparing traffic in a flow against the

TSpec for conformance. Reshaping consists of an attempt to restore the flow's traffic characteristics to conform to the *TSpec*. Reshaping mechanics delay the forwarding of datagrams until they are in conformance of the *TSpec*. As described in [IETF1997i], reshaping is done by combining a token bucket with a peak-rate regulator and buffering a flow's traffic until it can be forwarded in conformance with the token-bucket (r) and peak-rate (p) parameters. Such reshaping may be necessary because of small levels of distortion introduced by the packet-level of use of any transmission path. This packet-level *quantification* of flows is what is addressed by reshaping. In general, reshaping adds a small amount to the total delay, but it can reduce the overall jitter of the flow.

The mathematical computation and supporting calculations for implementing the guaranteed service mechanisms are documented in [IETF1997i], and thus mercifully are not duplicated here.

Traffic Control

The Integrated Services model defines four mechanisms that comprise the traffic-control functions at Layer 3 (the router) and above:

Packet scheduler. The scheduling mechanism may be represented as some exotic, non-FIFO queuing implementation and may be implementation specific. The packet scheduler also assumes the function of traffic policing, according to the Integrated Services model, because it must determine whether a particular flow can be admitted entry to the network.

Packet classifier. This maps each incoming packet to a specific class so that these classes may be acted on individually to deliver traffic differentiation.

Admission control. Determining whether a flow can be granted the requested QoS without affecting other established flows in the network.

Resource reservation. A resource reservation protocol (in this case, RSVP) is necessary to set up flow state in the requesting end-systems as well as each router along the end-to-end flow-transit path.

Figure 7.1 shows a reference model illustrating the relationship of these functions.

Resource-Sharing Requirements

It is important to understand that the allocation of network resources is accomplished on a flow-by-flow basis, and that although each flow is subject to admission-control criteria, many flows share the available resources on the network, which is described as *link sharing*. With link sharing, the aggregate bandwidth in the network is shared by various types of traffic. These types of traffic generally can be different network protocols (e.g., IP, IPX, SNA), different services within the same protocol suite (e.g., Telenet, FTP, SMTP), or simply different traffic flows that are segregated and classified by sender. It is important that different traffic types do not unfairly utilize more than their fair share of network resources, because that could result in a disruption of other traffic. The Integrated Services

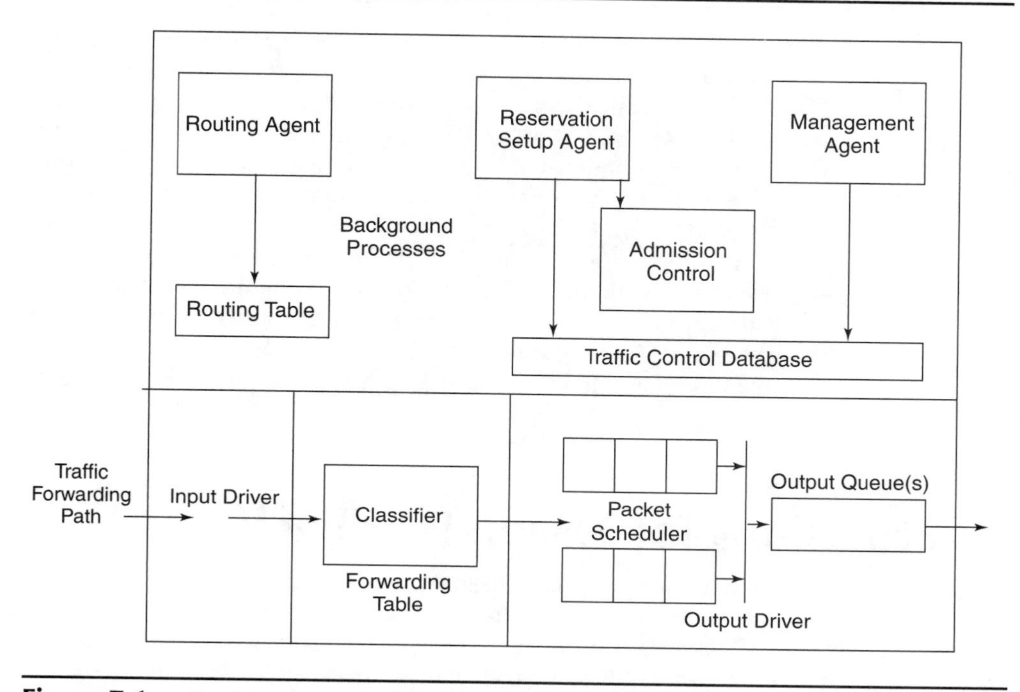

Figure 7.1: *Implementation reference model [IETF1994b].*

model also focuses on link sharing by aggregate flows and link sharing with an additional admission-control function—a fair queuing (i.e., Weighted Fair Queuing or WFQ) mechanism that provides proportional allocation of network resources.

Packet-Dropping Allowances

The Integrated Services model outlines different scenarios in which traffic control is implicitly provided by dropping packets. One concept is that some packets within a given flow may be *preemptable* or subject to drop. This concept is based on situations in which the network is in danger of reneging on established service commitments. A router simply could discard traffic by acting on a particular packet's *preemptability option* to avoid disrupting established commitments. Another approach classifies packets that are not subject to admission-control mechanisms.

Several other interesting approaches could be used, but naming each is beyond the scope of this book. Just remember that it is necessary to drop packets in some cases to control traffic in the network. Also, [IETF1997i] suggests that some guaranteed service implementers may want to use preemptive packet dropping as a substitute for traffic reshaping—if the result produces the same effect as reshaping at an intermediate node—by using a combined token bucket and peak-rate regulator to buffer traffic until it conforms to the *TSpec*. A preliminary proposal that provides guidelines for replacement services was published in [ID1997e].

TIP Packet dropping can be compared to the Random Early Detection (RED) mechanism for TCP (Transmission Control Protocol), as described in Chapter 4, "QoS and TCP/IP: Finding the Common Denominator." A common philosophy is that it is better to reduce congestion in a controlled fashion before the onset of resoure saturation until the congestion event is cleared, instead of waiting until all resources are fully consumed, resulting in complete discard. This leads to the observation that controlling the quality of a service often is the task of controlling the way in which a service degrades in the face of congestion. The point here it that it is more stable to degrade incrementally instead of waiting until the buffer resources are exhausted and they degrade to complete exhaustion in a single catastrophic collapse.

Provisions for Usage Feedback

Although it is commonly recognized that usage feedback (or accounting data, as it is more commonly called) is necessary to prevent abuses of the network resources, the IETF Integrated Services drafts do not go into a great amount of detail on this topic. In fact, [IETF1994b] says little about it at all, other than that usage feedback seems to be a highly contentious issue and occasionally an inflammatory one at that. The reason for this probably is that the primary need for accounting data is for subscriber billing, which is a matter of institutional business models and a local policy issue. Although technical documentation eventually will emerge that produces mechanisms that provide accounting data that could be used for these purposes, the technical mechanisms themselves deserve examination—not the policy that drives the use of such mechanisms.

RSVP: The Resource Reservation Model

As described earlier, the Integrated Services architecture provides a framework for applications to choose between multiple controlled levels of delivery services for their traffic flows. Two basic requirements exist to support this framework. The first requirement is for the nodes in the traffic path to support the QoS control mechanisms defined earlier—the controlled-load and guaranteed services. The second requirement is for a mechanism by which the applications can communicate their QoS requirements to the nodes along the transit path, as well as for the network nodes to communicate between one another the QoS requirements that must be provided for the particular traffic flows. This could be provided in a number of ways, but as fate would have it, it is provided by a resource reservation setup protocol called RSVP [IETF1997f].

The information presented here is intended to provide an synopsis of the internal mechanics of the RSVP protocol and is not intended to be a complete detailed description of the internal semantics, object formats, or syntactical construct of the protocol fields. You can find a more in-depth description in the RSVP version 1 protocol specification [IETF1997f].

As detailed in [IETF1997j], there is a logical separation between the Integrated Services QoS control services and RSVP. RSVP is designed to be used with a variety of QoS services, and the QoS control services are designed to be used with a variety of setup mechanisms. RSVP does not define the internal format of the protocol objects related to characterizing QoS control services; it treats these objects as opaque. In other words, RSVP is simply the signaling mechanism, and the QoS control information is the signal content. RSVP is analogous to other IP control protocols, such as ICMP (Internet Control Message Protocol), or one of the many IP routing protocols. RSVP is not a routing protocol, however, but is designed to interoperate with existing unicast and multicast IP routing protocols. RSVP uses the local routing table in routers to determine routes to the appropriate destinations. For multicast, a host sends IGMP (Internet Group Management Protocol) messages to join a multicast group and then sends RSVP messages to reserve resources along the delivery path(s) of that group.

In general terms, RSVP is used to provide QoS requests to all router nodes along the transit path of the traffic flows and to maintain the state necessary in the router required to actually provide the requested services. RSVP requests generally result in resources being reserved in each router in the transit path for each flow. RSVP establishes and maintains a *soft state* in nodes along the transit path of a reservation data path. A *hard state* is what other technologies provide when setting up virtual circuits for the duration of a data-transfer session; the connection is torn down after the transfer is completed. A *soft state* is maintained by periodic refresh messages sent along the data path to maintain the reservation and path state. In the absence of these periodic messages, which typically are sent every 30 seconds, the state is deleted as it times out. This soft state is necessary, because RSVP is essentially a QoS reservation protocol and does not associate the reservation with a specific static path through the network. As such, it is entirely possible that the path will change, so that the reservation state must be refreshed periodically.

RSVP also provides dynamic QoS; the resources requested may be changed at any given time for a number of reasons:

- An RSVP receiver may modify its requested QoS parameters at any time.
- An RSVP sender may modify its traffic-characterization parameters, defined by its *Sender TSpec* and cause the receiver to modify its reservation request.
- A new sender can start sending to a multicast group with a larger traffic specification than existing senders, thereby causing larger reservations to be requested by the appropriate receivers.
- A new receiver in a multicast group may make a reservation request that is larger than existing reservations.

The last two reasons are related inextricably to how reservations are merged in a multicast tree. Figure 7.2 shows a simplified version of RSVP in hosts and routers.

Method of Operation

RSVP requires the *receiver* to be responsible for requesting specific QoS services instead of the sender. This is an intentional design in the RSVP protocol that attempts to provide

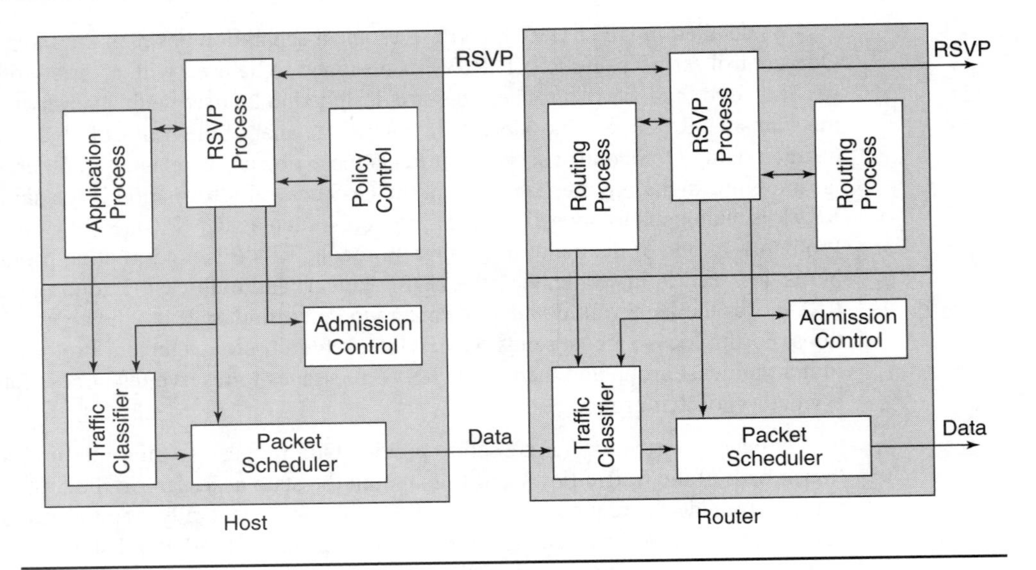

Figure 7.2: *RSVP in hosts and routers [IETF1997f].*

for efficient accommodation of large groups (generally, for multicast traffic), dynamic group membership (also for multicast), and diverse receiver requirements.

An RSVP sender sends *Path* messages *downstream* toward an RSVP receiver destination (Figure 7.3). Path messages are used to store path information in each node in the traffic path. Each node maintains this path state characterization for the specified sender's flow, as indicated in the sender's *TSpec* parameter. After receiving the Path message, the RSVP receiver sends *Resv* (reservation request) messages back *upstream* to the sender (Figure 7.4) along the same hop-by-hop traffic path the Path messages traversed when traveling toward the receiver. Because the receiver is responsible for requesting the desired QoS services, the *Resv* messages specify the desired QoS and set up the reservation state in each node in the traffic path. After successfully receiving the *Resv* message, the sender begins sending its data. If RSVP is being used in conjunction with multicast, the receiver first joins the appropriate multicast group, using IGMP, prior to the initiation of this process.

RSVP requests only unidirectional resources—resource reservation requests are made in one direction only. Although an application can act as a sender and receiver at the same time, RSVP treats the sender and receiver as logically distinct functions.

RSVP Reservation Styles

A Reservation (*Resv*) request contains a set of options which are collectively called the *reservation style*, which characterizes how reservations should be treated in relation to the sender(s). These reservation styles are particularly relevant in a multicast environment.

One option concerns the treatment of reservations for different senders within the same RSVP session. The option has two modes: establish a *distinct* reservation for each upstream sender or establish a *shared* reservation used for all packets of specified senders.

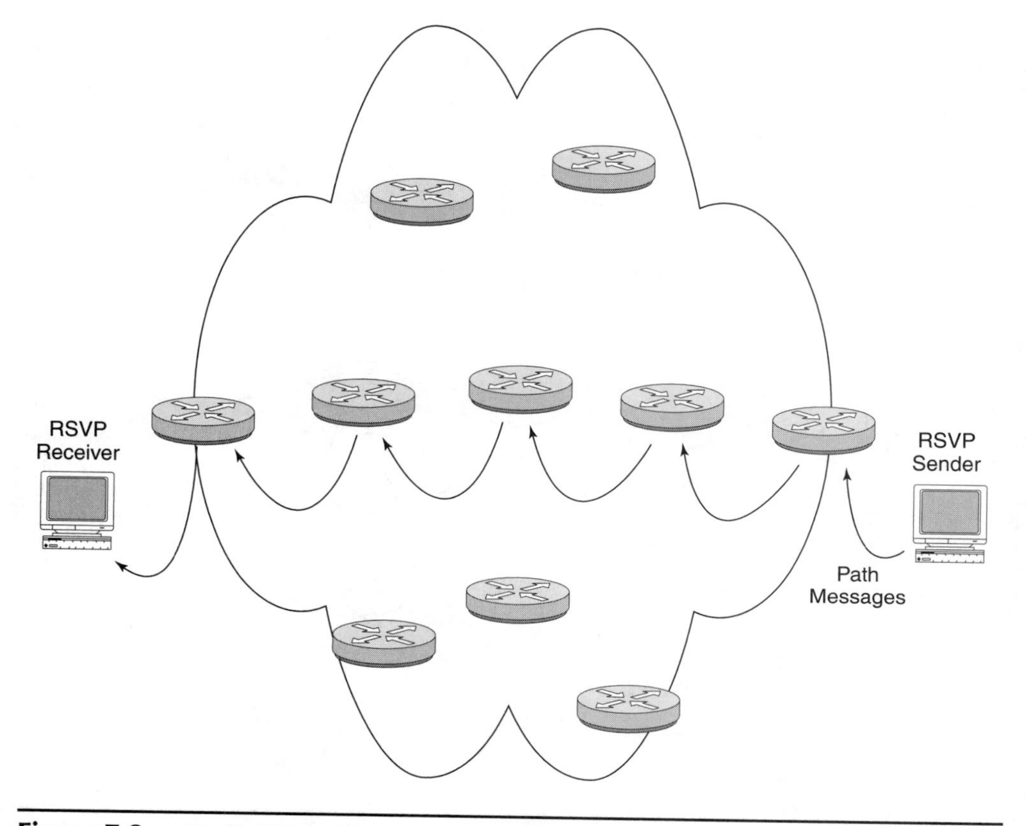

Figure 7.3: *Traffic flow of the RSVP Path message.*

Another option controls the selection of the senders. This option also has two modes: an *explicit* list of all selected senders or a *wildcard* specification that implicitly selects all senders for the session. In an explicit sender-selection reservation, each *Filter Spec* must match exactly one sender. In a wildcard sender selection, no *Filter Spec* is needed.

As depicted in Figure 7.5 and outlined earlier, the *Wildcard-Filter* (WF) style implies a *shared* reservation and a *wildcard* sender selection. A WF style reservation request creates a single reservation shared by all flows from all upstream senders. The *Fixed-Filter* (FF) style implies *distinct* reservations with *explicit* sender selection. The FF style reservation request creates a distinct reservation for a traffic flow from a specific sender, and the reservation is not shared with another sender's traffic for the same RSVP session. A *Shared-Explicit* (SE) style implies a *shared* reservation with *explicit* sender selection. An SE style reservation request creates a single reservation shared by selected upstream senders.

The RSVP specification does not allow the merging of shared and distinct style reservations, because these modes are incompatible. The specification also does not allow merging of explicit and wildcard style sender selection, because this most likely would produce unpredictable results for a receiver that may specify an explicit style sender selection. As a result, the WF, FF, and SE styles are all incompatible with one another.

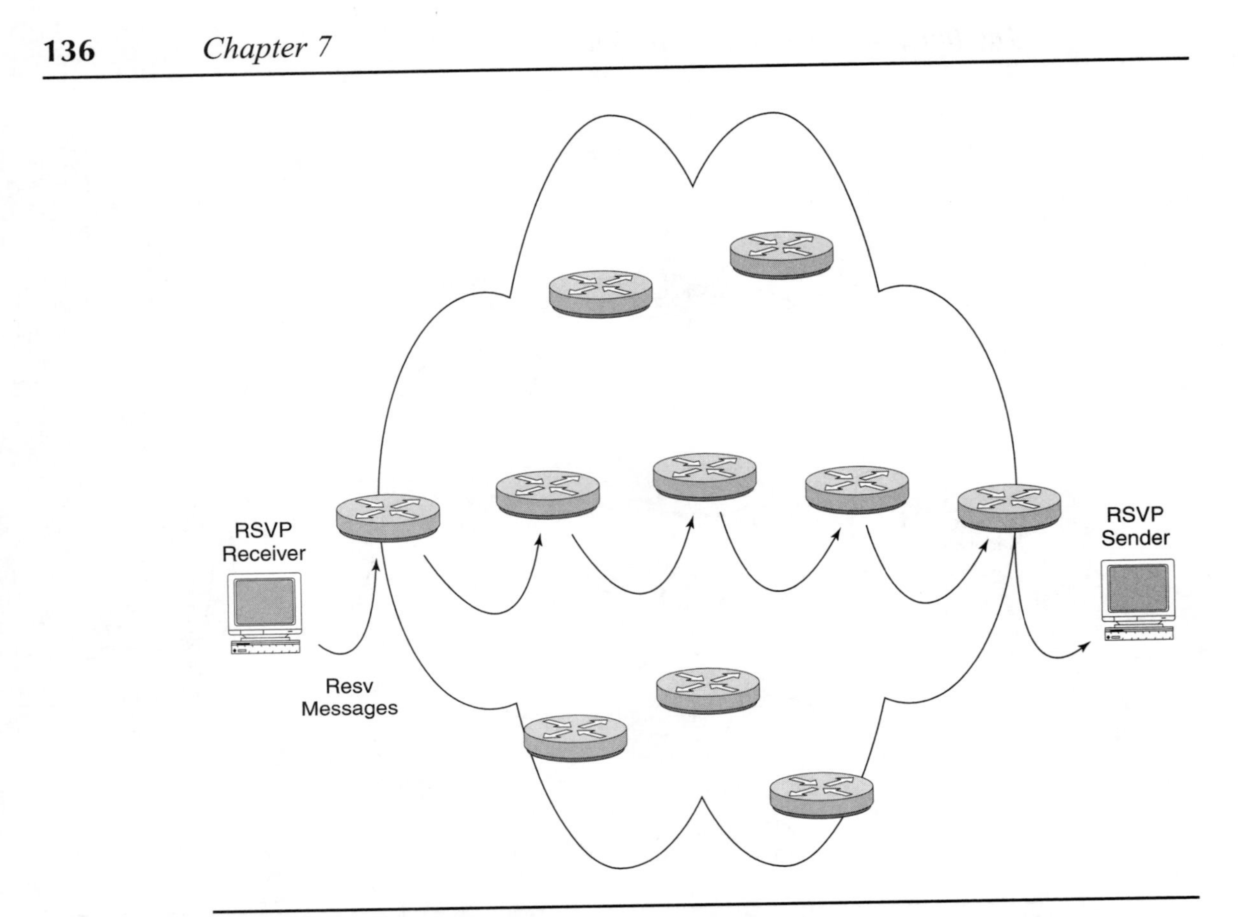

Figure 7.4: *Traffic flow of the RSVP Resv message.*

Sender Selection	Reservations	
	Distinct	Shared
Explicit	Fixed-Filter (FF) Style	Shared-Explicit (SE) Style
Wildcard	(No Style Defined)	Wildcard-Filter (WF) Style

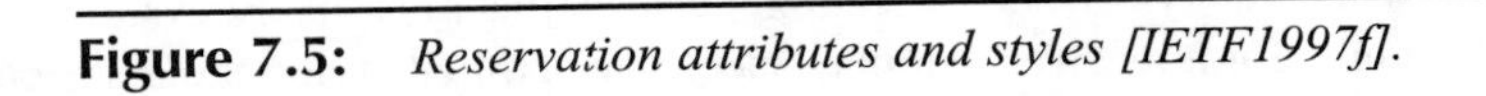

Figure 7.5: *Reservation attributes and styles [IETF1997f].*

RSVP Messages

An RSVP message contains a *message-type* field in the header that indicates the function of the message. Although seven types of RSVP messages exist, there are two fundamental RSVP message types: the *Resv* (reservation) and *Path* messages, which provide for the basic operation of RSVP. As mentioned earlier, an RSVP sender transmits *Path* messages downstream along the traffic path provided by a discrete routing protocol (e.g., OSPF). *Path* messages store path information in each node in the traffic path, which includes at a minimum the IP address of each Previous Hop (PHOP) in the traffic path. The IP address of the previous hop is used to determine the path in which the subsequent *Resv* messages will be forwarded. The *Resv* message is generated by the receiver and is transported back upstream toward the sender, creating and maintaining reservation state in each node along the traffic path, following the reverse path in which *Path* messages previously were sent and the same path data packets will subsequently use.

RSVP control messages are sent as raw IP datagrams using protocol number 46. Although raw IP datagrams are intended to be used between all end-systems and their next-hop intermediate node router, the RSVP specification allows for end-systems that cannot accommodate raw network I/O services to encapsulate RSVP messages in UDP (User Datagram Protocol) packets.

Path, *PathTear*, and *ResvConf* messages must be sent with the Router Alert option [IETF1997a] set in their IP headers. The Router Alert option signals nodes on the arrival of IP datagrams that need special processing. Therefore, nodes that implement high-performance forwarding designs can maximize forwarding rates in the face of normal traffic and be alerted to situations in which they may have to interrupt this high-performance forwarding mode to process special packets.

Path Messages

The RSVP *Path* message contains information in addition to the Previous Hop (PHOP) address, which characterizes the sender's traffic. These additional information elements are called the *Sender Template*, the *Sender TSpec*, and the *Adspec*.

The *Path* message is required to carry a *Sender Template*, which describes the format of data traffic the sender will originate. The *Sender Template* contains information called a *Filter Spec* (filter specification), which uniquely identifies the sender's flow from other flows present in the same RSVP session on the same link. The *Path* message is required to contain a *Sender TSpec*, which characterizes the traffic flow the sender will generate. The *TSpec* parameter characterizes the traffic the sender expects to generate; it is transported along the intermediate network nodes and received by the intended receiver(s). The *Sender TSpec* is not modified by the intermediate nodes.

Path messages may contain additional fragments of information contained in an *Adspec*. When an *Adspec* is received in a *Path* message by a node, it is passed to the local traffic-control process, which updates the *Adspec* with resource information and then passes it back to the RSVP process to be forwarded to the next downstream hop. The *Adspec* contains information required by receivers that allow them to choose a QoS control service and determine the appropriate reservation parameters. The *Adspec* also allows the receiver to determine whether a non-RSVP capable node (a router) lies in the transit path or whether a specific QoS control service is available at each router in the transit path. The *Adspec* also provides default or service-specific information for the characterization parameters for the guaranteed service class.

Information also can be generated or modified within the network and used by the receivers to make reservation decisions. This information may include specifics on available resources, delay and bandwidth estimates, and various parameters used by specific QoS control services. This information also is carried in the *Adspec* object and is collected from the various nodes as it makes its way toward the receiver(s). The information in the Adspec represents a cumulative summary, computed and updated each time the Adspec passes through a node. The RSVP sender also generates an initial Adspec object that characterizes its QoS control capabilities. This forms the starting point for the accumulation of the path properties; the *Adspec* is added to the RSVP *Path* message created and transmitted by the sender.

As mentioned earlier, the information contained in the Adspec is divided into fragments; each fragment is associated with a specific control service. This allows the *Adspec* to carry information about multiple services and allows the addition of new service classes in the future without modification to the mechanisms used to transport them. The size of the *Adspec* depends on the number and size of individual per-service fragments included, as well as the presence of nondefault parameters.

At each node, the *Adspec* is passed from the RSVP process to the traffic-control module. The traffic-control process updates the *Adspec* by identifying the services specified in the *Adspec* and calling each process to update its respective portion of the *Adspec* as necessary. If the traffic-control process discovers a QoS service specified in the *Adspec* that is unsupported by the node, a flag is set to report this to the receiver. The updated *Adspec* then is passed from the traffic-control process back to the RSVP process for delivery to the next node in the traffic path. After the RSVP *Path* message is received by the receiver, the *Sender TSpec* and the *Adspec* are passed up to the RAPI (RSVP Application Programming Interface).

The *Adspec* carries flag bits that indicate that a non-IS-aware (or non-RSVP-aware) router lies in the traffic path between the sender and receiver. These bits are called break bits and correspond to the NON_IS_HOP characterization parameter described earlier. A set break bit indicates that at least one node in the traffic path did not fully process the *Adspec*, so the remainder of the information in the *Adspec* is considered unreliable.

Resv Messages

The *Resv* message contains information about the reservation style, the appropriate *Flowspec* object, and the *Filter Spec* that identify the sender(s). The pairing of the *Flowspec* and the *Filter Spec* is referred to as the *Flow Descriptor*. The *Flowspec* is used to set parameters in a node's packet-scheduling process, and the *Filter Spec* is used to set parameters in the packet-classifier process. Data that does not match any of the *Filter Specs* is treated as best-effort traffic.

Resv messages are sent periodically to maintain the reservation state along a particular traffic path. This is referred to as *soft state*, because the reservation state is maintained by using these periodic refresh messages.

Various bits of information must be communicated between the receiver(s) and intermediate nodes to appropriately invoke QoS control services. Among the data types that need to be communicated between applications and nodes is the information generated by each receiver that describes the QoS control service desired, a description of the traffic flow to which the resource reservation should apply (*Receiver TSpec*), and the necessary parameters required to invoke the QoS service (*Receiver RSpec*). This information is contained in the *Flowspec* (flow specification) object carried in the *Resv* messages. The information contained in the *Flowspec* object may be modified at any intermediate node in the traffic path because of reservation merging and other factors.

The format of the *Flowspec* is different depending on whether the sender is requesting controlled-load or guaranteed service. When a receiver requests controlled-load service, only a *TSpec* is contained in the *Flowspec*. When requesting guaranteed service, both a *TSpec* and an *RSpec* are contained in the *Flowspec* object. (The *RSpec* element was described earlier in relation to the guaranteed service QoS class.)

In RSVP version 1, all receivers in a particular RSVP session are required to choose the same QoS control service. This restriction is due to the difficulty of merging reservations that request different QoS control services and the lack of a service-replacement mechanism. This restriction may be removed in future revisions of the RSVP specification.

At each RSVP-capable router in the transit path, the *Sender TSpecs* arriving in *Path* messages and the *Flowspecs* arriving in *Resv* messages are used to request the appropriate resources from the appropriate QoS control service. State merging, message forwarding, and error handling proceed according to the rules defined in the RSVP specification. Also, the merged *Flowspec* objects arriving at each RSVP sender are delivered to the application, informing the sender of the merged reservation request and the properties of the data path.

Additional Message Types

Aside from the *Resv* and *Path* messages, the remaining RSVP message types concern path and reservation errors (*PathErr* and *ResvErr*), path and reservation teardown (*PathTear* and *ResvTear*), and confirmation for a requested reservation (*ResvConf*).

The *PathErr* and *ResvErr* messages simply are sent upstream to the sender that created the error and do not modify the path state in the nodes through which they pass. A *PathErr* message indicates an error in the processing of *Path* messages and are sent back to the sender. *ResvErr* messages indicate an error in the processing of *Resv* messages and are sent to the receiver(s).

RSVP *teardown* messages remove path or reservation state from nodes as soon as they are received. It is not always necessary to explicitly tear down an old reservation, however, because the reservation eventually times out if periodic refresh messages are not received after a certain period of time. *PathTear* messages are generated explicitly by senders or by the time-out of path state in any node along the traffic path and are sent to all receivers. An explicit *PathTear* message is forwarded downstream from the node that generated it; this message deletes path state and reservation state that may rely on it in each node in the traffic path. A *ResvTear* message is generated explicitly by receivers or any node in which the reservation state has timed out and is sent to all pertinent senders. Basically, a *ResvTear* message has the opposite effect of a *Resv* message.

ResvConf

A *ResvConf* message is sent by each node in the transit path that receives a *Resv* message containing a reservation confirmation object. When a receiver wants to obtain a confirmation for its reservation request, it can include a confirmation request (RESV_CONFIRM) object in a *Resv* message. A reservation request with a *Flowspec* larger than any already in place for a session normally results in a *ResvErr* or a *ResvConf* message being generated and sent back to the receiver. Thus, the *ResvConf* message acts as an end-to-end reservation confirmation.

Merging

The concept of *merging* is necessary for the interaction of multicast traffic and RSVP. Merging of RSVP reservations is required because of the method multicast uses for delivering packets—replicating packets that must be delivered to different next-hop nodes. At each replication point, RSVP must merge reservation requests and compute the maximum of their *Flowspecs*.

Flowspecs are merged when *Resv* messages, each originating from different RSVP receivers and initially traversing diverse traffic paths, converge at a *merge point* node and are merged prior to being forwarded to the next RSVP node in the traffic path (Figure 7.6). The largest *Flowspec* from all merged *Flowspecs*—the one that requests the most stringent QoS reservation state—is used to define the single merged *Flowspec*, which is forwarded to the next hop node. Because *Flowspecs* are opaque data elements to RSVP, the methods for comparing them are defined outside of the base RSVP specification.

As mentioned earlier, different reservation styles cannot be merged, because they are fundamentally incompatible.

You can find specific ordering and merging guidelines for message parameters within the scope of the controlled-load and guaranteed service classes in [IETF1997h] and [IETF1997i], respectively.

Sender Selection	Reservations	
	Distinct	Shared
Explicit	Fixed-Filter (FF) Style	Shared-Explicit (SE) Style
Wildcard	(No Style Defined)	Wildcard-Filter (WF) Style

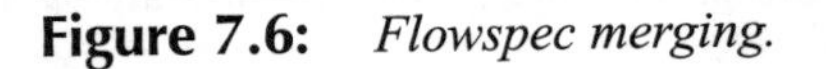

Figure 7.6: *Flowspec merging.*

Non-RSVP Clouds

RSVP still can function across intermediate nodes that are not RSVP-capable. End-to-end resource reservations cannot be made, however, because non-RSVP-capable devices in the traffic path cannot maintain reservation or path state in response to the appropriate RSVP messages. Although intermediate nodes that do not run RSVP cannot provide these functions, they may have sufficient capacity to be useful in accommodating tolerant real-time applications.

Because RSVP relies on a discrete routing infrastructure to forward RSVP messages between nodes, the forwarding of *Path* messages by non-RSVP-capable intermediate nodes is unaffected. Recall that the *Path* message carries the IP address of the Previous Hop (PHOP) RSVP-capable node as it travels toward the receiver. As the *Path* message arrives at the next RSVP-capable node after traversing an arbitrary non-RSVP cloud, it carries with it the IP address of the previous RSVP-capable node. Therefore, the *Resv* message then can be forwarded directly back to the next RSVP-capable node in the path.

Although RSVP functions in this manner, its use may severely distort the QoS request by the receiver.

Low-Speed Links

Low-speed links, such as analog telephone lines, ISDN connections, and sub-T1 rate lines present unique problems with regard to providing QoS, especially when multiple flows are present. It is problematic for a user to receive consistent performance, for example, when different applications are active at the same time, such as a Web browser, an FTP (File Transfer Protocol) transfer, and a streaming-audio application. Although the Integrated Services model is designed implicitly for situations in which some network traffic can be treated preferentially, it does not provide tailored service for low-speed links such as those described earlier.

At least one proposal has been submitted to the IETF's Integrated Services over Specific Link Layers (ISSLL) working group [ID1997j] that proposes the combination of enhanced, compressed, real-time transport protocol encapsulation; optimized header compression; and extensions to the PPP (Point-to-Point Protocol) to permit fragmentation and a method to suspend the transfer of large packets in favor of packets belonging to flows that require QoS services. The interaction of this proposal and the IETF Integrated Services model is outlined in [ID1997k].

Integrated Services and RSVP-over-ATM

The issues discussed in this section are based on information from a collection of documents currently being drafted in the IETF (each is referenced individually in this section). This discussion is based on the ATM Forum Traffic Management Specification Version 4.0 [AF1996a].

A functional disconnect exists between IP and ATM services, the least of which are their principal modes of operation: IP is non-connection oriented and ATM is connection oriented. An obvious contrast exists in how each delivers traffic. IP is a best-effort delivery service, whereas ATM has underlying technical mechanisms to provide differentiated levels of QoS for traffic on virtual connections. ATM uses point-to-point and point-to-multipoint VCs. Point-to-multipoint VCs allow nodes to be added and removed from VCs, providing a mechanism for supporting IP multicast.

Although several models exist for running IP-over-ATM networks [IETF1996a], any one of these methods will function as long as RSVP control messages (IP protocol 46) and data packets follow the same data path through the network. The RSVP *Path* messages must follow the same path as data traffic so that path state may be installed and maintained along the appropriate traffic path. With ATM, this means that the ingress and egress points in the network must be the same in both directions (remember that RSVP is only unidirectional) for RSVP control messages and data.

Supporting documentation discussing the interaction of ATM and upper-layer protocols can be complex and confusing, to say the least. When additional issues are factored in concerning ATM support for Integrated Services and RSVP, the result can be overwhelming.

Background

The technical specifications for running "Classical" IP-over-ATM is detailed in [IETF1994c]. It is based on the concept of an LIS (Logical IP Subnetwork), where hosts within an LIS communicate via the ATM network, and communication with hosts that reside outside the LIS must be through an intermediate router. Classical IP-over-ATM also provides a method for resolving IP host addresses to native ATM addresses called an ATM ARP (Address Resolution Protocol) server. The ATM Forum provides similar methods for supporting IP-over-ATM in its MPOA (Multi-Protocol Over ATM) [AF1997a] and LANE (LAN Emulation) [AF1995b] specifications. By the same token, IP multicast traffic and ATM interaction can be accommodated by a Multicast Address Resolution Server (MARS) [IETF1996b].

The technical specifications for LANE , Classical IP, and NHRP (Next Hop Resolution Protocol) [ID1997p] discuss methods of mapping best-effort IP traffic onto ATM SVCs (Switched Virtual Connections). However, when QoS requirements are introduced, the mapping of IP traffic becomes somewhat complex. Therefore, the industry recognizes that ongoing examination and research is necessary to provide for the complete integration of RSVP and ATM.

Using RSVP over ATM PVCs is rather straightforward. ATM PVCs emulate dedicated point-to-point circuits in a network, so the operation of RSVP is no different than when implemented on any point-to-point network model using leased lines. The QoS of the PVCs, however, must be consistent with the Integrated Services classes being implemented to ensure that RSVP reservations are handled appropriately in the ATM network. Therefore, there is no apparent reason why RSVP cannot be successfully implemented in an ATM network today that solely uses PVCs.

Using SVCs in the ATM network is more problematic. The complexity, cost, and efficiency to set up SVCs can impact their benefit when used in conjunction with RSVP. Additionally, scaling issues can be introduced when a single VC is used for each RSVP flow. The number of VCs in any ATM network is limited. Therefore, the number of RSVP flows that can be accommodated by any one device is limited strictly to the number of VCs available to a device.

The IP and ATM interworking requirements are compounded further by VC management issues introduced in multicast environments. A primary concern in this regard is how to integrate the many-to-many connectionless features of IP multicast and RSVP into the one-to-many, point-to-multipoint, connection-oriented realm of ATM.

ATM point-to-multipoint VCs provide an adequate mechanism for dealing with multicast traffic. With the introduction of ATM Forum 4.0, a new concept has been introduced called *Leaf Initiated Join* (LIJ), which allows an ATM end-system to join an existing point-to-multipoint VC without necessarily contacting the source of the VC. This reduces the resource burden on the ATM source as far as setting up new branches, and it more closely resembles the receiver-based model of RSVP and IP multicast. However, several scaling issues still exist, and new branches added to an existing point-to-multipoint VC will end up using the existing QoS parameters as the existing branches, posing yet another problem. Therefore, a method must be defined to provide better handling of heterogeneous RSVP and multicast receivers with ATM SVCs.

By the same token, a major difference exists in how ATM and RSVP QoS negotiation is accomplished. ATM is sender oriented and RSVP is receiver oriented. At first glance, this might appear to be a major discrepancy. However, RSVP receivers actually determine the QoS required by the parameters included in the sender's *TSpec*, which is included in received *Path* messages. Therefore, whereas the resources in the network are reserved in response to receiver-generated *Resv* messages, the resource reservations actually are initiated by the sender. This means that senders will establish ATM QoS VCs and receivers must accept incoming ATM QoS VCs. This is consistent with how RSVP operates and allows senders to use different RSVP flow-to-VC mappings for initiating RSVP sessions.

Several issues discussed in previous chapters concerned attempts to provide QoS by using traditional IP and the underlying ATM mechanics, and some of these same concerns have driven efforts in the IETF and elsewhere to develop methods for using RSVP and Integrated Services with ATM to augment the existing model. Two primary areas in which the integration of ATM and the Integrated Services model is important are QoS translation or mapping between Integrated Services and ATM, and VC management, which deals with VC establishment and which traffic flows are forwarded over the VCs. An ATM edge device (router) in an IP network must provide IP and ATM interworking functions, servicing the requirements of each network (Figure 7.7). In the case of RSVP, it must be able to process RSVP control messages, reserve resources, maintain soft state, and provide packet scheduling and classification services. It also must be able to initiate, accept, or refuse ATM connections via UNI signaling. Combining these capabilities, the edge device also must translate RSVP reservation semantics to the appropriate ATM VC establishment parameters.

As stated in [ID1997o], the task of providing a translation mechanism between the Integrated Services controlled-load and guaranteed services and appropriate ATM QoS parameters is "a complex problem with many facets." This document provides a proposal for a mapping of both the controlled-load and guaranteed service classes as well as best-effort traffic to the appropriate ATM QoS services.

As depicted in Figure 7.8, the mapping of the Integrated Services QoS classes to ATM service categories would appear to be straightforward. The ATM CBR and rt-VBR service

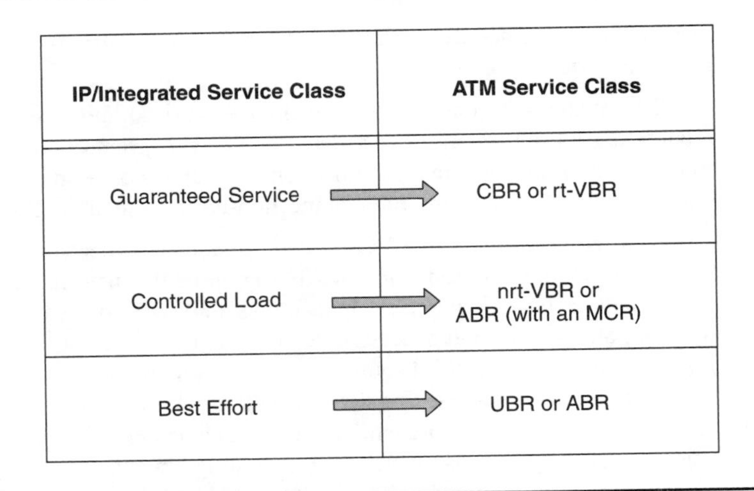

IP/Integrated Service Class	ATM Service Class
Guaranteed Service	CBR or rt-VBR
Controlled Load	nrt-VBR or ABR (with an MCR)
Best Effort	UBR or ABR

Figure 7.7: *ATM edge functions [ID1997o].*

categories possess characteristics that make them prospective candidates for guaranteed service, whereas the nrt-VBR and ABR (albeit with an MCR) service categories provide characteristics that are the most compatible with the controlled-load service. Best-effort traffic fits well into the UBR service class.

The practice of tagging nonconformant cells with CLP = 1, which designates cells as lower priority, can have a special use with ATM and RSVP. As outlined previously, you can determine whether cells are tagged as conformant by using a GCRA (Generic Cell Rate Algorithm) leaky-bucket algorithm. Also recall that traffic in excess of controlled-load or guaranteed service specifications must be transported as best-effort traffic. Therefore, the practice of dropping cells with the CLP bit set should be exercised with excess guaranteed

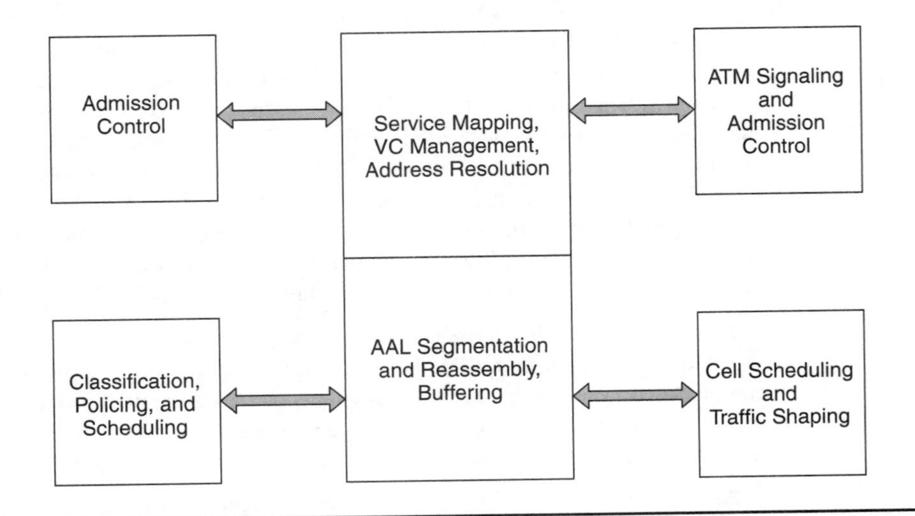

Figure 7.8: *Integrated Services and ATM QoS mapping [ID1997o].*

service or controlled-load traffic. Of course, this is an additional nuance you should consider in an ATM/RSVP interworking implementation.

Several ATM QoS parameters exist for which there are no IP layer equivalents. Therefore, these parameters must be configured manually as a matter of local policy. Among these parameters are CLR (Cell Loss Ratio), CDV (Cell Delay Variation), SECBR (Severely Errored Cell Block Ratio), and CTD (Cell Transfer Delay).

The following section briefly outlines the mapping of the Integrated Services classes to ATM service categories. This is discussed in much more detail in [ID1997o].

Guaranteed Service and ATM Interaction

The guaranteed service class requires reliable delivery of traffic with as little delay as possible. Thus, an ATM service category that accommodates applications with end-to-end time synchronization also should accommodate guaranteed service traffic, because the end-to-end delay should be closely monitored and controlled. The nrt-VBR, ABR, and UBR service categories, therefore, are not good candidates for guaranteed service traffic. The CBR service category provides ideal characteristics in this regard. However, CBR does not adapt to changing data rates and can leave large portions of bandwidth underutilized. In most scenarios, CBR proves to be a highly inefficient use of available network resources.

Therefore, rt-VBR is the most appropriate ATM service category for guaranteed service traffic because of its inherent adaptive characteristics. The selection of the rt-VBR service category, however, requires two specified rates to be quantified: the SCR (Sustained Cell Rate) and PCR (Peak Cell Rate). These two parameters provide a burst and average tolerance profile for traffic with bursty characteristics. The rt-VBR service also should specify a low enough CLR for guaranteed service traffic so that cell loss is avoided as much as possible.

When mapping guaranteed service onto an rt-VBR VC, [ID1997o] suggests that the ATM traffic descriptor values for PCR, SCR, and MBS (Maximum Burst Size) should be set within the following bounds:

$R <= \text{PCR} <= \text{minimum } p \text{ or minimum line rate}$
$r <= \text{SCR} <= \text{PCR}$
$0 <= \text{MBS} <= b$

where $R =$ *RSpec*
$p =$ peak rate
$r =$ *Receiver TSpec*
$b =$ bucket depth

In other words, the *RSpec* should be less than or equal to the PCR, which in turn should be less than or equal to the minimum peak rate, or alternatively, the minimum line rate. The *Receiver TSpec* (assumed here to be identical to the *Sender TSpec*) should be less than or equal to the SCR, which in turn should be less than or equal to the PCR. The MBS should be greater than or equal to zero (generally, greater than zero, of course) but less than or equal to the leaky-bucket depth defined for traffic shaping.

Controlled-Load Service and ATM Interaction

Of the three remaining ATM service categories (nrt-VBR, ABR, and UBR), only nrt-VBR and ABR are viable candidates for the controlled-load service. The UBR service category does not possess traffic capabilities strict enough for the controlled-load service. UBR does not provide any mechanism to allocate network resources, which is the goal of the controlled-load service class. Recall that traffic appropriate for the controlled-load service is characterized as tolerant real-time applications, with the applications somewhat tolerant of packet loss but still requiring reasonably efficient performance.

The ABR service category best aligns with the model for controlled-load service, which is characterized as being somewhere between best-effort and a requiring service guarantees. Therefore, if the ABR service class is used for controlled-load traffic, it requires that an MCR (Minimum Cell Rate) be specified to provide a lower bound for the data rate. The *TSpec* rate should be used to determine the MCR.

The nrt-VBR service category also can be used for controlled-load traffic. However, the maxCTD (Maximum Cell Transfer Delay) and CDV (Cell Delay Variation) parameters must be chosen for the edge ATM device and is done manually as a matter of policy.

When mapping controlled-load service onto an nrt-VBR VC, [ID1997o] suggests that the ATM traffic descriptor values for PCR and MBS should be set within the following bounds:

$r <= \text{SCR} <= \text{PCR} <=$ minimum p or minimum line rate
$0 <= \text{MBS} <= b$

where $p =$ peak rate
$r =$ *Receiver TSpec*
$b =$ bucket depth

Integrated Services, Multicast, and ATM

For RSVP to be useful in multicast environments, flows must be distributed appropriately to multiple destinations from a given source. Both the Integrated Services model and RSVP support the idea of heterogeneous receivers. Conceptually, not all receivers of a particular multicast flow are required to ask for the same QoS parameters from the network. An example of why this concept is important is that when multiple senders exist on a shared-reservation flow, enough resources can be reserved by a single receiver to accommodate all senders in the shared flow. This is problematic in conjunction with ATM. If a single maximum QoS is defined for a point-to-multipoint VC, network resources could be wasted on links where the reassembled packets eventually will be dropped. Additionally, a maximum QoS may impose a degradation in the service provided to the best-effort branches. In ATM networks, additional end-points of a point-to-multipoint VC must be set up explicitly. Because ATM does not currently provide support for this situation, any RSVP-over-ATM implementations must make special provisions to handle heterogeneous receivers.

It has been suggested that RSVP heterogeneity can be supported over ATM by mapping RSVP reservations onto ATM VCs by using one of four methods proposed in [ID1997m].

In the *full heterogeneity* model, a separate VC is provided for each distinct multicast QoS level requested, including requests for best-effort traffic. In the *limited heterogeneity* model, each ATM device participating in an RSVP session would require two VCs: one point-to-multipoint VC for best-effort traffic and one point-to-multipoint QoS VC for RSVP reservations. Both these approaches require what could be considered inefficient quantities of network resources. The *full heterogeneity* model can provide users with the QoS they require but makes the most inefficient use of available network resources. The *limited heterogeneity* model requires substantially less network resources. However, it is still somewhat inefficient, because packets must be duplicated at the network layer and sent on two VCs.

The third model is a *modified homogeneous* model. In a homogeneous model, all receivers—including best-effort receivers—on a multicast session use a single QoS VC that provides a maximum QoS service that can accommodate all RSVP requests. This model most closely matches the method in which the RSVP specification handles heterogeneous requests for resources. However, although this method is the simplest to implement, it may introduce a couple of problems. One such problem is that users expecting to making a small or no reservation may end up not receiving any data at all, to include best-effort traffic; their request may be rejected because of insufficient resources in the network. The modified homogeneous model proposes that special handling be added for the situation in which a best-effort receiver cannot be added to the QoS VC by generating an error condition that triggers a request to establish a best-effort VC for the appropriate receivers.

The fourth model, the *aggregation* model, proposes that a single, large point-to-multipoint VC be used for multiple RSVP reservations. This model is attractive for a number of reasons, primarily because it solves the inefficiency problems associated with full heterogeneity, because concerns about induced latency imposed by setting up an individual VC for each flow are negated. The primary problem with the aggregation model is that it may be difficult to determine the maximum QoS for the aggregate VC.

The term *variegated VCs* [ID1997m] has been created to describe point-to-multipoint VCs that allow a different QoS on each branch. However, cell-drop mechanisms require further research to retain the best-effort delivery characterization for nonconformant packets that traverse certain branch topologies. Implementations of Early Packet Discard (EPD) should be deployed in these situations so that all cells belonging to the same packet can be discarded—instead of discarding only a few arbitrary cells from several packets, making them useless to their receivers.

Another issue of concern is that IP-over-ATM currently uses a multicast server or reflector that can accept calls from multiple senders and redirect them to a set of senders through the use of point-to-multipoint VCs. This moves the scaling issue from the ATM network to the multicast server. However, the multicast server needs to know how to interpret RSVP messages to enable VC establishment with the appropriate QoS parameters.

Integrated Services over Local Area Media

The Integrated Services discussions up to this point have focused on two basic network entities: the host and the intermediate nodes or routers. There may be many cases in which an intermediate router does not lie in the end-to-end path of an RSVP sender and receiver, or perhaps several link-layer bridges or switches may lie in the data path between the RSVP sender or receiver and the first intermediate *IS-aware* router. Because LAN technologies, such as Ethernet and token ring, typically constitute the *last-hop* link-layer media between the host and the wide-area network, it is interesting to note that, currently, no standard mechanisms exist for providing service guarantees on any of these LAN technologies. Given this consideration, link-layer devices such as bridges and LAN switches may need a mechanism that provides customized admission-control services to provide support for traffic that requests Integrated Services QoS services. Otherwise, delay bounds specified by guaranteed services end-systems may be impacted adversely by intermediate link-layer devices that are not *IS aware*, and controlled-load services may not prove to be *better than best effort*.

The concept of a subnetwork *Bandwidth Manager* first was described in [ID1997q] and provides a mechanism to accomplish several things on a LAN subnet that otherwise would be unavailable. Among these are admission control, traffic policing, flow segregation, packet scheduling, and the capability to reserve resources (to include maintaining soft state) on the subnet.

The conceptual model of operation is fairly straightforward. The Bandwidth Manager model consists of two distinctive components: a Requester Module (RM) and a Bandwidth Allocator (BA), as illustrated in Figure 7.9. The RM resides in every end-system that resides on the subnetwork and provides an interface between the higher-layer application (which can

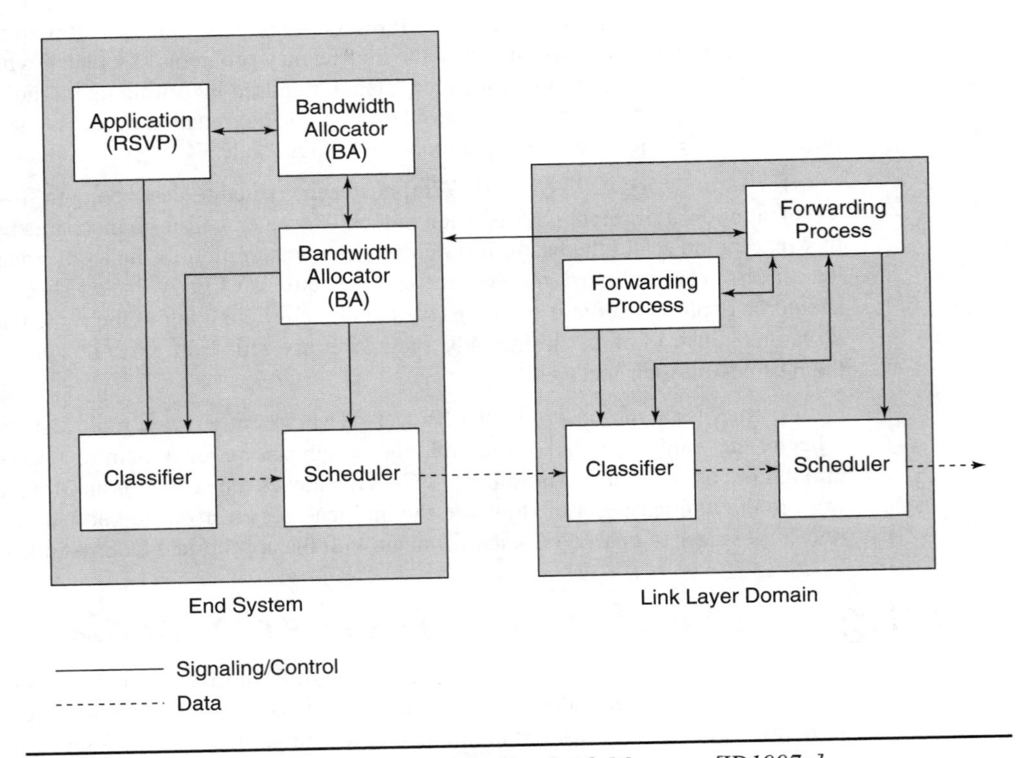

Figure 7.9: *Conceptual operation of the Bandwith Manager [ID1997q].*

be assumed to be RSVP) and the Bandwidth Manager. For the end-system to initiate a resource reservation, the RM is provided with the service desired (guaranteed or controlled-load), the traffic descriptors in the *TSpec*, and the *RSpec* defining the amount of resources requested. This information is extracted from RSVP *Path* and *Resv* messages.

As a demonstration of the merit of this idea, the Bandwidth Manager concept is expanded and described in more detail in the following section.

The Subnet Bandwidth Manager (SBM)

The concept of the *Subnet Bandwidth Manager* (SBM) is defined further in [ID1997r], which is a proposal that provides a standardized signaling protocol for LAN-based admission control for RSVP flows on IEEE 802-style LANs. The SBM proposal suggests that this mechanism, when combined with per-flow policing on the end-systems and traffic control and priority queuing at the link-layer, will provide a close approximation of the controlled-load and guaranteed services. However, in the absence of any link-layer traffic controls or priority-queuing mechanisms in the LAN infrastructure (for example, on a shared media LAN), the SBM mechanism limits only the total amount of traffic load imposed by RSVP-associated flows. In environments of this nature, no mechanism is available to separate RSVP flows from best-effort traffic. This brings into question the usefulness of using the SBM model in a LAN infrastructure that does not support the capability to forward packets tagged with an IEEE 802.1p priority level.

In each SBM-managed segment, a single SBM is designated to be the DSBM (Designated Subnet Bandwidth Manager) for the managed LAN segment. The DSBM is configured with information about the maximum bandwidth that can be reserved on each managed segment under its control. Although this information most likely will be statically configured, future methods may dynamically discover this information. When the DSBM clients come online, they attempt to discover whether a DSBM exists on each of the segments to which they may be attached. This is done through a dynamic DSBM discovery-and-election algorithm, which is described in Appendix A of [ID1997r]. If the client itself is capable of serving as a DSBM, it may choose to participate in the election process.

When a DSBM client sends or forwards a *Path* message over an interface attached to a managed segment, it sends the message to its DSBM instead of to the RSVP session destination address, as is done in conventional RSVP message processing. After processing and possibly updating the *Adspec*, the DSBM forwards the *Path* message to its destination address. As part of its processing, the DSBM builds and maintains a *Path* state for the session and notes the Previous Hop (PHOP) of the node that sent the message. When a DSBM client wants to make a reservation for an RSVP session, it follows the standard RSVP message-processing rules and sends an RSVP *Resv* message to the corresponding PHOP address specified in a received *Path* message. The DSBM processes received *Resv* messages based on the bandwidth available and return *ResvErr* messages to the requester if the request cannot be granted. If sufficient resources are available, and the reservation request is granted, the DSBM forwards the *Resv* message to the PHOP based on the local *Path* state for the session. The DSBM also merges and orders reservation requests in accordance with traditional RSVP message-processing rules.

In the example in Figure 7.10, an "intelligent" LAN switch is designated as the DSBM for the managed segment. The "intelligence" is only abstract, because all that is required is

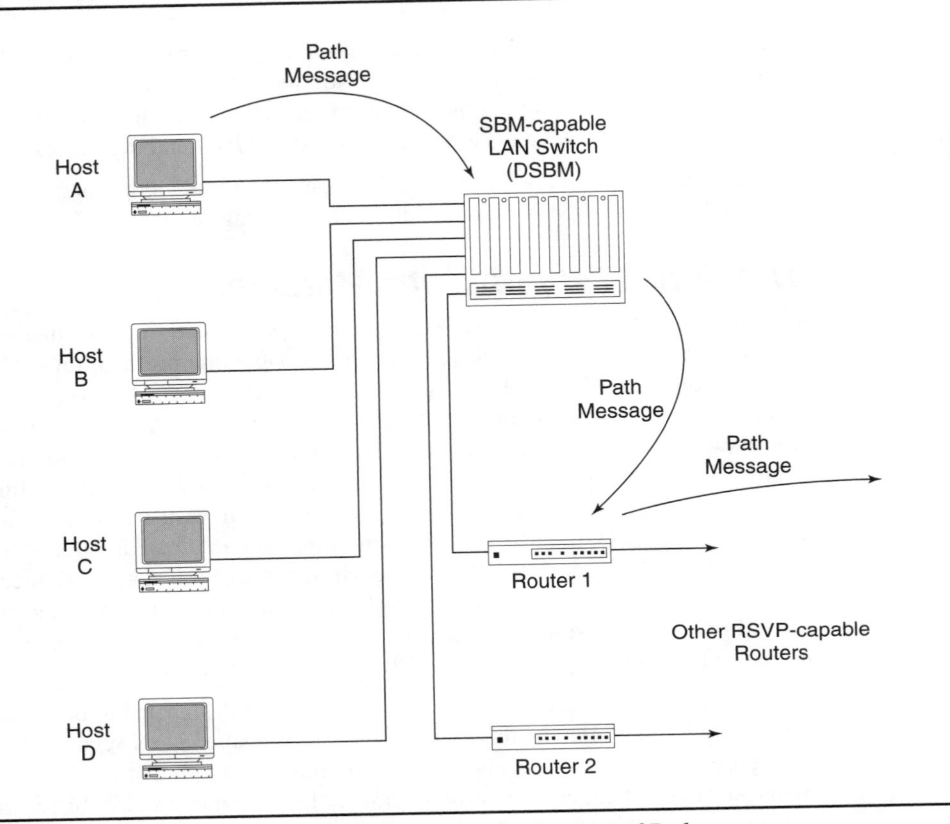

Figure 7.10: *SBM-managed LAN segment: forwarding of Path messages.*

that the switch implement the SBM mechanics as defined in [ID1997r]. As the DSBM client, host A, sends a *Path* message upstream, it is forwarded to the DSBM, which in turn forwards it to its destination session address, which lies outside the link-layer domain of the managed segment. In Figure 7.11, the *Resv* message processing occurs in exactly the same order, following the *Path* state constructed by the previously processed *Path* message.

The addition of the DSBM for admission control for managed segments results in some additions to the RSVP message-processing rules at a DSBM client. In cases in which a DSBM needs to forward a *Path* message to an egress router for further processing, the DSBM may not have the Layer 3 routing information available to make the necessary forwarding decision when multiple egress routers exist on the same segment. Therefore, new RSVP objects have been proposed, called LAN_NHOP (LAN Next Hop) objects, which keep track of the Layer 3 hop as the *Path* message traverses a Layer 2 domain between two Layer 3 devices.

When a DSBM client sends out a *Path* message to its DSBM, it must include LAN_NHOP information in the message. In the case of unicast traffic, the LAN_HOP address indicates the destination address or the IP address of the next hop router in the path to the destination. As a result, when a DSBM receives a *Path* message, it can look at the

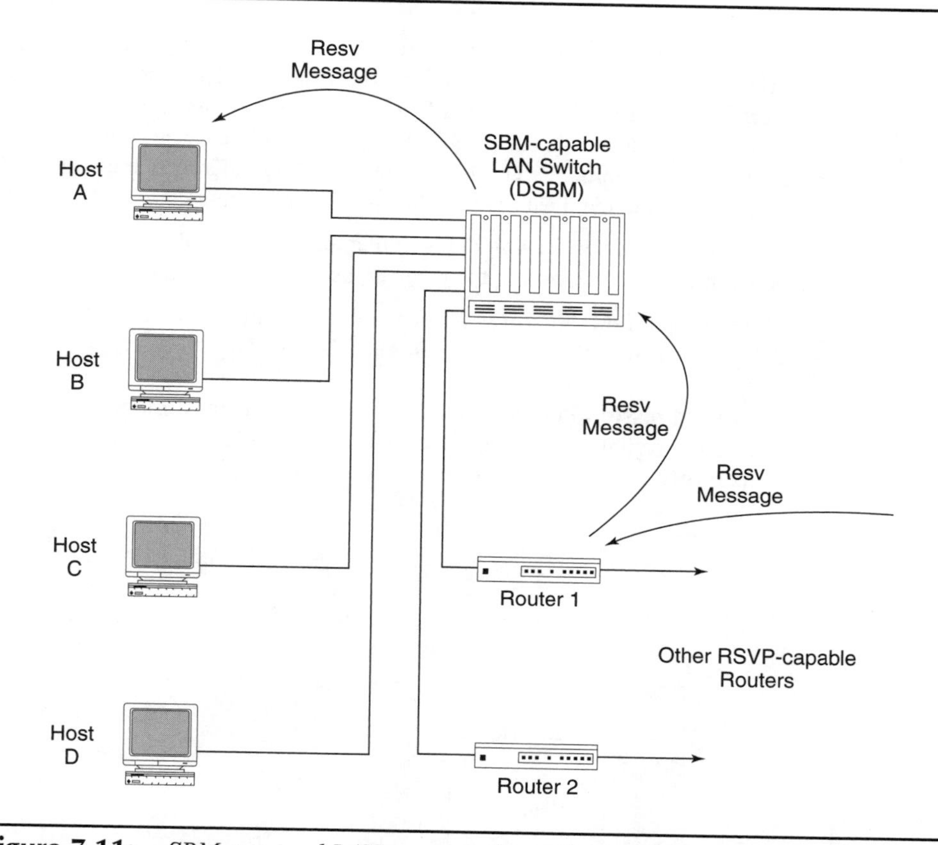

Figure 7.11: *SBM-managed LAN segment: forwarding of Resv messages.*

address specified in the LAN_NHOP object and appropriately forward the message to the appropriate egress router. However, because the link-layer devices (LAN switches) must act as DSBMs, the level of "intelligence" of these devices may not include an ARP capability that enables it to resolve MAC (Media Access Control) addresses to IP addresses. For this reason, [ID1997r] requires that LAN_HOP information contain both the IP address (LAN_NHOP_L3) and corresponding MAC address (LAN_NHOP_L2) for the next Layer 3 device.

Because the DSBM may not be able to resolve IP addresses to MAC addresses, a mechanism is needed to dispense with this translation requirement when processing *Resv* messages. Therefore, the RSVP_HOP_L2 object is used to indicate the Layer 2 MAC address of the previous hop. This provides a mechanism for SBM-capable devices to maintain the *Path* state necessary to accommodate forwarding *Resv* messages along link-layer paths that cannot provide IP-address-to-MAC-address resolution.

There is at least one additional proposed new RSVP object, called the TCLASS (Traffic Class) object. The TCLASS object is used with IEEE 802.1p *user-priority* values, which can be used by Layer 2 devices to discriminate traffic based on these priority values. These values are discussed in more detail in the following section. The priority value assigned to each packet is carried in the new extended frame format defined by IEEE 802.1Q [IEEE-5]. As an SBM Layer 2 switch, which also functions as an 802.1p device, receives a *Path* message, it inserts a TCLASS object. When a Layer 3 device (a router) receives *Path* messages, it retrieves and stores the TCLASS object as part of the process of building *Path* state for the session. When the same Layer 3 device needs to forward a *Resv* message back toward the sender, it must include the TCLASS object in the *Resv* message.

The Integrated Services model is implemented via an SBM client in the sender, as depicted in Figure 7.12, and in the receiver, as depicted in Figure 7.13.

Figure 7.14 shows the SBM implementation in a LAN switch. The components of this model are defined in the following summary [ID1997s]:

Local admission control. One local admission control module on each switch port manages available bandwidth on the link attached to that port. For half-duplex links, this involves accounting for resources allocated to both transmit and receive flows.

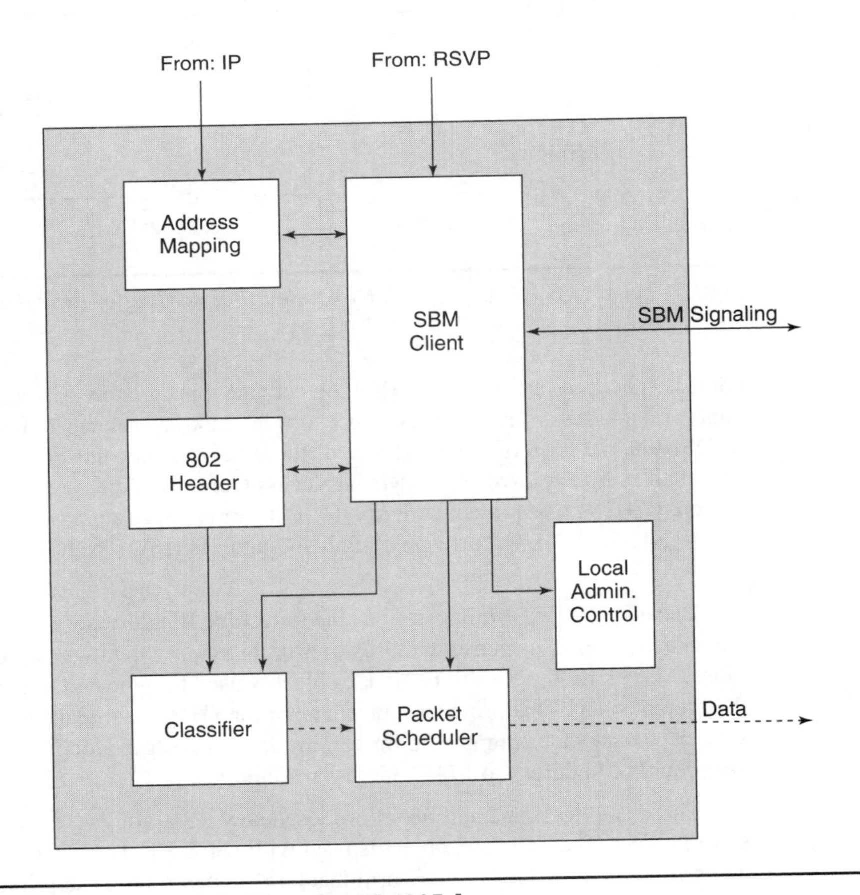

Figure 7.12: *SBM in a LAN sender [ID1997s].*

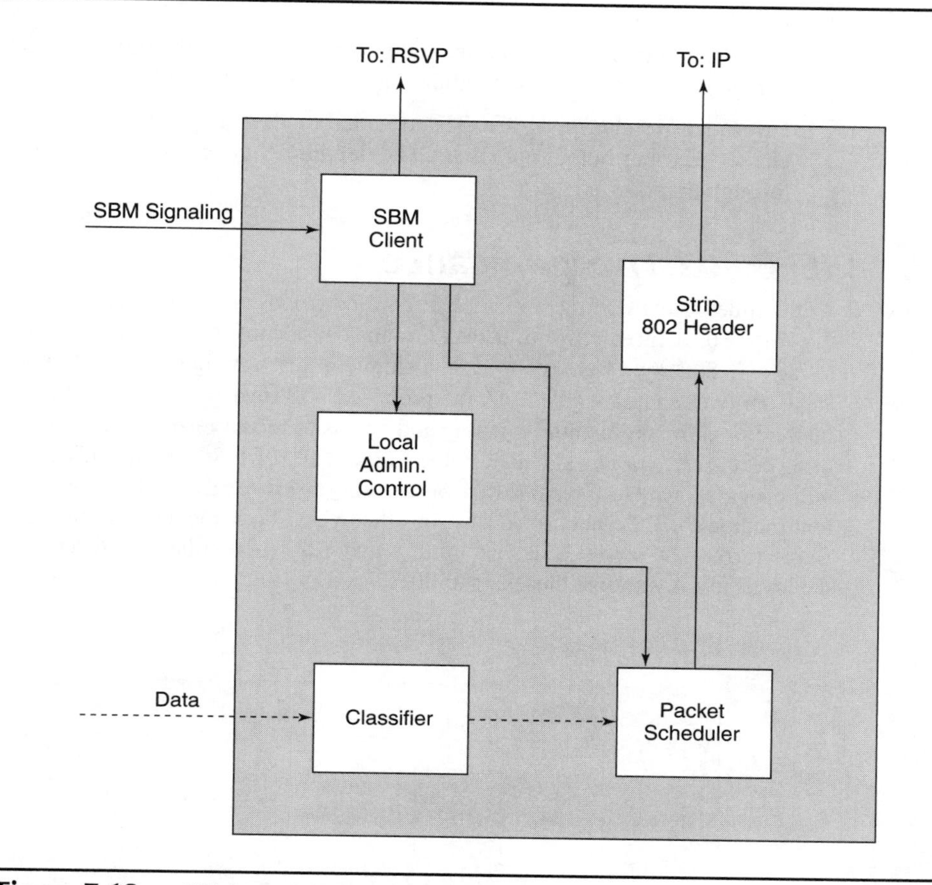

Figure 7.13: *SBM in a LAN receiver [ID1997s].*

Input SBM module. One instance per port. This module performs the *network* portion of the client-network peering relationship. This module also contains information about the mapping of Integrated Service classes to IEEE 802.1p *user_priority*, if applicable.

SBM propagation. Relays requests that have passed admission control at the input port to the relevant output port's SBM module(s). As indicated in Figure 7.14, this requires access to the switch's forwarding table and port spanning-tree states.

Output SBM module. Forwards messages to the next Layer 2 or Layer 3 network hop.

Classifier, Queuing, and scheduler. The classifier function identifies the relevant QoS information from incoming packets and uses this, with information contained in the normal bridge forwarding database, to determine which queue of the appropriate output port to direct the packet for transmission. The queuing and scheduling functions manage the output queues and provide the algorithmic calculation for servicing the queues to provide the promised service (controlled-load or guaranteed service).

Ingress traffic class mapper and policing. This optional module may check on whether the data in the traffic classes conforms to specified behavior. The switch may police this traffic and remap to another class or discard the traffic altogether. The default behavior should be to allow traffic through unmodified.

Egress traffic class mapper. This optional module may apply remapping of traffic classes on a per-output port basis. The default behavior should be to allow traffic through unmodified.

IEEE 802.1p Significance

At the time of this writing, an interesting set of proposed enhancements is being reviewed by the IEEE 802.1 Internetworking Task Group. These enhancements would provide a method to identify 802-style frames based on a simple priority. A supplement to the original IEEE MAC Bridges standard [IEEE-1], the proposed 802.1p specification [IEEE-2] provides a method to allow preferential queuing and access to media resources by traffic class on the basis of a *user_priority* signaled in the frame. The IEEE 802.1p specification, if adopted, will provide a way to transport this value (*user_priority*) across the subnetwork in a consistent method for Ethernet, token ring, or other MAC-layer media types using an extended frame format. Of course, this also implies that 802.1p-compliant hardware may have to be deployed to fully realize these capabilities.

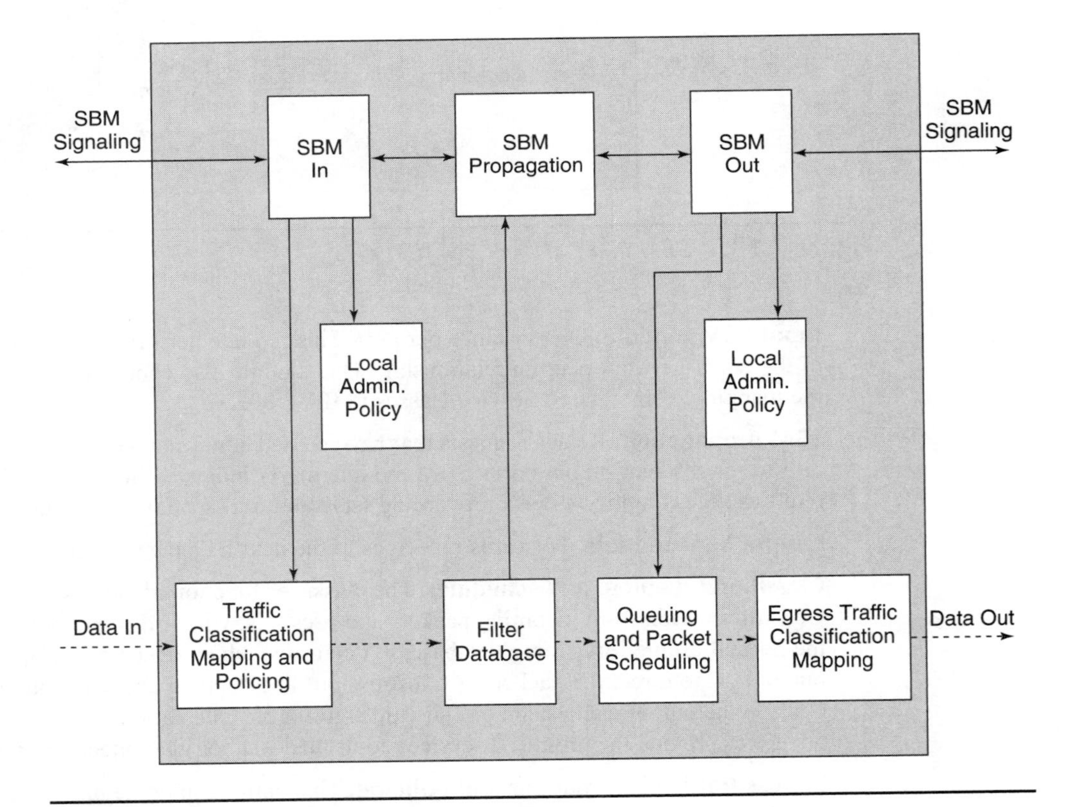

Figure 7.14: *SBM in a LAN switch [ID1997s].*

The current 802.1p draft defines the *user_priority* field as a 3-bit value, resulting in a variable range of values between 0 and 7, with 7 indicating high priority and 0 indicating low priority. The IEEE 802 specifications do not make any suggestions on how the *user_priority* should be used by the end-system by network elements. It only suggests that packets may be queued by LAN devices based on their *user_priority* values.

A proposal submitted recently in the Integrated Services over Specific Link Layers (ISSLL) working group of the IETF provides a suggestion on how to use the IEEE 802.1p *user_priority* value to an Integrated Services class [ID1997s]. Because no practical experience exists for mapping these parameters, the suggestions are somewhat arbitrary and provide only a framework for further study. As shown in Table 7.1, two of the *user_priority* values provide separate classifications for guaranteed services traffic with different delay requirements. The *less-than-best-effort* category could be used by devices that tag packets that are in nonconformance of a traffic commitment and may be dropped elsewhere in the network.

Because no explicit traffic class or *user_priority* field exists in Ethernet 802.3 [IEEE-3] packets, the *user_priority* value must be regenerated at a downstream node or LAN switch by some predefined default criteria, or by looking further into the higher-layer protocol fields in the packet and matching some parameters to a predefined criteria. Another option is to use the IEEE 802.1Q encapsulation proposal [IEEE-5] tailored for VLANs (Virtual Local Area Networks), which may be used to provide an explicit traffic class field on top of the basic MAC format.

The token-ring standard [IEEE-4] does provide a priority mechanism, however, that can be used to control the queuing of packets and access to the shared media. This priority mechanism is implemented using bits from the Access Control (AC) and Frame Control (FC) fields of an LLC frame. The first three bits (the token priority bits) and the last three bits (the reservation bits) of the AC field dictate which stations get access to the ring. A token-ring station theoretically is capable of separating traffic belonging to each of the eight levels of requested priority and transmitting frames in the order of indicated priority. The last three bits of the FC field (the user priority bits) are obtained from the higher layer in the *user_priority* parameter when it requests transmission of a packet. This parameter also

Table 7.1: *IEEE 802.1p user_priority Mapping to Integrated Services Classes*

User Priority	Service
0	Less than best effort
1	Best effort
2	Reserved
3	Reserved
4	Controlled load
5	Guaranteed service, 100 ms bound
6	Guaranteed service, 10 ms bound
7	Reserved

establishes the access priority used by the MAC. This value usually is preserved as the frame is passed through token-ring bridges; thus, the *user_priority* can be transported end-to-end unmolested.

In any event, the implementation of 802.1p-capable devices in a LAN provides an additional level of granularity, and when used in conjunction with SBM, it presents the possibility to push the Integrated Services mechanics further into the network infrastructure.

Observations

It has been suggested that the Integrated Services architecture and RSVP are excessively complex and possess poor scaling properties. This suggestion is undoubtedly prompted by the existence of the underlying complexity of the signaling requirements. However, it also can be suggested that RSVP is no more complex than some of the more advanced routing protocols, such as BGP (Border Gateway Protocol). An alternative viewpoint might suggest that the underlying complexity is required because of the inherent difficulty in establishing and maintaining path and reservation state information along the transit path of data traffic. The suggestion that RSVP has poor scaling properties deserves additional examination, however, because deployment of RSVP has not been widespread enough to determine the scope of this assumption.

As discussed in [IETF1997k], there are several areas of concern about the wide-scale deployment of RSVP. With regard to concerns of RSVP scaleability, the resource requirements (computational processing and memory consumption) for running RSVP on routers increase in direct proportion to the number of separate RSVP reservations, or sessions, accommodated. Therefore, supporting a large number of RSVP reservations could introduce a significant negative impact on router performance. By the same token, router-forwarding performance may be impacted adversely by the packet-classification and scheduling mechanisms intended to provide differentiated services for reserved flows. These scaling concerns tend to suggest that organizations with large, high-speed networks will be reluctant to deploy RSVP in the foreseeable future, at least until these concerns are addressed. The underlying implications of this concern also suggest that without deployment by Internet service providers, who own and maintain the high-speed backbone networks in the Internet, the deployment of pervasive RSVP services in the Internet will not be forthcoming.

At least one interesting proposal has been submitted to the IETF [ID1997l] that suggests a rationale for grouping similar guaranteed service flows to reduce the bandwidth requirements that guaranteed service flows might consume individually. This proposal does not suggest an explicit implementation method to provide this grouping but instead provides the reasoning for identifying identical guaranteed service flows in an effort to group them. The proposal does suggest offhand that some sort of tunneling mechanism could be used to transport flow groups from one intermediate node to another, which could conceivably reduce the amount of bandwidth required in the nodes through which a flow group tunnel passes. Although it is well intentioned, the obvious flaw in this proposal is that it only partially addresses the scaling problems introduced by the Integrated Services model. The flow group still must be policed at each intermediate node to provide traffic-conformance monitoring, and path and reservation state still must be maintained at each intermediate node for individual flows. This proposal is in the embryonic stages of review and investigation, however, and it is unknown at this time how plausible the proposal might be in practice.

Another important concern expressed in [IETF1997k] deals with policy-control issues and RSVP. Policy control addresses the issue of who is authorized to make reservations and provisions to support access control and accounting. Although the current RSVP specification defines a mechanism for transporting policy information, it does not define the policies themselves, because the policy object is treated as an opaque element. Some vendors have indicated that they will use this policy object to provide proprietary mechanisms for policy control. At the time of this writing, the IETF RSVP working group has been chartered to develop a simple policy-control mechanism to be used in conjunction with RSVP. There is ongoing work on this issue in the IETF. Several mechanisms already have been proposed to deal with policy issues [ID1997h, ID1997I, ID1997v], in addition to the aforementioned vendor-proprietary policy-control mechanisms. It is unclear at this time, however, whether any of these proposals will be implemented or adopted as a standard.

The key recommendation contained in [IETF1997k] is that given the current form of the RSVP specification, multimedia applications run within smaller, private networks are the most likely to benefit from the deployment of RSVP. The inadequacies of RSVP scaling and lack of policy control may be more manageable within the confines of a smaller, more controlled network environment than in the expanse of the global Internet. It certainly is possible that RSVP may provide genuine value and find legitimate deployment uses in smaller networks, both in the peripheral Internet networks and in the private arena, where these issues of scale are far less important. Therein lies the key to successfully delivering quality of service using RSVP. After all, the purpose of the Integrated Services architecture and RSVP is to provide a method to offer quality of service, not to degrade the service quality.

chapter 8

QoS and Dial Access

This chapter deals with the delivery of Quality of Service (QoS) mechanisms at the demand dial-access level. Some would argue that discernible quality of any IP service is impossible to obtain when using a telephone modem. Most of the millions of people who use the Internet daily are visible only at the other end of a modem connection, however, and QoS and dial access is a very real user requirement.

So what environment are we talking about here? Typically, this issue concerns dial-in connections to an underlying switched network, commonly a Public Switched Telephone Network (PSTN) or an Integrated Services Digital Network (ISDN), that is used as the access mechanism to an Internet Service Provider (ISP). The end-user's system is connected by a dynamically activated circuit to the provider's Internet network.

There are a number of variations on this theme, as indicated in Figure 8.1. The end-user environment may be a single system (Panel A), or it may be a local network with a gateway that controls the circuit dynamically (Panel B). The dynamics of the connectivity may be a modem call made across the telephone network or an ISDN data call made across an ISDN network. This is fast becoming a relatively cosmetic difference; integration of ISDN and analog call-answer services finally has reached the stage where the service provider can configure a single unit to answer both incoming ISDN and analog calls, so no significant differences exist between analog modem banks and ISDN Primary Rate Interface (PRI) systems. Finally, the logical connection between the user's environment and the ISP may be layered directly on the access connection. Alternatively, the connectivity may use IP tunneling to a remote ISP so that the access service and Internet service are operated by distinct entities in different locations on the network (Panel C).

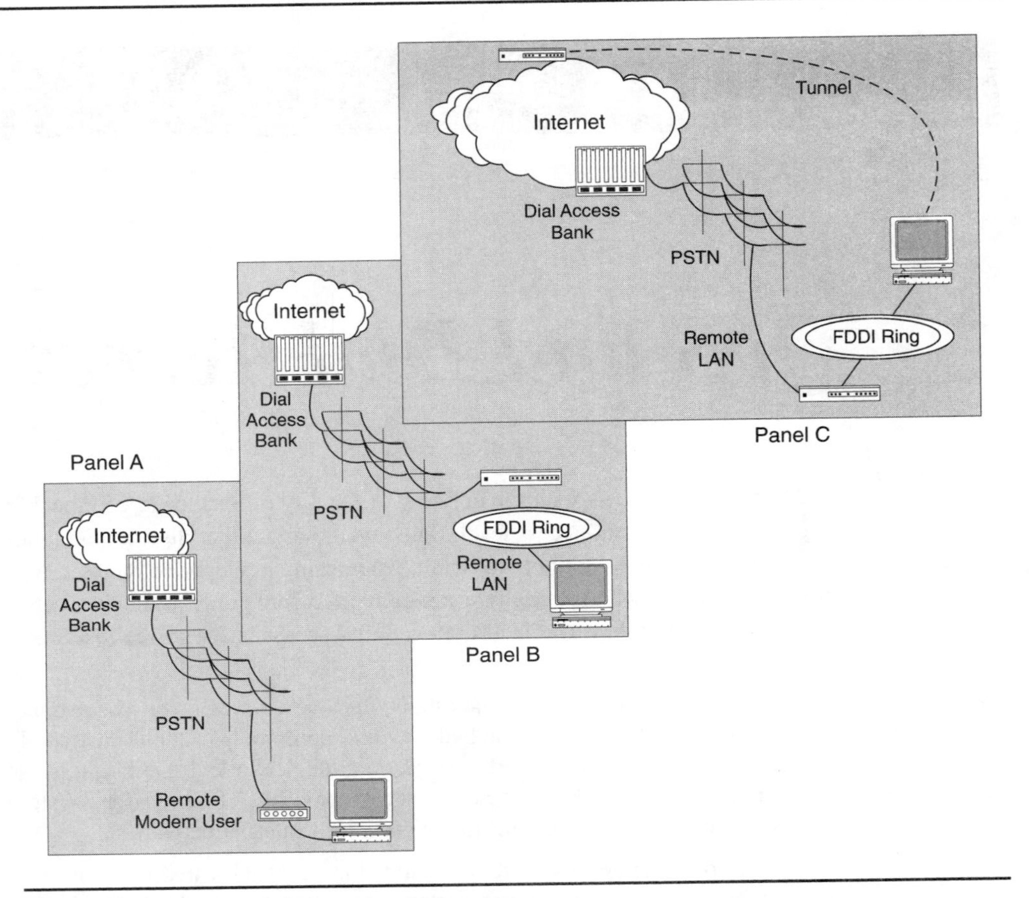

Figure 8.1: *Typical dial-access environments.*

Answering the Dial-Access Call

Answering a call in the traditional telephony world does not appear to be a particularly challenging problem. When the phone rings, you pick up the handset and answer the call. Of course, it is possible to replicate this simple one-to-one model for dial access. If the ISP is willing to provide dedicated access equipment for the exclusive use of each subscriber, whenever the client creates a connection to the service, the dedicated equipment answers the call and completes the connection.

Of course, when the client is not connected, the equipment remains idle. It is this state of equipment idleness, however, that makes this exclusive provisioning strategy a high-cost option for the service provider and a relatively expensive service for the subscriber. Competitive market pressures in the service-provider arena typically dictate a different approach to the provisioning of access ports to achieve a more efficient balance between operating cost and service-access levels.

Dial-Access Pools

A refinement of this exclusivity model is sharing a pool of access ports among a set of clients. As long as an adequate balance exists among the number of clients, the number of access ports, the average connection period, and times when the connection is established, the probability of a client not being able to make a connection because of exhaustion of the pool of access ports can be managed to acceptably low levels. The per-client service costs are reduced, because the cost of operation of each access port can be defrayed over more than one client, and the resulting service business model can sustain lower subscription fees. Accordingly, such a refinement of the access model has a direct impact on the competitive positioning of the service provider.

Considering Cost Structure

Unlike a more traditional telephony model of service economics, where the capital cost of infrastructure can be deferred over many years of operation, the Internet has a more aggressive model of capital investment with operational lifetimes of less than 18 months, particularly in the rapidly moving dial-access technology market. Accordingly, the considerations of the efficiency of utilization of the equipment plant are a major factor in the overall cost structure for access service operators.

Typically, in such environments, a pool of access ports is provided for a user community, and each user dials a common access number. The access call is routed to the next available service port on the ISP's equipment (using a rotary access number), the call is answered, and the initial authentication phase of the connection is completed (Figure 8.2).

In such a shared-pool configuration, the intent is to have adequate port availability to meet peak demands (Figure 8.3) while ensuring that the access port bank is not excessively provisioned. In this environment, the users compete against each other for the exclusive use of an access port, because the number of available access ports typically is lower than the number of users.

Such full servicing during peak periods is not always possible when using the shared-pool approach. Although it may be possible to continually provision a system to meet all peak demands within the available pool, this is an unnecessarily expensive mode of operation. More typically, service providers attempt to provision the pool size so that there are short periods in the week when the client may receive a busy signal when attempting to access the pool. This signals that the access port pool is exhausted and the subscriber simply should redial.

Queuing Analysis of Dial-Access Pools

Management of the dial-access pool form of congestion control is well understood, and the discipline of queuing theory is a precise match to this problem. Using queuing theory as a tool, it is possible to model the client behavior and dimension a pool of access ports so that

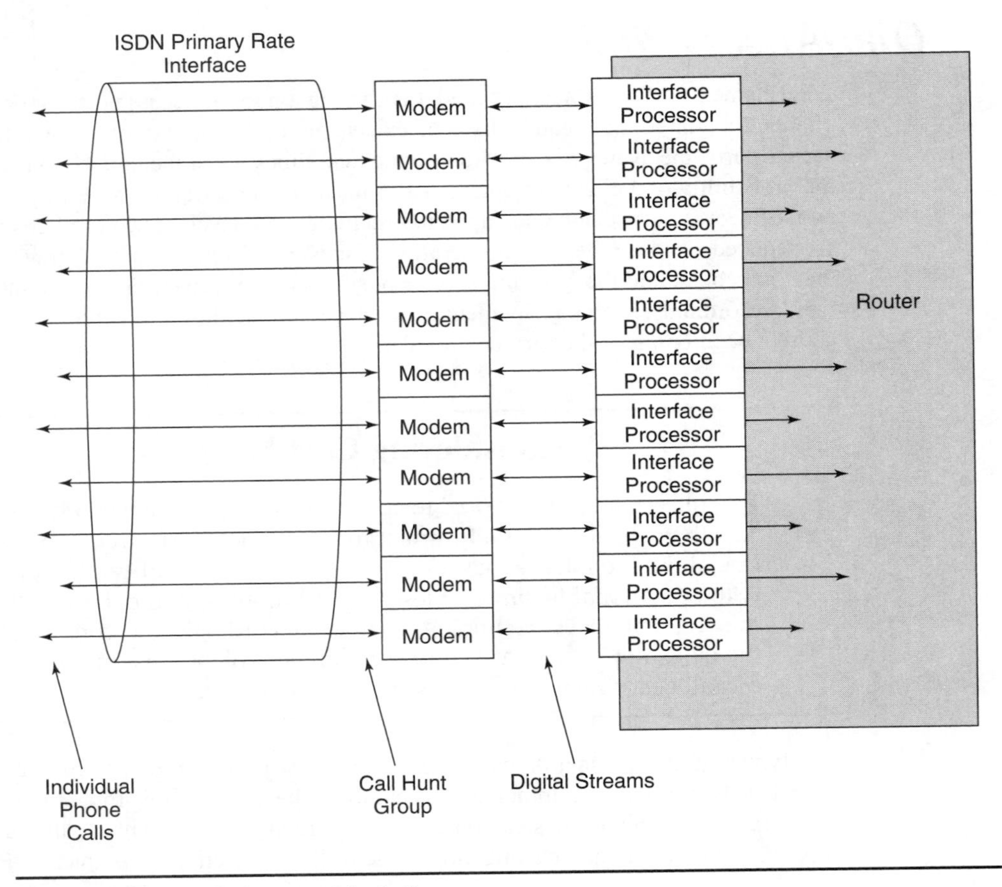

Figure 8.2: *A dial-access block diagram.*

the delay to access an available port is bounded by time, and the delay can be selected by dimensioning the number of individual access servers.

Such engineering techniques are intended to bound the overall quality of the access service for all clients. The periods during which the access bank returns a busy signal can be determined statistically, and the length of time before a client can establish a connection during such periods also can be statistically determined from a queuing model and can be observed by monitoring the occupancy level of the modem bank to a sufficiently fine level of granularity. You should note that the queuing theory will predict an exponential probability distribution of access-server saturation if the call-arrival rate and call-duration intervals are both distributed in a Markovian distribution model. You can find a complete treatment of queuing theory in [Kleinrock1976], and an analysis of a connection environment is available in [Tannenbaum1988].

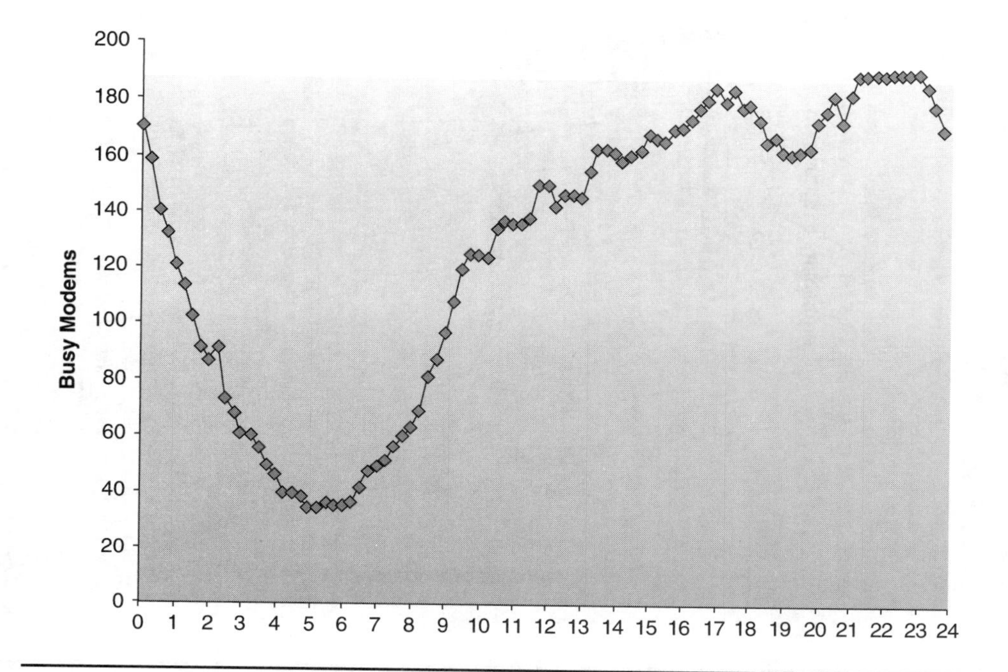

Figure 8.3: *Modem-pool use over 24 hours.*

TIP A.A. Markov published a paper in 1907 in which he defined and examined various behaviors and properties that create interdependencies and relationships among certain random variables, forming a stochastic process. These processes are now known as *Markovian processes*. More detailed information on Markovian theory can be found in *Queuing Systems, Volume I: Theory*, by Leonard Kleinrock, and published by John Wiley & Sons (1975).

If you are interested in queueing theory, its practice, and its application, Myron Hylkka's Queueing Theory Page at www2.uwindsor.ca/~hlynka/queue.html contains a wealth of information on these topics.

The probability that an access-port configuration with n ports will saturate and an incoming call will fail can be expressed by this relationship:

$$P_n = \frac{pn/n!}{\sum_{k=0}^{n} pk/k!}$$

Where the traffic intensity, r, is a function of the call rate λ and the service interval μ:

$$r = \frac{l}{m}$$

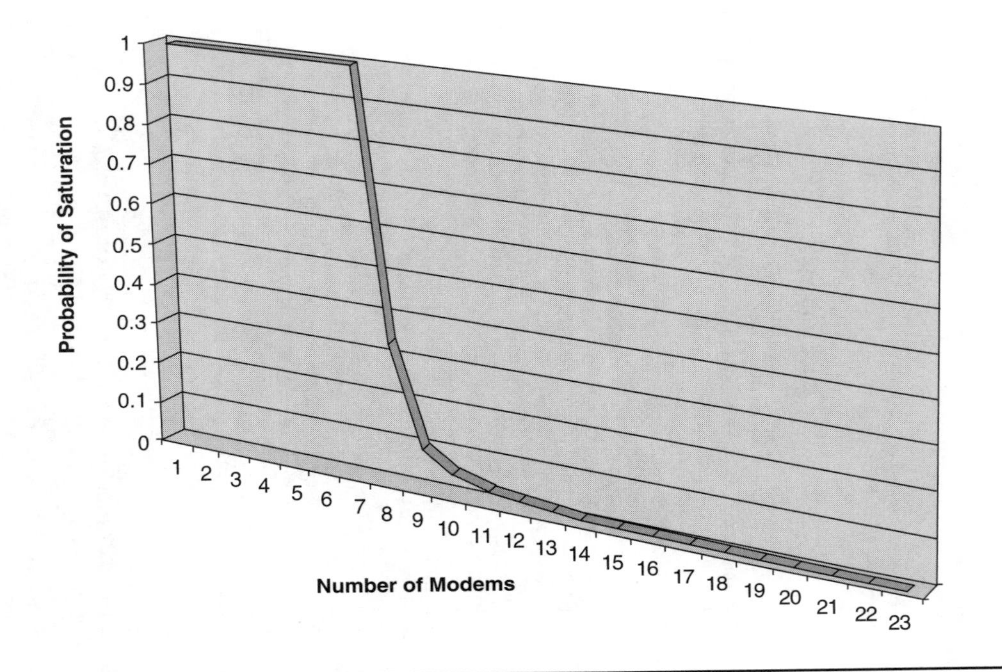

Figure 8.4: *Queueing theory model of modem-pool availability.*

An analysis of this model, putting in values for λ of 0.6 and μ of 0.59, and assuming that each call will be held for a minimum of 5 units, appears in Figure 8.4.

The conclusion from this very brief theoretical presentation is that in sufficiently large client population sets and access pools, the behavior of the pool can be modeled quite accurately by using queuing theory. This theoretical analysis yields a threshold function of pool provisioning—there is a point at which the law of diminishing returns takes over and the addition of another access port to the pool makes a very minor change to the probability of denial of service at the busy period. The consequent observation is that the "last mile" of provisioning an access pool that attempts to eliminate denial of service access is the most expensive component of the access rack, because the utilization rates for such access ports drop off sharply after the provisioning threshold is reached.

Alternative Access—Pool Management Practices

Instead of provisioning more access ports to the pool, which would be busy only during very limited peak-usage periods (and hence would generate usage-based revenue for very limited periods each day), another option is to change the access-port management policies. An approach along such lines could introduce prioritized allocation of access ports to increase the overall utilization rate of each modem and still provide a high-availability rate of access ports, but only to a subset of the total client population that wants a higher quality of service.

How can you introduce a scheme that prioritizes the allocation of access ports so that the contention can be reduced for a subclass of the serviced client base? Does the opportunity exist to differentiate clients, allowing preemptive allocation of access ports to meet the requirements of a differentiated section of the client pool?

The current state of the art as far as dial access is concerned ends at this point. It is simply the case that an integration of the two parts of dial access—the call-switching environment and the Internet-access environment—does not exist to the extent necessary to implement such QoS structures at the access level. However, such observations should not prevent you from examining what you need to implement differentiated classes of access service and the resulting efficiencies in the operation of any such scheme.

Managing Call-Hunt Groups

The basic access model makes very few assumptions about the switching mechanisms of the underlying access network substrate. If a free port of a compatible type exists within the access bank, the port is assigned to the incoming call, and the call is answered. For an analog modem configuration, this is a call-hunt group, in which a set of service numbers is mapped by the PSTN switch to a primary hunt-group number. When an incoming call to the hunt-group number is detected, the switch maps the next available service number to the call, and the call is routed to the corresponding modem. For ISDN, this is part of the functionality of Primary Rate Access (PRA), where incoming switched calls made to the PRA address are passed to the next available B-channel processor.

Addressing the Problems of Call-Hunt Groups

The call-hunt group approach has two problems. First, if the modem or channel processor fails without notifying the hunt-group switch of a modem fault, and if the hunt group is noncyclic, all subsequent calls are trapped to the faulty modem, and no further calls can be accepted until the modem condition is cleared. Second, no mechanism exists for preemption of access. When the access bank is fully subscribed, no further incoming calls can be answered until a line hang-up occurs on any of the currently active ports.

The problem of modem faults causing call blocking can be addressed with a simple change to the behavior of the hunt-group switch, which searches for a free port at the next port in sequence from where the last search terminated. If a client is connected to a faulty port, hanging up the connection and redialing the hunt-group number causes the hunt-group switch to commence the search for a free port at the modem following in sequence from the faulty port. Although this is an effective workaround, it is still far from ideal. The missing piece of the puzzle here is a feedback signal from the modem to the call-hunt group, so that the modem manager agent can poll all modems for operational integrity and pull out of the call-hunt group any modems that fail such an integrity self-test.

The second problem, the lack of preemption of service, is not so readily handled by the access equipment acting in isolation. The mechanism to allow such preemptive access lies in a tighter integration of the call-management environment and the access equipment. One potential model uses two hunt-group numbers: a *basic* and a *premium* access number overlaid on a single access port bank. If the basic access number is called, ports are allocated and incoming calls are answered up to a high-water mark allocation point of the total access-port pool. All subsequent access calls to this basic access number are not answered until the total amount of the busy port numbers falls below a low-water allocation mark. All calls to a premium access service can be allocated on any available port.

In this way, the pool of ports available to the differentiated premium access service is the total number of available ports reserving a pool (equal to the total pool size minus the high-water mark point) for the differentiated access service, as shown in Figure 8.5. With a differential number of effective servers for the two populations, where the smaller population is accessing (in effect) a larger pool of access servers, queuing analysis of the system results in significantly shorter busy wait times for the elevated priority client group. To implement preemptive access, you must pass the information related to the incoming call (caller's number and number called) to the access equipment to allow the access equipment to apply configuration rules to the two numbers to determine whether to accept or block the call. Recording the called number for the session-accounting access record allows the service provider to bill the client on a differential fee structure for the higher-access service class.

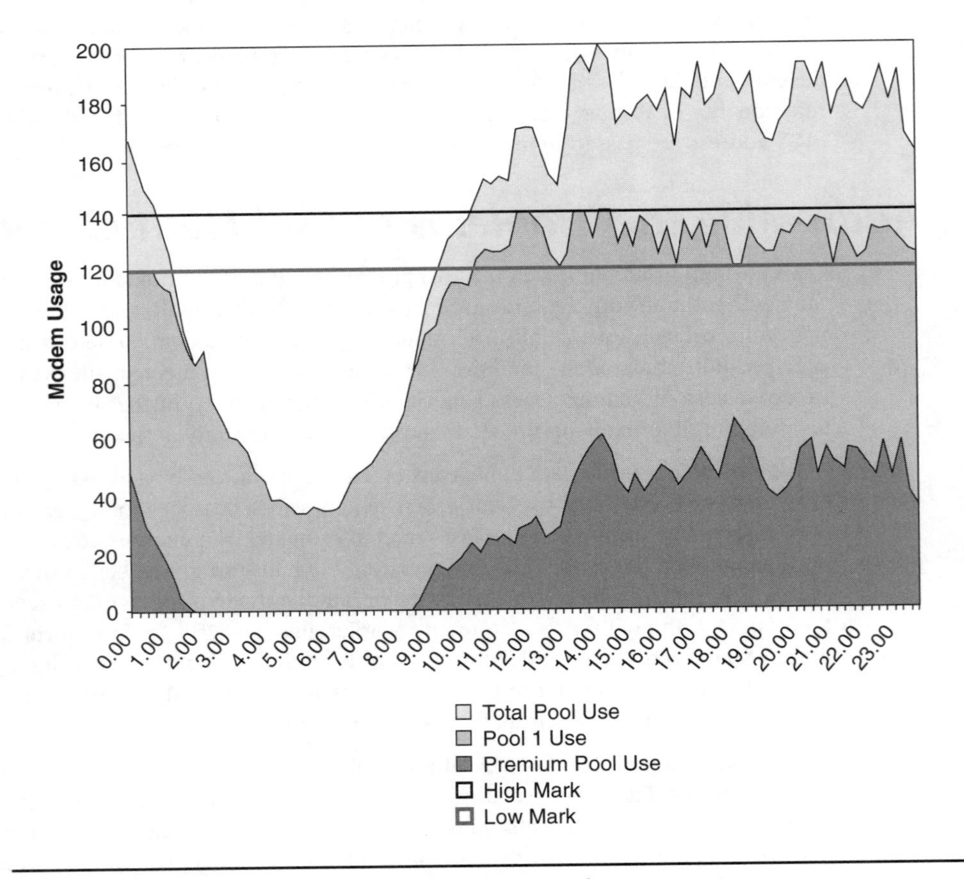

Figure 8.5: *Modem use: high- and low-water marks.*

Further refinements are possible if the signaling between the call switch and the access server is sufficiently robust enough to allow passback of the call. In such an environment, the switch can implement overflow to other access groups in the event of saturation of the local access port, by the access server sending a local congestion signal back to the switch, which allows the call to be rerouted to an access server pool with available ports.

From the perspective of the client, there are two levels of service access: a *basic rate-access* number, which provides a high probability of access in off-peak periods but with reduced service levels during peak-use periods, and a *premium-access* number, which provides high availability at all times, with presumably a premium fee to match.

The intended result of each of these schemes is to provide a differentiated service. This service would provide mechanisms that would allow the service operator to increase the average equipment utilization rate of the access equipment. This in turn would offer a number of higher-quality, differentiated access services and these, presumably, would also be differentiated according to subscriber-fee structure and business model.

Differentiated Access Services at the Network Layer

After the access call is assigned to an access port and data-link communications is negotiated, the next step is to create an IP connection environment for the call. The most critical component of this step is to authenticate the identity of the client in a secure fashion. Several technologies have been developed to support this phase of creating a remote-access session, most notably TACACS (and its subsequent refinements) [IETF1993a], Kerberos [IETF1993b], CHAP [IETF1996d], PAP [IETF1993c], and RADIUS [IETF1997d]. For the purpose of examining QoS structures in remote access, the most interesting of these approaches is RADIUS, which not only can undertake the authentication of the user's identity, but also can upload a custom profile to the access port based on the user's identity. This second component allows the access server to configure itself to deliver the appropriate service to the user for the duration of the user's access session.

TIP Predominately, these technologies consist of two remote access-control mechanisms called TACACS and RADIUS. See RFC1492, "An Access Control Protocol, Sometimes Called TACACS," C. Finseth, July 1993 and RFC2138, "Remote Authentication Dial-In User Service (RADIUS)," C. Rigney, A. Rubens, W. Simpson, S. Willens, April 1997. An updated variation of the TACACS protocol was introduced by Cisco Systems, Inc., called TACACS+. However, RADIUS is still more widely used in the networking community and is considered by many to be the predominate and preferred method of remote-access control for authentication, authorization, and accounting for dial-up services.

Remote Authentication Dial-In User Service (RADIUS)

RADIUS uses a distributed client/server architecture in which the Network Access Server (NAS) is the client and the RADIUS server can hold and deliver the authentication and profile information directly or act as a proxy to other information servers. Subscriber authentication typically uses the provided username and an associated password. Variants of this approach can be configured to use *calling number* information or *called number*, which can be useful in applications that configure closed environment information servers within a larger Internet environment. Authentication may involve a further user challenge/response phase at this stage. The final response from the RADIUS server is to *Access-Reject* or *Access-Accept*. The *Access-Accept* response contains a profile to be loaded into the NAS and, from the perspective of supporting QoS, is most relevant here.

Figure 8.6 shows the precise format of the RADIUS *Access-Accept* packet. The packet has identification headers and a list of attribute values encoded in standard TLV (Type, Length, Value) triplets. The RADIUS specification [IETF1997d] specifies attribute types 1 through 63, noting that the remainder of the types are available for future use and reserved for experimentation and implementation-specific use.

RADIUS allows the network service provider to determine a service profile specific to the user for the duration of the access session. The *Access-Accept* packet allows the RADIUS server to return a number of attribute-value pairs; the precise number and interpretation is left to the RADIUS access server and client to negotiate. The implicit assumption is that default values for these attributes are configured into the NAS, and the RADIUS *Access-Accept* message has the capability to set values for attributes that are not defaulted.

Within the defined attribute types, the attributes that can be used to provide QoS structure are described in the following sections.

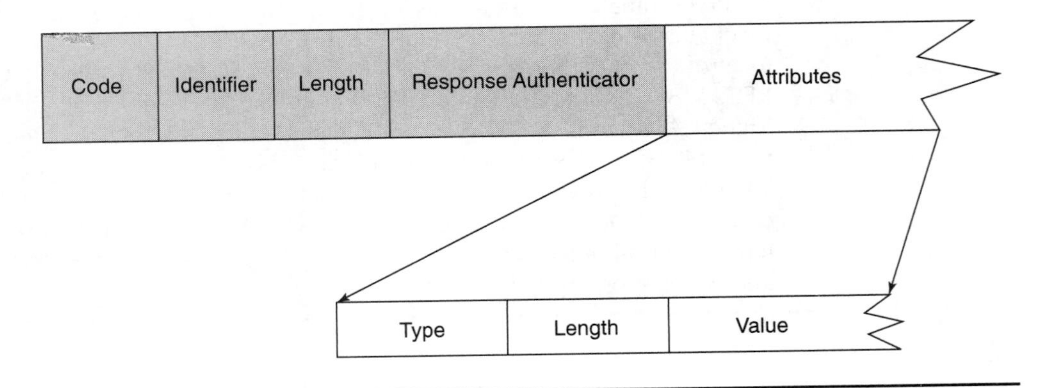

Figure 8.6: *RADIUS Access-Accept packet format.*

Session-Timeout

The *Session-Timeout* attribute defines the maximum session duration before the NAS initiates session termination. Within a QoS environment, service differentiation can include the capability to set longer (or unbounded) session times.

Idle-Timeout

The *Idle-Timeout* attribute sets the maximum idle period before the NAS initiates session termination. Again, within a QoS environment, this attribute can vary according to the service class.

Port-Limit

Using the PPP (Point-to-Point Protocol) Multilink protocol [IETF1996e], it is possible to logically configure a number of parallel access circuits into a common virtual circuit. This attribute is sent to the NAS to place an upper limit on the number of ports that can be configured into a PPP Multilink session. Within the context of QoS, this attribute can be used to define an upper bound on the aggregate access bandwidth available to a single user. As the traffic rate to (or from) the user reaches a point where the line is saturated for a determined interval, the user's equipment can activate additional circuits into the PPP Multilink group, relieving the congestion condition. This ordinarily is used in ISDN access configurations, where the second B-channel of a Basic Rate Interface (BRI) can be activated on a dial-on-demand basis.

Other RADIUS Attributes

The RADIUS specification also allows extensions to the RADIUS attribute set that are implementation specific or experimental values. Interestingly enough, there is no further negotiation between the RADIUS server and client, so if an attribute value is specified that is unknown to the NAS, no mechanism exists for informing the server or the client that the attribute has not been set to the specified value.

Extensions to the RADIUS Profile

A number of interesting potential extensions to the RADIUS profile set can be used to define a QoS mechanism relevant to the dial-access environment.

The first possibility is to set the IP precedence field bit settings of all IP packets associated with the access session to a specified value as they enter the network. This could allow QoS mechanisms residing in the network to provide a level of service with which the settings correlate. As noted previously, such settings may be used to activate precedence-queuing operations, weighting of RED (Random Early Detection) congestion-avoidance mechanisms, and similar service-differentiation functions. Obviously, such a QoS profile can be set as an action that is marked on all user packets as they enter the network. An alternative interpretation of such an attribute is to use the field settings as a mask that defines the maximum QoS profile the user can request, allowing the user's application to request particular QoS profiles on a user-determined basis.

The immediately obvious application of QoS profiles is on ingress traffic to the network. Of possibly higher importance to QoS is the setting of relative levels of precedence to egress traffic flows as they are directed to the dial-up circuit. For a dial-connected client making a number of simultaneous requests to remote servers, the critical common point of flow-control management of these flows is the output queue of the NAS driving the dial-access circuit to the user.

A 1500-byte packet on a 28.8 Kbps modem link, for example, makes the link unavailable for 400 ms. Any other packets from other flows arriving for delivery during this interval must be queued at the output interface at the NAS. If the remote server is well connected to the network (well connected in terms of high availability of bandwidth between the NAS and the server at a level that exceeds the bandwidth of the dial-access circuit), as the server attempts to open up the data-transfer rate to the user's client, the bandwidth rate attempts to exceed the dial-access circuit rate, causing the NAS output queue to build. Continued pressure on the access bandwidth, brought about by multiple simultaneous transfers attempting to undertake the same rate increase, leads to queue saturation and nonselective tail-drop discard from the output queue. If QoS mechanisms are used by the server and supported by the NAS, this behavior can be modified by the server; it can set high precedence to effectively saturate the output circuit. However, if QoS is a negotiated attribute in which the dial-access client also plays an active role, this server-side effective assertion of control over the output queue still is somewhat inadequate.

An additional mechanism to apply QoS profiles is to allow the service provider to set the policy profile of the output circuit according to the user's RADIUS profile. This can allow the user profile to define a number of filters and associated access-rate settings to be set on the NAS output queue, allowing the behavior of the output circuit to respond consistently with the user-associated policy described in the RADIUS profile. Accordingly, the user's RADIUS profile can include within the filter set whether or not the output queue management is sensitive to packets with QoS headings. Equally, the profile can specify source addresses, source ports, relative queuing priorities, and/or committed bandwidth rates associated with each filter.

The general feature of dial access in an environment with limited bandwidth at the access periphery of the network with a high-bandwidth network core is that the control mechanism for QoS structures focuses on the outbound queue feeding the access line. Although the interior of the network may exhibit congestion events, such congestion events are intended to be transitory (with an emphasis on *intended* here), while the access line presents to the application flow a constant upper limit on the data-transfer rate. Accordingly, it is necessary in the dial-access environment to focus QoS-management structures to work on providing reliable control over the queue management of the dial circuit outbound queue at the NAS.

These two potential extensions into the dial-access environment are highly capable in controlling the effective relative quality of service given to various classes of traffic passing to and from the access client. These extensions also exhibit relatively good scaling properties, given that the packet rate of dial-access clients is relatively low, and the subsequent per-access port-processing resources required by the NAS to implement these structures is relatively modest. This is a good example of the general architectural principle of pushing QoS traffic-shaping mechanisms to the edge of the network and, in doing so, avoiding the scaling issues of attempting to impose extensive per-flow QoS control conditions into the interior high-speed components of the network.

Layer 2 Tunnels

The model assumed in the discussion of this topic so far as that the dial-access service and the network-access service are bundled into a single environment, and the logical boundary between the network-access service and the user's dial-access circuit is within the NAS device. Within the Internet today, this model has posed a number of scaling issues related to the provisioning of very large-scale analog or ISDN access services to the service providers' locations and has lead to dramatic changes in the switching load imposed on the underlying carrier network in routing a large number of access calls from local call areas to a single site. This model can enable you to significantly alter the provisioning models used in the traditional PSTN environment, which in turn calls for rapid expansion of PSTN capacity to service this new load pattern, or to examine more efficient means to provide dial-access services within a model that is architecturally closer to the existing PSTN model. Given that the latter approach can offer increased scaling of the dial-access environment without the need for associated capital investment in increased PSTN switching capacity, it is no surprise that this is being proposed as a model to support future ubiquitous dial-access environments.

Such an approach attempts to unbundle the dial-access service and the network-access service, allowing multiple network-access services to be associated with a single NAS. The incoming call to one of a set of network-access service call numbers is terminated on the closest common NAS that supports access to the called number, where the PSTN uses intelligent PSTN network features to determine what is closest in terms of PSTN topology and NAS access-port availability. The user-authentication phase then can use the called number to determine the selected access service, and the NAS's RADIUS server can invoke, via a RADIUS proxy call, the RADIUS user-authentication process of the called network access service. The user profile then can be specified by the remote network access and downloaded into the local NAS. The desired functionality of this model is one in which the network-access services can be a mixture of both competing public Internet dial-access services and Virtual Private Network (VPN) access services, so that the NAS must support multiple access domains associated with multiple network access services. This is known as a Virtual Private Dial Network (VPDN).

The implementation of this functionality requires the use of Layer 2 tunneling features to complete the access call, because the end-to-end logical connection profile being established in this model is between the client device and the boundary of the network dial-access service's infrastructure. The user's IP platform that initiates the call also must initiate a Layer 2 peering with a virtual access port provided by the selected network-access service for the access session to be logically coherent in terms of IP routing structures. All IP packets from the dial-access client must be passed without alteration to the remote network access server, and all IP packets from the remote network access server must be passed to the NAS for delivery across the dial-access circuit to the client. The local NAS is placed in the role of an intermediary Layer 2 repeater or bridge, translating from the dial-access Layer 2 frames over the dial circuit to Layer 2 frames over some form of Layer 2 tunnel. This process is illustrated in Figure 8.7. Such a tunneling environment is described in the Layer 2 Tunneling Protocol (L2TP) [ID1997a1].

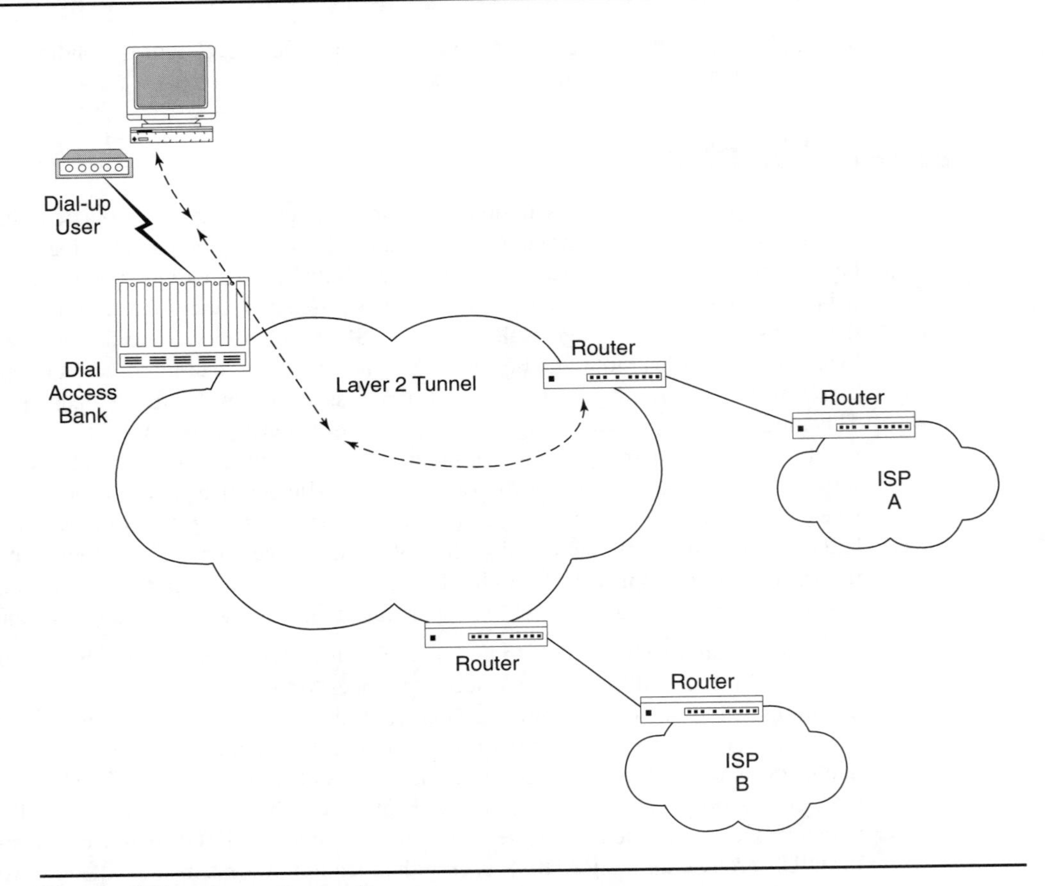

Figure 8.7: *Layer 2 tunneling.*

As a tunneling protocol, L2TP does not specify any particular transport substrate for the tunnel, allowing the tunnel to be implemented over Frame Relay PVCs or as IP-encapsulated tunnels, for example. The IP encapsulation mode of tunnel operation is discussed here, with particular reference to QoS mechanisms that can be undertaken on the tunnel itself.

The general nature of the problem is whether the IP transport level should signal per-packet QoS actions to the underlying IP transport path elements and, if so, how the data-link transport system can signal its dynamic capability to match the desired QoS to the actual delivery.

Tunnel-encapsulation specifications using IP as the tunnel substrate [IETF1991a, IETF1994d] typically map the clear channel QoS fields of the IP header into the QoS fields of the tunnel IP header, allowing the tunnel to be supported in a semitransparent mode, where the tunnel path attempts to handle tunneled packets using QoS mechanisms activated indirectly by the encapsulated packet. In doing so, the behavior of the tunnel transport may differ from many *native-mode* data-link transport technologies, because the order of packets placed into the tunnel is not necessarily the order of packets leaving the tunnel. Packets placed into the tunnel with a higher preference QoS may *overtake* packets placed into the tunnel with a lower QoS within the IP tunnel itself, and sequence-sensitive end-to-end Layer 2 actions, such as header compression, may be adversely affected by such an outcome. This leads to a corollary of the QoS attribute setting: To ensure predictability of behavior if a data flow requests distinguished service, all packets within that flow should request the same service level. Thus, with reference to TCP flows, the QoS bits should remain the same from the initial SYN (TCP Synchronize Sequence Numbers Flag in a TCP session establishment negotiation) packets to the final FIN ACK (Finish Flag in a TCP Acknowledgment, indicating the acknowledged end of a session) packets, and both ends of the connection should use the same QoS settings.

As the environment of distributed dial-access services becomes more prolific, you can anticipate further refinement of the tunneling model. It may be the case that tunnel encapsulation would map to an RSVP-managed flow between the NAS and the remote network access server and, instead of making the tunnel QoS transparent and relying on best-effort QoS management in the tunnel path, the tunnel characteristics could be specified at setup time and the clear data frames would be placed into a specific RSVP-managed flow tunnel depending on the QoS characteristics of the frame. The use of this type of traffic differentiation for voice, video, and non-real-time data applications is immediately obvious. Although this mechanism does imply a proliferation of tunnels and their associated RSVP flows across the network, the attraction of attempting to multiplex traditional telephony traffic with other data streams is, in industry terms, extraordinarily overwhelming. The major determinant of capability here is QoS structures.

RSVP and Dial Access

As you can see, a need seems to exist for guaranteed and controlled-load flows across low-speed links. The problem that must be addressed here is that large data packets can take significant data-transmission times, and some form of preemption must be specified to support such flows across the dial-access link.

One approach is to use a reduced maximum packet size on the link to bound the packet-transmission latency. Such an approach filters back through the network to the server, via MTU (Maximum Transmission Unit) discovery, and increases the overall network packet-switching load. An alternative approach has been advocated to introduce fragmentation within the PPP Multilink protocol using a *multiclass extension* to the PPP Multilink protocol [ID1997a2]. This approach attempts to allow the transmission of larger packet to be fragmented, so as to defer to a flow a packet with defined latency and jitter characteristics. Extensions to the HDLC (High-Level Data Link Control) framing protocol also could provide this preemption capability, although with an associated cost of byte-level scanning and frame-context blocks.

However, there is a real issue with attempting to offer guaranteed access rates across analog modem connections: The data-bit rate of the modem circuit is variable. First, the carrier rate is adjusted dynamically by the modems themselves as they perform adaptive carrier-rate detection throughout a session. Second, the extent of the compressibility of the data causes the data-bit rate to vary across a constant carrier rate. The outcome of this observation is that across the dial circuit, it is reasonable to expect, at most, two or three RSVP-mediated flows to be established in a stable fashion at any time. As a consequence, it still is a somewhat open issue whether RSVP or outbound traffic queue management is the most effective means of managing QoS across dial-access circuits.

Like so many other parts of the QoS story on the Internet, the best that can be said today is "watch this space."

chapter 9

QoS and Future Possibilities

There are, of course, several ongoing avenues of research and development on technologies that may provide methods to deliver Quality of Service in the future. Current methods of delivering QoS are wholly at the mercy of the underlying routing system, whereas *QoS Routing* (QoSR) technologies offer the possibility of integrating QoS distinctions into the network layer routing infrastructure. The second category is *MPLS* (Multi Protocol Label Switching), which has been an ongoing effort in the IETF (Internet Engineering Task Force) to develop an integrated Layer 2/Layer 3 forwarding paradigm. Another category examined is a collection of proposed *RSVP extensions* that would allow an integration of RSVP and the routing system. This chapter also examines the potential of introducing QoS mechanisms into the IP multicast structure. Additionally, you'll look at the possibilities proposed by the adoption of congestion pricing structures and the potential implementation issues of end-to-end QoS within the global Internet. And finally, you'll look briefly at the IPv6 protocol, which contains a couple of interesting possibilities of delivering QoS capabilities in the future.

QoS Routing Technologies

QoS Routing (QoSR) is "a routing mechanism within which paths for flows are determined based on some knowledge of resource availability in the network, as well as the QoS requirements of flows" [ID1997t]. QoSR is considered by many to be the missing piece of the puzzle in the effort to deliver *real* quality of service in data networks. To this end, a new working group in the IETF, the QoSR working group, has been formed to design a framework and examine possible methods for developing QoSR mechanisms in the Internet.

> **TIP** You can find the IETF QoSR working group charter and related documents at www.ietf.org/html.charters/qosr-charter.html.

QoSR presents several interesting problems, the least of which is determining whether the QoS requirements of a flow can be accommodated on a particular link or along a particular end-to-end path. Some might argue that this specific issue basically has been reduced to a solved problem with the tools provided by the IETF Integrated Services architecture and RSVP. However, no integral relationship exists between the routing system and RSVP, which can be considered a shortcoming in this particular QoS mechanism.

Because the current routing paradigm is destination-based, you can assume that any reasonably robust QoSR mechanism must provide routing information based on both source and destination, flow identification, and some form of flow profile. RSVP side-steps the destination-based routing problem by making resource reservations in one direction only. Although an RSVP host may be both a sender and receiver simultaneously, RSVP senders and receivers are logically discrete entities. Path state is established and maintained along a path from sender to receiver, and subsequently, reservation state is established and maintained in the reverse direction from the receiver to the sender. Once this tedious process is complete, data is transmitted from sender to receiver. The efficiency concerns surrounding RSVP are compounded by its dependency on the stability of the routing infrastructure, however.

When end-to-end paths change because of the underlying routing system, the path and reservation state must be refreshed and reestablished for any given number of flows. The result can be a horribly inefficient expenditure of time and resources. Despite this cost, there is much to be said for this simple approach in managing the routing environment in very large networks. Current routing architectures use a unicast routing paradigm in which unicast traffic flows in the opposite direction of routing information. To make a path symmetrical, the unicast routing information flow also must be symmetrical. If destination A is announced to source B via path P, for example, destination B must be announced to source A via the same path. Given the proliferation of asymmetric routing policies within the global Internet, asymmetric paths are a common feature of long-haul paths. Accordingly, when examining the imposition of a generic routing overlay, which attempts to provide QoS support, it is perhaps essential for the technology to address the requirement for asymmetry, where the QoS paths are strictly unidirectional.

Intradomain QoSR

A reasonably robust QoSR mechanism must provide a method to calculate and select the most appropriate path based on a collection of metrics. These metrics should include information about the bandwidth resources available along each segment in the available path, end-to-end delay information, and resource availability and forwarding characteristics of each node in the end-to-end path.

Resource Expenditure

As the degree of granularity with which a routing decision can be made increases, so does the cost in performance. Although traditional destination-based routing protocols do not provide robust methods of selecting the most appropriate paths based on available path and intermediate node resources, for example, they do impact less on forwarding performance, impose the least amount of computational overhead, and consume less memory than more elaborate schemes. As the routing scheme increases in complexity and granularity, so does the amount of resource expenditure.

The following holds true when considering resources consumed by a given routing paradigm:

A < B < C < D

Here, A represents traditional destination-based routing. B is routing that computes the best path based on source and destination information. C is a flow-based routing scheme. D is a flow-based QoSR scheme that also computes the most appropriate path based on available per-path bandwidth and delay characteristics.

Resource consumption notwithstanding, [ID1997t] reinforces the consensus that path computation based on certain combinations of metrics to include delay and jitter is difficult. Therefore, it generally is acceptable to make certain concessions with regard to optimum path computation in exchange for decreased computational complexity. Some metrics may be considered primary, others secondary, and other available metrics ranked in a similar fashion, for example. By implementing a series of ordered, sequential computational comparisons, it may be sufficient to consider the result of comparing of two or more primary metrics instead of a computationally intensive comparison of all available metrics. If recursive computation is required, comparisons of secondary metrics may then be performed.

Don't forget that the principal objective of QoSR is that, given traffic with a clearly defined set of QoS parameters, routers must calculate an appropriate path based on available QoS path metrics without degrading the overall performance of the network. Therefore, administrative controls should be implemented to reduce the amount of traffic as much as possible for which these types of path computations must be made. Of course, this leaves room for creative interpretation. Various administrative mechanisms that provide admission control, prejudicial packet drop with IP precedence, and partitioning of available bandwidth may be candidates in this regard.

A modified link-state routing protocol would appear to be an ideal candidate for QoSR, because the changes in the network topology state are quickly detected and communicated to neighbors in the routing domain. However, given the increased amount of information that would be contained in LSAs (Link-State Advertisements), the additional propagation overhead may be inefficient. This particular aspect may be an architectural design issue, however, and may require the number of peers within a routing domain to be fewer as a result.

An immediate feedback loop exists between path selection and traffic flows. If a path is selected because of available resources and traffic then is passed on this path, the available resource level of the path deteriorates, potentially causing the next iteration of the QoSR protocol to select an alternate path, which may in turn act as an attractor to the traffic flows. As established in the ARPAnet many years ago, where the routing protocol uses measured Round Trip Time (RTT) as the link metric, such routing algorithms with similar feedback loops from traffic flows to the routing metric often suffer from the inability to converge to a stable state. The problem of similar QoSR protocols is the management aspect of this feedback loop to allow some level of workable stability.

QoSR should be scaleable, and the side-effect of having to reduce the number of peers in a routing area or subdomain is somewhat disconcerting. Therefore, any QoSR approach also should provide a method for hierarchical aggregation, so that scaling properties are not significantly diminished by constraints imposed in smaller subdomains. However, introducing aggregation into this equation may produce a problem in maintaining accuracy in the path state information when senders and receivers reside in different subdomains.

Intradomain QoSR Requirements

An *intradomain* QoSR scheme must satisfy a set of minimum requirements [ID1997t]. The QoSR protocol must route a flow along a specific path that can accommodate its QoS requirements, or it must provide a mechanism to indicate why the flow has not been admitted. The protocol also must indicate disturbances in the route of a flow because of topological changes. The QoSR protocol must accommodate best-effort traffic without requiring a resource reservation. The routing protocol should support QoS-based multicast traffic with receiver heterogeneity and shared reservation styles (when a resource reservation protocol is used). It also is recommended that the QoSR make a reasonable effort to minimize computational and resource overhead and provide some mechanisms to allow for administrative admission control.

Depending on the mechanism used for path selection, there are two possibilities for route processing at a *first-hop router* (the first router in the traffic path). A first-hop router simply may forward traffic to the most appropriate next-hop router, as is traditionally done in hop-by-hop routing schemes, or it may select each and every intermediate node through which the traffic will traverse in the end-to-end path to its destination, also called *explicit routing*.

PNNI (Private Network-to-Network Interface) may offer some useful insights into how to address this topic, because it provides a form of QoS routing for Virtual Connection (VC) establishment in ATM networks. Although PNNI does not actually perform routing calculations as would a more traditional Layer 3 routing protocol such as OSPF (Open Shortest Path First), admirable QoS parameters (e.g., available node and path resources, delay, and jitter) certainly are used in its path calculations.

Retrofitting OSPF for QoS

There have been at least two proposals [ID1996f, ID1997u] that have suggested the addition of QoS extensions to the OSPF routing protocol. OSPF appears to be an ideal choice in this regard. OSPF is a standardized intradomain routing protocol that provides a fast recalculation mechanism to determine changes in topology state, and it also provides a method to quickly notify neighboring nodes of topology state changes. The recalculation mechanism is not so ironically called SPF (Shortest-Path First) and is based on the Dijkstra algorithm. SPF provides a mechanism to determine the shortest-path between two end-points when multiple paths exist, as opposed to simply choosing the path with the shortest hop count. OSPF accomplishes this by calculating the sum of all metrics of links in each path and comparing them to determine the least-cost path.

Edsger W. Dijkstra

Edsger W. Dijkstra began programming in 1952, became a professor of mathematics at Eindhoven University of Technology in 1962, and added work as a research fellow with Burroughs Corporation in 1973. In 1984, Dijkstra left the Netherlands to accept the Schlumberger Centennial Chair in the Department of Computer Sciences at the University of Texas at Austin. Dijkstra wrote about all aspects of the computer programming field —mostly while he was a research fellow for the Burroughs Corporation in the Netherlands. Most notably (for network engineers, at any rate), Dijkstra formulated what now is commonly referred to as the *Dijkstra algorithm,* which is a single-source, shortest-path algorithm that computes all shortest paths from a single point of reference based on a collection of link metrics.

Both of the aforementioned proposals have made concessions in "optimality" in an effort to keep the complexity at a minimum. In other words, path selection is based on a simplified set of criteria, which does not calculate all possible combinations of QoS metrics when making a path selection (i.e., sequenced calculations) to facilitate realistic deployment of QoS routing. Both approaches also deal only with unicast traffic.

The approach outlined in [ID1997u] assumes that a flow with QoS requirements will convey these requirements to the QoSR protocol (OSPF, in this case) via parameters contained in an RSVP *Path* message; in other words, the *Sender TSpec* is conveyed to the routing protocol along with the destination address.

Before you look at this proposal in more detail, you should examine the OSPF Options field and TOS semantics.

OSPF Options and TOS Semantics

The OSPF Options field enables OSPF routers to support optional capabilities and to communicate their capability level to other OSPF routers. When used in OSPF Hello packets, the 8-bit Options field allows a router to reject a neighbor because of a capability mismatch. The *T-bit*, depicted in Figure 9.1, describes the router's TOS capability. Because you looked at the concept of IP TOS in Chapter 4, "QoS and TCP/IP: Finding the Common Denominator," it is not revisited here in depth. If the *T-bit* is set to 0, this is an indication that the router supports only a single TOS (TOS 0). This also indicates that a router is incapable of TOS-routing and is referred to as a *TOS-0-only* router. The absence

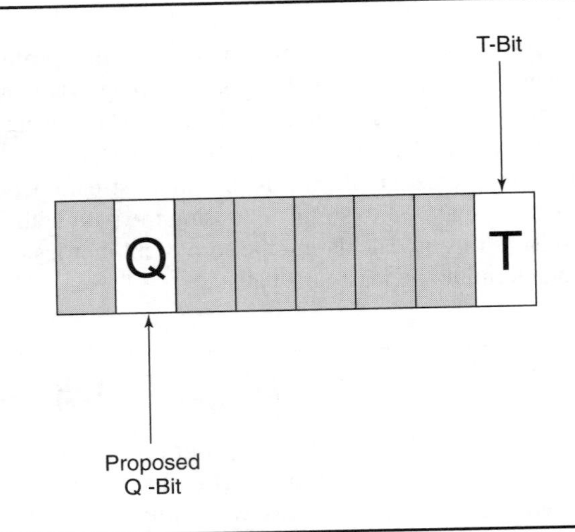

Figure 9.1: *The OSPF Options field.*

of the *T-bit* in a router-links advertisement (Type 1 LSA) causes the router to be skipped when building a nonzero TOS shortest-path topology calculation. In other words, routers incapable of TOS routing are avoided as much as possible when forwarding traffic requesting a nonzero TOS.

> **TIP** Avoiding routers that are incapable of TOS routing may not be a very wise decision. Many network architectures use a very high-speed backbone network, with the design parameter of using the fastest possible routing platforms available in order to ensure that the core of the network exhibts no congestion. Such very high-speed routers may offer no TOS-based switching capability simply because it adds additional delay to the router and impairs performance. Hence, routers that support a single TOS may be doing so to offer a minimal latency service to all traffic classes at very high speed.

The absence of the *T-bit* in a summary link advertisement (Type 3 or 4 LSA) or an AS (Autonomous System) external link advertisement (Type 5 LSA) indicates that the advertisement is describing a TOS 0 route only (and not routes for nonzero TOS).

All OSPF LSAs, with the exception of Type 2 LSAs (network-link advertisements), specify metrics. In Type 1 LSAs, the metrics indicate the cost of the links associated with each router interface. In Type 3, 4, and 5 LSAs, the metric indicates the cost of the end-to-end path. In each of these LSAs, a separate metric can be specified for each IP TOS. Table 9.1 specifies the encoding of TOS in OSPF LSAs and correlates the OSPF TOS encoding to the TOS field in the IP packet header (as defined in RFC1349). The OSPF encoding is expressed as a decimal value, and the IP packet header's TOS field is expressed in the binary TOS values used in RFC1349.

Table 9.1: *Representing TOS in OSPF [ID1995a]*

OSPF Encoding	*RFC1349 TOS Values*
0	0000 normal service
2	0001 minimize monetary cost
4	0010 maximize reliability
6	0011
8	0100 maximize throughput
10	0101
12	0110
14	0111
16	1000 minimize delay
18	1001
20	1010
22	1011
24	1100
26	1101
28	1110
30	1111

In router links advertisements (Type 1 LSAs), the T-bit is set in the advertisement's Option field if and only if the router is able to calculate a separate set of routes for each IP TOS. The # TOS field contains the number of different TOS metrics given for this link, not counting the required metric for TOS 0. If no additional TOS metrics are given, this field should be set to 0, for example. The TOS 0 metric field indicates the cost of using this particular router link for TOS 0. For each link, separate metrics may be specified for each Type of Service (TOS); however, the metric for TOS 0 always must be included. The TOS field indicates the IP TOS this metric refers to. The description of the encoding of TOS in OSPF LSA follows. The Metric field indicates the cost of using this outbound router link for traffic of the specified TOS.

You can find a detailed description for the remainder of the fields in Figure 9.2 in RFC1583 [IETF1993a].

The method outlined in [ID1997u] also proposes the addition of a Q-bit to the OSPF Options field (also shown in Figure 9.1) to allow routers that support this option to recognize OSPF Hello packets, OSPF database description packets, and packets containing information flagged as QoS-capable (e.g., LSAs). Because the OSPF Options field already is present in each of these packets, it is assumed that the addition of this bit is not a major architectural modification of the OSPF protocol but an extension that may selectively be supported by routers.

The Q-bit would be set for all routers and links that support QoS routing. A *set* Q-bit in an OSPF packet indicates that the associated network described in the advertisement is QoS-

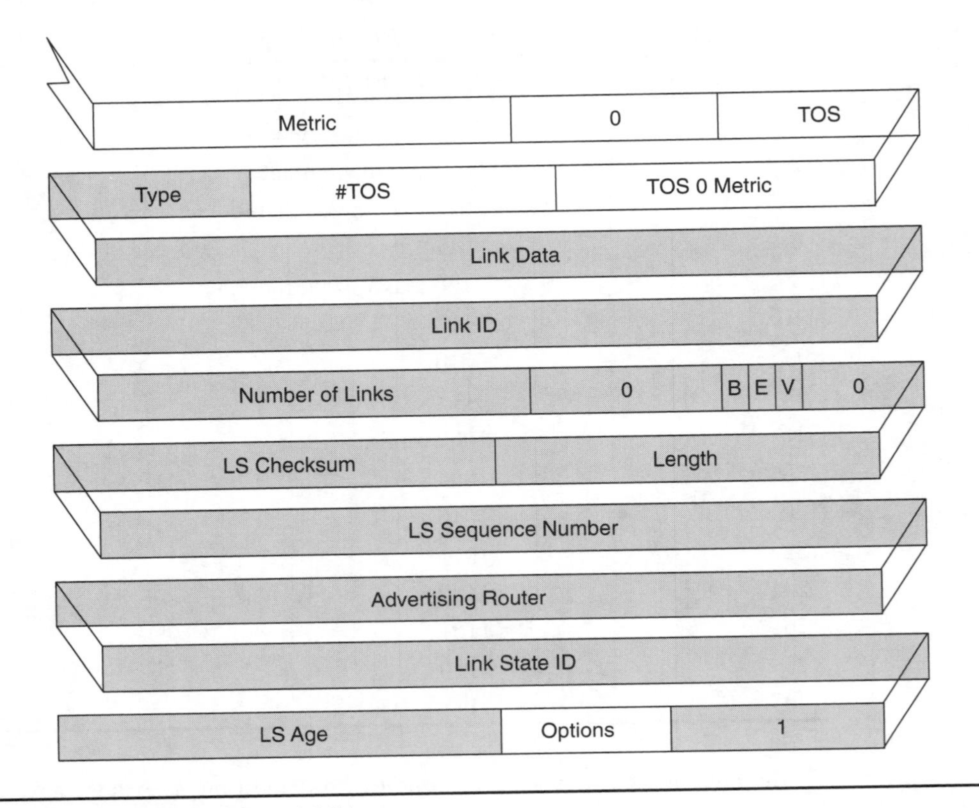

Figure 9.2: *OSPF type 1 LSA.*

capable and that additional QoS fields must be processed in the packet. Because the TOS field in OSPF packets are 8 bits and the existing TOS specification [IETF1992b] only specifies values of 4 bits, this allows a substantial expansion to the encoding of TOS values that may be expressed in the OSPF TOS field. Therefore, by using an additional fifth bit, [ID1997u] proposes an additional 16 TOS values for a TOS field, as depicted in Table 9.2.

Table 9.2: *Representing QoS in OSPF TOS [ID1997u]*

OSPF Encoding	*QoS Encoding Values*
32	10000
34	10001
36	10010
38	10011
40	10100 bandwidth
42	10101
44	10110

OSPF Encoding	QoS Encoding Values
46	10111
48	11000 delay
50	11001
52	11010
54	11011
56	11100
58	11101
60	11110
62	11111

Notice that the two specified parameters for *bandwidth* and *delay* (decimal 40 and 48, respectively) overlay seamlessly onto the existing TOS values for *maximize throughput* and *minimize delay* (decimal 8 and 16, respectively), therefore avoiding conflicts with existing values.

The *bandwidth* and *delay* parameters in [ID1997u] are expressed in the Metric field. However, because the Metric field is only 16 bits in length, and gigabit-per-second links soon will become a reality, a linear expression of the available bandwidth is not practical. Therefore, [ID1997u] proposes to use a 3-bit exponent and a 13-bit mantissa. The delay value is encoded in a similar format and is expressed in microseconds.

It also is appropriate to define the concept of *route pinning*, which loosely defined means that a particular hop-by-hop route is held in place for the flow duration so that changes in the routing topology or network load do not cause consistent rerouting of the flow. This proposal uses *path pinning* instead of *route pinning*, because what actually is "pinned" is the path for a given flow—not a route computed by the routing protocol. The proposed scheme asserts that the pinning is tied to the RSVP soft state and relies on RSVP message refreshes and time-out mechanisms to "pin" and "unpin" paths selected by the routing protocol.

Give this set of translation mechanics, it may indeed be possible to translate QoS requirements from an incoming flow into useful information that can be used to determine routing paths for various classes of traffic. However, its usefulness and efficiency remain to be seen.

Interdomain QoSR

Interdomain routing can effectively be described as the exchange of routing information between two dissimilar administrative routing domains or autonomous systems (ASs). Likewise, the basic definition of *interdomain QoSR* factors in the exchange of routing information, which includes QoS metrics. The principle concern in interdomain routing of any sort is stability. This continues to be a crucial issue in the Internet, because stability is a key factor in scalability. Instability affects the service quality of subscriber traffic.

For this reason, a link-state routing protocol may not be ideal property in an interdomain QoSR protocol. First, LSAs are very useful within a single routing domain to quickly communicate topology changes to other routers in the routing area. Flooding of LSAs into another, adjacent routing domain (AS) most likely will inject instability into routing exchanges between neighboring domains. By the same token, an AS may not want to advertise details of its interior topology to a neighboring AS.

Also, link-state routing is desirable in intradomain routing because of the speed at which topology changes are computed and the granularity of information that can be disseminated quickly to peers within the routing domain. However, the utility of this mechanism would be greatly diminished by the aggregation of state and flow information, which normally is done in interdomain routing schemes as the routing information is passed between ASs. It also would be prudent to assume that limiting the rate of information exchanged between ASs is a good thing, because interdomain routing scalability is directly related to the frequency at which information is exchanged between routing domains. Therefore, a mechanism used to provide interdomain QoSR should provide only infrequent, periodic updates of QoS routing information instead of frequently flooding information needlessly into neighboring ASs.

It is unclear at this time what mechanism would provide dynamic, interdomain QoSR routing. Given the pervasive currently deployed base of the BGP protocol, however, it is not difficult to imagine that a defined set of BGP (Border Gateway Patrol) extensions certainly could be deployed to provide this functionality. By the same token, it may be unnecessary to run a dynamic QoS-based routing protocol between different routing domains in lieu of a common method to differentiate IP packets based on the values carried in their TOS or IP precedence fields of the IP packet header.

There is, however, at least one proposal [ID1996gk] that might provide a mechanism with which to communicate QoS requirements between ASs. The proposal, outlined briefly in the following section, proposes a mechanism with which BGP might use RSVP to calculate QoS-capable paths.

Multi-Protocol Label Switching (MPLS)

The MPLS moniker describes a united effort in the IETF to blend the best of several similar concepts into a standardized framework and protocol suite. The IETF draft framework document [ID1997w] says that MPLS as a " . . . base technology (label swapping) is expected to improve the price/performance of network layer routing, improve the scalability of the network layer, and provide greater flexibility in the delivery of (new) routing services (by allowing new routing services to be added without a change to the forwarding paradigm)." MPLS does hold some interesting possibilities in the realm of QoS, however.

> **TIP** You can find the IETF MPLS Working Group home page, along with any associated documents at www.ietf.org/html.charters/mpls-charter.html.

The terms *label switching* or *label swapping* are an attempt to convey an amazingly simple concept, although you do need to know a bit of the historical context and background information to understand why the basic routing and forwarding paradigm of today is somewhat less than ideal. To appreciate the label-swapping concept, you must examine the current *longest-match* routing and forwarding paradigm used today. The current method of routing and forwarding is referred to as *longest match* because a router references a routing table of variable-length prefixes and installs the "longest," or most specific, prefix as the preference for subsequent forwarding mechanisms. Consider an example of a router that receives a packet destined for a 199.1.1.1 host address. Suppose that the router has routing table entries for both 199.1.0.0/24 and 199.1.0.0/16. Assuming that no administrative controls are in place that would interfere with the basic behavior of dynamic routing, the router will (based on an algorithmic lookup) choose to forward the packet on an output interface toward the next-hop router from which it received an announcement for 199.1.0.0/24, because it is a "longer" and more specific prefix than 199.1.0.0/16.

At first, the significance of this paradigm may not seem important. However, given the number of prefixes in the global Internet (currently hovering in the neighborhood of about 50,000 variable-length unique prefixes) and the realization that the growth trend most likely will continue to increase, it is important to note that the amount of time and computational resources required to make path calculations is directly proportional to the number of prefixes and possible paths. One of the expected results of label swapping is to reduce the amount of time and computational resources required to make these decisions. Given that at any routing point within the network, the number of paths always is far fewer than the number of address prefixes, the intent here is to reduce the cumulative switching decision in an n node configuration from a calculation in a space of order (n times the number of unique prefixes) to a calculation in a space that can be bounded approximately by order ($n!$). For closed systems in which the number of unique end-to-end paths is well constrained (which implies relatively small values of n), this labeling technique can facilitate a highly efficient forwarding process within the router.

The concept of label swapping involves replacing the need to do longest-match, to a certain extent, by inserting a fixed-length *label* between the network-layer header (Layer 3) and the link-layer header (Layer 2), which can be used to make path and forwarding decisions (Figure 9.3). Determining a best path based on a set of fixed-length values, as opposed to the same number of variable-length values, is computationally less intensive and thus takes less time.

MPLS has jokingly been referred to as *Layer 2.5*, because it is neither Layer 3 nor Layer 2 but is inserted somewhere in the middle between the network-layer and the link-layer. The network-layer could be virtually any of the various network protocols in use today, such as IP, IPX, AppleTalk, and so on—thus, the significance of *multiprotocol* in the MPLS technology framework. Because of the huge deployed base in the global Internet and the growing base of IP networks elsewhere, however, the first and foremost concern and application for this technology is to accommodate the IP protocol.

Alternatively, MPLS can be implemented natively in ATM switching hardware, where the labels are substituted for (and situated in) the VP/VC Identifiers.

Conceptually, the labels are distributed in the MPLS network by a dynamic label-distribution protocol, and the labels are bound to a prefix or set of prefixes. The association and prefix-to-tag binding are done by an MPLS network edge node—a router that interfaces other nodes that are not MPLS-capable. The MPLS edge node exchanges routing information

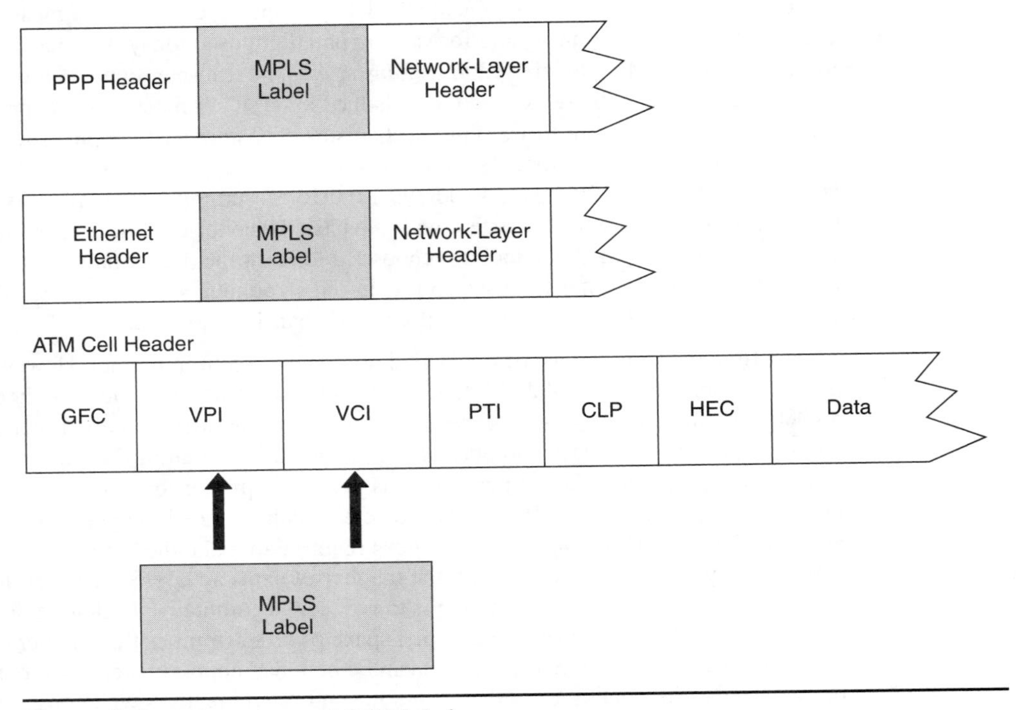

Figure 9.3: *Inserting an MPLS label.*

with non-MPLS-capable nodes, locally associates and binds prefixes learned via Layer 3 routing to MPLS labels, and distributes the labels to MPLS peers (Figure 9.4).

QoS and MPLS

QoS has several possibilities with MPLS. One of the most straightforward is a direct mapping of the 3 bits carried in the IP precedence of the incoming IP packet headers to a Label CoS field, as proposed in Cisco Systems' contribution to the MPLS standardization process, *Tag Switching*. For all intents and purposes, the terms *tag* and *label* can be considered interchangeable. As IP packets enter an MPLS domain, the edge MPLS router is responsible for—in addition to the functions mentioned earlier—mapping the bit settings in the IP packet header into the CoS field in the MPLS header, as shown in Figure 9.5.

One of the most compelling uses for MPLS is the capability to build explicit *label-switched paths* from one end of an MPLS network domain to another. It is envisioned that label-switched paths could be determined, and perhaps modeled, with a collection of traffic engineering tools that perhaps reside on a workstation and then downloaded to network devices. This is an important aspect for traffic-engineering purposes; it gives network administrators the capability to define explicit paths through an MPLS cloud based on any arbitrary criteria. Traffic-engineering functions such as this may contribute significantly to enhanced *service quality*. As a result, some method to offer differentiated services may be possible.

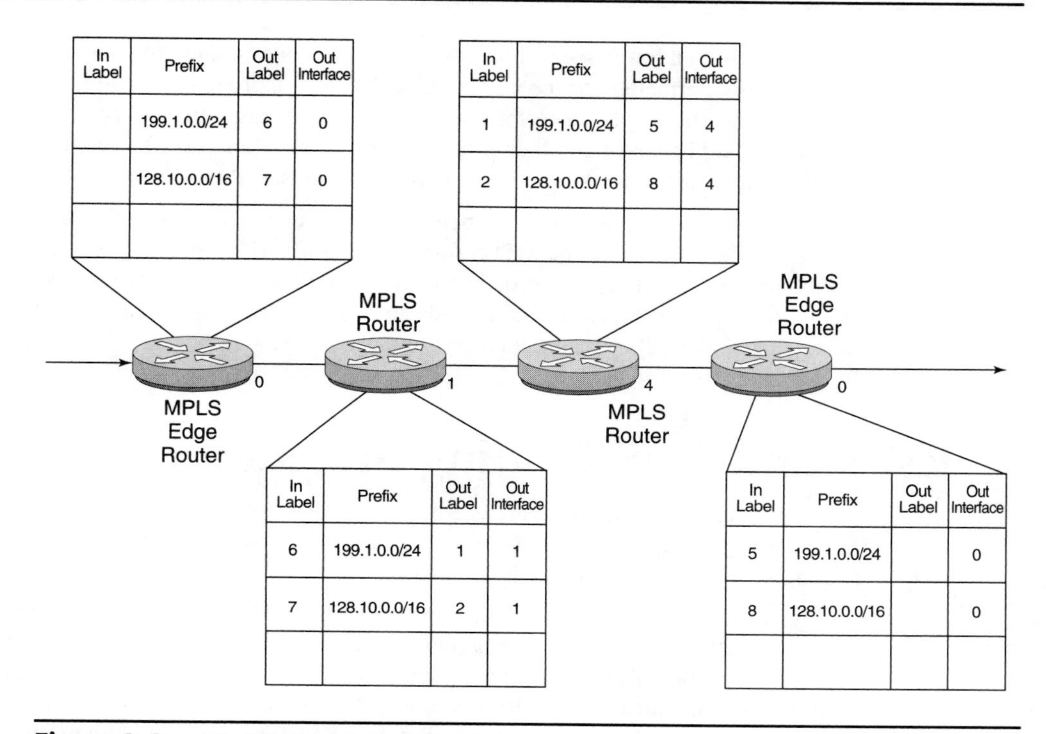

Figure 9.4: *The label-swapping process.*

Several parallel paths may exist from one end of an MPLS network domain to another, for example, each of varying bandwidth and utilizations. It certainly is possible that explicit ingress-to-egress paths could be chosen for each specific CoS type, each path offering a distinct differentiated characteristic. Traffic labeled with higher CoS values could be forwarded along a higher-speed, lower-delay path, whereas traffic labeled with a lower CoS value could

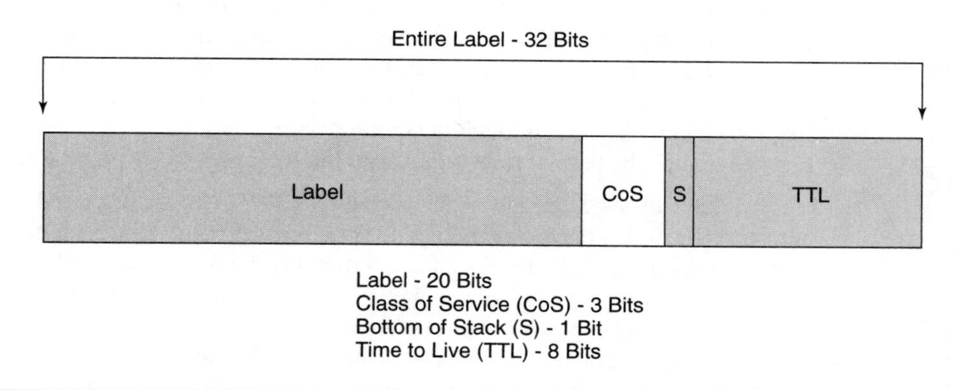

Figure 9.5: *CoS bits carried in the MPLS label.*

be forwarded on a lower-speed, higher-delay path. This example illustrates one method of providing differentiated service with MPLS; different approaches certainly exist. In fact, traffic engineering could be performed without consideration of the CoS designation altogether. It could be based on other criteria, such as source address, destination address, or per-flow characteristics; or it could be in response to RSVP messages.

MPLS traffic engineering in response to RSVP messages is one that presents an interesting possibility, because the MPLS framework draft [ID1997w] suggests that any MPLS-compliant implementation *must* be interoperable with RSVP. Although this mandate does not explicitly mean that MPLS would provide an explicit forwarding path through the network in response to RSVP messaging, it certainly is an interesting possibility and an avenue for further study.

RSVP and QoS Routing Integration

As mentioned several times, there is a noticeable disconnect between RSVP and the underlying routing system. Without some sort of interface between the two, the utility of a QoS routing scheme or RSVP alone is less than ideal.

At least one proposal has been submitted to the IETF community that provides a method for RSVP to request information and services from the local routing protocol process [ID1996h]. This proposal recommends an RSVP-to-routing interface called RSRR (Routing Support for Resource Reservations). The proposal describes the role of the RSRR interface: providing for communication between RSVP and an underlying routing protocol similar in operation to any other type of API (Application Programming Interface). This task is accomplished via an exchange of asynchronous queries and replies, with the RSRR protocol as a conduit. Recall that traditionally, RSVP does not perform its own routing; instead, it relies on the underlying routing system for path selection. As outlined in this proposal, RSVP could obtain routing entries via the RSRR interface, which then would allow it to send RSVP control messages (e.g., *Path, Resv, PathTear, ResvTear*) hop-by-hop, as well as request notifications of route changes.

Using the RSRR interface, RSVP may obtain routing information by sending a *Route_Query* to the routing process; RSVP uses the returned route entries (*Route_Response*) to activate the transmission of Path messages. Additionally, RSVP may ask the routing process to notify it explicitly of routing changes through the use of a notification flag (*Notify_Flag*) in the queries, which results in the routing protocol immediately communicating route changes to RSVP (*Route_Change*). The result of this interaction allows RSVP to adapt to changes in paths through the routing system by sending periodic *Path* or *Resv* messages to refresh the path or reservation state for a flow.

Additionally, an extension to the routing-to-RSVP interface described earlier has been proposed [ID1997x] that provides support for a broader range of routing mechanisms—namely, support for explicit paths and QoS routing. The support for explicit paths is significant. Traditional hop-by-hop path computation necessitates each node to locally determine the best path for a packet based on the destination specified in it. The packet then is forwarded to the appropriate next-hop router, where the process is repeated. The liability in hop-by-hop path determination is that although a node may compute what it perceives to be the optimal path to a particular destination, the next-hop router may deviate from the previously determined path and locally determine to forward the packet to a next-hop that was not in the previously computed feasible path. By contrast, explicit routing is the practice of a single router computing the end-to-end, hop-by-hop path a packet will traverse, as well as each subsequent router in the path, honoring this path without locally computing a path for it itself.

This particular draft [ID1997x], along with another proposal [ID1997z], details how RSVP could set up reservations based on explicit route support. Between these two proposals, a suggested new object, *Explicit_Route*, could be added to the RSVP specification, which contains specific information to support explicit routes. The proposed *Explicit_Route* object contains a pointer, source IP address, destination IP address, and a series of IP addresses that, when processed sequentially, indicate the explicit hop-by-hop path with which the *Path* message should be forwarded. Because RSVP objects are handled as opaque, RSVP transports only the *Explicit_Route* objects in Path messages. The routing protocol then uses the information contained in the object to determine the next hop to which the *Path* message should be forwarded by using the *Route_Query* and *Route_Response* mechanisms. At each intermediate node, this identification is done by extracting the *Explicit_Route* object and examining the pointer, which is updated at each hop to the next sequential address in the object. This pointer determines what point in the sequence of addresses is locally relevant and where to forward the message. After making this determination, the pointer is updated, and the *Path* message is sent on its way to the specified next hop.

The extended RSVP-to-routing interface mentioned earlier [ID1997x] also proposes that the RSVP *Sender TSpec* information be contained in the *Route_Query* so that the necessary QoS information be made available to a QoS-based routing protocol and can be used in determining an appropriate path for a flow.

Path Management and Path Pinning

It is important to acknowledge the significance of the role QoS path management plays in any QoS-based routing scheme. QoS path management recalculates, adjusts, and maintains paths in the network based on changes in the network topology, traffic QoS requirements, and network load. Additionally, an adequate QoS path-management scheme must ensure that established QoS flows are not disrupted by the dynamics of the routing system or the management scheme itself. This requires that after a path for a flow is determined and the flow established, the path for the flow is "pinned" for the duration of the flow.

At this point, it is important to restate the concept of *path pinning* as it relates to QoS path management and RSVP. This aspect of QoS path management is tricky. Simply stated, QoS-based routing computes the path, and RSVP manages *Path* state via its normal method of message processing. Paths are pinned during the processing of *Path* messages. RSVP issues a *Route_Query* to the QoS routing process to determine the appropriate next-hop for which to send the *Path* message. Paths may become unpinned under several conditions. Paths may be unpinned when *Path* state for the flow is removed, via a *PathTear* or refresh time-out. Paths also may become unpinned when *TSpec* parameters in a *Path* message change, resulting in the receiver reinitiating a reservation. Paths may become unpinned when a local admission-control failure is detected after receipt of a *Resv* message, when a *PathErr* message is received, or when a local link failure is detected. As mentioned earlier, if *Route_Querys* are issued with the *Notify_Flag* set, the QoS routing process notifies RSVP of specific route changes or link failures.

Given this set of criteria and characteristics, support for path pinning requires some basic modifications to the way in which RSVP control messages are processed [ID1997y]. In particular, when an RSVP receives the first initial *Path* message, it issues a *Route_Query* to the QoS routing protocol, obtains the best path for expected traffic, and then stores the next hop as part of the *Path* message with a pinned flag set. After *Path* refreshes are received, RSVP checks the *Path* state for changes in the *Previous Hop* (PHOP), the IP Time to Live (TTL), and the status of the pinned flag for the next hop. If there are no changes in the *Path* state and the next hop is flagged as pinned, the indicated next hop is used for forwarding the *Path* message. If the *Path* state changes and the next hop is not flagged as pinned, RSVP again issues a *Route_Query* to determine the best path and again sets the pinned flag for the returned next hop within the subsequent *Path* refresh. Similarly, when a *Route_Change* is received by RSVP, it immediately unpins the next hop.

RSVP and QoS Routing Observations

Given the importance of QoS path management, it is clear that relying on the dynamics of an underlying routing protocol (to include a QoS-based protocol) can be detrimental if left to its own devices. The dynamics of the link-state QoS-based routing can interfere with RSVP state and flow establishment when the network topology changes or when network utilization changes. Therefore, the support for explicit routes and path pinning cannot only be considered desirable but mandatory. Even with this support, the interaction between traffic that is so constrained and the flow levels of best-effort routed traffic must be considered. The inescapable conclusion is that path-constrained flows should be associated with some form of router-by-router resource-reservation scheme.

Remember that the concepts and proposals discussed here are, for the most part, simply that—conceptual design plans and proposed mechanisms to enhance the delivery of QoS in the network. It therefore is unclear at this point whether these mechanism will be adopted en masse, portions adopted ad hoc, or none adopted at all. It is encouraging, however, in that the Internet community is beginning to discuss these issues, because a large portion of the same community feels that mechanisms similar to the ones described here are necessary to allow for a more granular, controlled level of QoS delivery.

QoS and IP Multicast

The discussion of IP multicast is placed in this section on "futures" for a good reason. Currently, the capability to impose any form of QoS structure on multicast traffic flows is not well understood. Fred Baker, the current chair of the IETF, was heard to say of IP multicast, "Multicast makes any problem harder. If you think you understand a problem, repeat the problem statement using the word 'multicast.'" Certainly this observation appears to be well applied to QoS.

The multicast model is one of group communication, where any member of the group can initiate data that is delivered to all other members of the group. To achieve this, the wide-area network-transmission mode is one of *controlled flooding*, in which a single packet may be replicated and forwarded onto multiple output interfaces simultaneously. Several routing protocols are designed to facilitate this controlled flooding, including DVRMP, a simple Distance Vector Routing Multicast Protocol, a sparse and dense mode Protocol-Independent Multicast (PIM), and most recently, a proposal for interdomain multicast routing with extensions to the BGP (Border Gateway Protocol). This model of multiple asynchronous receivers obviously makes any form of end-to-end flow control very challenging, so QoS multicast flow management must be undertaken by the network as an imposed profile on the multicast traffic flow.

Because multicast does not follow an adaptive end-to-end flow control model, selective discard control structures, such as RED (Random Early Detection), have no significant impact on multicast applications. To date, the major control facility is admission control, using a leaky-bucket rate-control structure to impose a rate limitation on a multicast path within a network. Other QoS control mechanisms that can be triggered by the TOS and precedence fields of the packet header could be deployed by the router. Note that any precedence queuing must be applied to all output interfaces uniformly if the QoS structure is to be applied consistently to the entire multicast group.

Resource-reservation models and QoS routing in a multicast environment are challenging problems, but ones that are increasingly important. Currently, one of the most attractive multicast applications is *groupware* with shared audio and video streams complemented by some form of common workspace. For this real-time multicast flow to be effective, it is likely you will see resource-reservation models that attempt to place a controlled-load or guaranteed-service specification into the network. The twist is that, instead of using the current model of a unidirectional path reservation, the end-points may be the multicast host at one end and the network *state* of the multicast group at the other. The resource-discovery problem becomes selecting the point at which the host is connected to the multicast group, as well as a path between the host and the multicast connection point. It is not clear whether this selection is as a pair of unidirectional paths and connection points or whether it is a bidirectional path selection.

As you can see, at this stage, the IP multicast QoS subject is largely speculative. A significant body of further work must be undertaken to provide a well-structured platform to support QoS traffic profiles.

QoS as a Congestion Pricing Algorithm

Jeff MacKie-Mason and Hal Varian [MV1994] have suggested that pricing mechanisms are a viable method of relieving congestion. One potential refinement of this approach is to look at a model of QoS activation based on the user's choice as a mechanism of activating a congestion fee structure.

The basis of QoS as a congestion-avoidance mechanism is that you can provide permanent QoS measures undertaken by the network on behalf of the customer or implement QoS structures that can be activated dynamically by traffic passed to the network by the customer.

Network-Activated QoS on the Customer's Behalf

You can compare the first structure—a network-activated QoS mechanism activated on behalf of the customer—to a form of performance insurance. Traffic that matches the QoS criteria is marked at the network ingress device as being of an elevated precedence. All such traffic would attract a premium tariff, so this could be considered as a premium service associated with pricing. In an unloaded network, this action triggers negligible actions by the network, given that the elevated precedence does not displace any other traffic within the queues, which are located on the customer's end-to-end paths within the network.

The consequent observation is that within such conditions of an unloaded network, the difference between the throughput of traffic flows that are marked with an elevated precedence and traffic flows that can be considered *normal* is negligible. In the event of congestion along a higher-precedence traffic path, notable comparative differences in throughput levels will result. Of course, congestion is a localized phenomenon, because each router is not synchronized to the flow state of its neighbors. Because the queue structure on any given router is not linked to the queue structure of a neighboring router, the differences activated by the entry-level elevated precedence are readily distinguishable only on long-lived flows that traverse identical paths and similarly pass across the same congestion point.

The outcome of this entry function is that the difference between traffic that matches the entry conditions of elevated precedence and normal best-effort traffic can be characterized not by the presence of positive metrics related to flow performance but the selective absence of negative metrics. This is an interesting result: The less the incidence of congestion within the underlying network, the less the visible perception to the user of the value of this QoS precedence elevation. This leads to the observation that this style of QoS setting is a "just in case" option; the eventuality being insured against with the permanent QoS setting is small-scale isolated burst congestion events or a more systemic resource constraint within the network, which is activated by peak load patterns. It is still unclear how this somewhat negative image of network-activated QoS can be marketed selectively to a customer base that already is paying for normal best-effort service. Although some segments of the customer base may be prepared to pay a premium price for such an elevated service structure, bear in mind that the service is described accurately as *better effort*, whereas the customer is more motivated to see a *service performance guarantee* in exchange for this form of price premium.

Customer-Activated Precedence Traffic

A second structure is customer-activated precedence traffic, where the network implements structures that enable precedence, precisely the same as required in the first structure discussed earlier. However, the principal difference is that the network ingress actions that set the precedence level within the packet header are removed. In their place, the customer-defined precedence levels in the packet header are honored. This may appear to be a minor distinction, but it has significant implications related to the pricing structures that may be associated with this customer capability. If the pricing structure is done on a per-packet basis, or a volume-based premium based on the packet precedence and packet length, a congestion-based pricing structure can be implemented. If the customer observes some form of performance degradation, the option is available to increase the precedence level of transmitted packets to avoid continued degradation, which is associated with an incremental cost. A simple congestion-based premium pricing structure would use a single elevated precedence, whereas more complex models can be constructed with a sequence of precedence levels and an associated sequence of pricing levels.

A number of potential structures can assist the customer in making an informed decision as to whether the option of precedence setting is a viable action. One is the use of a precedence-discovery protocol, which is valuable when precedence structures may not be deployed uniformly across the full extent of the network's (or networks') path. This is possible by using a variant of the MTU discovery protocol, where ICMP (Internet Control Message Protocol) packets are used to probe the network path, and the required change is the return of an ICMP error packet with a precedence not honored if a router on the path will not (or cannot) accommodate the desired precedence setting. Another approach is perhaps more ambitious and implements more closely the congestion pricing model of MacKie-Mason and Varian. Here, the packet accounting is undertaken on egress from the network, but it is accounted in correlation to the source address as a precursor. The additional requirement is that a single bit field of the header is reserved for precedence activation (which could be 1 bit of the IP precedence field). The bit is cleared on ingress to the network and is set by any router that exercises the precedence structure; this usurps resources that would have been allocated to another packet. Egress accounting is performed if the bit is set, which indicates that the precedence function has been exercised within the packet's transit within the network.

The result of this more sophisticated network structure is that the customer can indicate a willingness to pay a congestion premium on entry to the network, and the network will complete the pricing transaction only if the network had to exercise the precedence because of congestion in the packet's path across the network. The complete congestion pricing structure can be implemented with multiple precedence levels.

The pragmatic observation is that in transitioning to a QoS-defined service, you probably will see a number of increasingly sophisticated implementation phases. The initial phase is most likely to be a simple model of ingress precedence setting by the network, associated with internal network-precedence mechanisms, and a uniform pricing premium for all such traffic on a customer-by-customer basis. Successive refinements of this model could include egress accounting, customer setting of the precedence levels, precedence discovery, and precedence flagging by intermediate network nodes. It may well be that the resultant QoS framework is one in which QoS structures are activated by the user as an indication of an option for the network to exercise QoS differentiation, and that pricing is associated with the level of differentiation requested and whether the network exercises the option in transit.

Interprovider Structures and End-to-End QoS

It would appear that QoS is going to be deployed in two fundamentally different environments: the *private* network domain, where QoS is tied to particular performance objectives for certain applications related fundamentally to business needs and drivers, and the *public* Internet, where QoS is tied to competitive service offerings.

The public Internet deployment is particularly challenging. For QoS to be truly effective, the mechanisms used across a multiprovider transit path require all transit elements to accept the originating QoS signaling. The minimum condition for QoS to work is that the QoS signals preserve the semantics of the originating customer, so that the QoS field value in the customer's packet is interpreted consistently as a request, for example, for precedence elevation. The situation that should be avoided is one in which the neighboring provider in the traffic path accepts the packet's field value without modification but places an entirely different semantic interpretation on the value. As an example, here the situation to avoid is where provider A interprets a field value as a request for elevated queuing and scheduling priority, while the next provider (B) interprets the same value as a discard eligibility flag. Accordingly, the first requirement for interdomain QoS deployment is an operational consensus of the QoS signaling semantics within the deployed global Internet.

It is a realistic prediction that QoS will not be deployed uniformly on the public Internet, and accordingly, there will be situations in which QoS signaling will be ignored within an end-to-end network path. In stateful mechanisms, this is addressed explicitly within the implementation of path setup and maintenance, while in a stateless QoS environment, where QoS signaling is undertaken by use of the packet header field values, this is an open issue. Should an ingress network node that does not implement QoS services respond to all QoS-labeled packets with ICMP destination unreachable messages? This interprets the QoS fields as a mandatory condition, where if a network cannot honor the request for elevated or distinguished service, the network must signal the originator of this inability. Alternatively, the ingress node may choose to accept the packet, effectively ignoring the implicit QoS request contained in the header. There are arguments for both types of responses, although the prevailing philosophy of a liberal acceptance philosophy tends to see the second response as the most appropriate here.

Another situation in which QoS signaling may or may not be honored in the end-to-end network path is when the neighboring peer network implements QoS structures and conforms to the semantics of the originating network. If the packet is accepted by the peer network, it transits the network with elevated priority and is passed on with the QoS fields intact. Although this is technically feasible, it is readily apparent that the major issues here relate to the capability of the commercial agreement between the two providers to accept and honor QoS settings from peer networks. If the commercial agreement is not sufficiently robust enough to accommodate QoS interaction, the provider has little choice but to clear the QoS header values to ensure that these are not activated within the local transit and that the end-to-end QoS signaling mechanism is cleared at an intermediate position within the transit path.

Current indications are that local QoS structures will indeed appear within the local Internet with some limited-pricing premium and some limited applicability to transit paths within each participating ISP's customer domain, so that end-to-end QoS will be achievable, but only when both ends are attached to the same provider. This initial deployment no doubt will use ingress filtering so that interprovider QoS structures will not be a default feature of the initial deployment models. It therefore is likely that you will see some refinement of the existing interprovider commercial agreements to admit bilateral and ultimately transit QoS mechanisms, where the initial QoS agreement between the customer and the local ISP can be reflected by these interprovider agreements to extend the reachability of the QoS domain from the customer's perspective. As to whether this "bottom-up" method of QoS deployment will yield truly uniform end-to-end QoS deployment in the global Internet remains very much a subject of speculation. Interestingly enough, however, the nature of this speculation is more about the robustness of the commercial models of interaction, the associated issues of ISP business practices, and the policies of interprovider agreements than speculation about the capabilities of QoS technology.

Although the telecommunications world has been able to craft an environment that deploys a relatively uniform quality level across a multiprovider global network, one result is a relatively consistent view of end-customer retail pricing models. Whether the pressure to implement consistent interprovider commercial agreements that encompass meaningful QoS levels instead of a highly variable perspective on what the *better* of *better effort* quantitatively translates to, and whether it will result in increasing uniformity of the retail pricing structure for the global Internet, remains to be seen.

Should QoS Be Bidirectional?

In much of the discussion about QoS mechanisms, it has been implied that QoS is a unidirectional mechanism and that it applies to transmission flows that are injected into the network. RSVP goes one step further and makes this "unidirectionality" explicit. The question is whether this is adequate as a QoS mechanism, whether it will yield tangible results in terms of elevated service, or whether bidirectionality is a necessary component of the QoS picture. Of course, this discussion is predicated on the observation that it is relevant only to end-to-end controlled traffic flows that are clocked by the network and most readily is applicable only to TCP traffic. Externally clocked, UDP unidirectional flows explicitly fall outside consideration within this topic.

The intent of elevating the precedence of packets within a QoS structure is directed toward removing jitter and loss the network may impose on a traffic flow. The intent is to allow an end-to-end flow-control algorithm to develop an accurate and consistent view of the network-propagation delay between the sender and transmitter. If this is obtained by the flow, the optimal operating condition of the network-clocked flow control can be achieved, and as the receiver peels data packets off the network, the sender pushes another packet into the network. This is achieved by timing the sender's data-transmission events against the receipt of acknowledgment (ACK) packets.

In this optimal steady state, each time the sender receives an ACK, it advertises the availability of a receive window, which then allows the sender to push another packet into the network. Of course, the initial state is that the sender is unaware of the RTT delay and the maximum available capacity of the path to the receiver, so the TCP flow-control algorithm uses the slow-start mechanism to determine these parameters. Starting with one packet, the number of packets the sender pushes into the network doubles with every RTT, using the pacing of received ACKs to send two data packets for each received ACK. This algorithm continues until either the buffer space of the sender is exhausted (so that the flow is running at the maximum pace the sender can withstand) or the network signals that maximum available capacity for the flow has been achieved. The latter signaling is done by reception of an ACK, which indicates that the network has lost data. Within the network, the condition that created this signal is caused by the output queue of an intermediate node reaching a critical capacity state (saturation), and the data packet being discarded. The packet discard is caused as a byproduct of queue exhaustion, or possibly through the actions of a RED mechanism active on the router in question. Thereafter, the sender moves into a state of congestion avoidance, where the amount of capacity in the path is probed more *gently* by attempting to increase the number of packets in flight within the network by one segment per RTT interval. Again, packet loss signaled within the reverse ACK stream causes the sender to back off the rate to the previous start point of the congestion-avoidance behavior.

In the face of continued loss, the sender backs off further and attempts to reestablish the new flow rate by again using slow-start probing. Elevating the priority of the sending packets within this flow causes the queue on the intermediate node that is experiencing congestion (or selective packet drop) to delay the advent of packet discard event for this flow, because the elevated precedence is a signal to the router to attempt to discard packets of some other *nonelevated* priority flow in preference. This has the effect of allowing the elevated priority flow to take a greater proportion of the bandwidth resources at the intermediate node in question, because as the data packets of the flow continue to be transmitted, the intent of the sender is to continue to increase the transmission rate, either aggressively (if it is in slow-start mode) or more temperately (as in congestion-avoidance mode).

The ACK packets may not necessarily take a symmetrical path back from the sender to the receiver. Taking into account the explicit or implicit assumption that QoS structures are imposed unidirectionally on packet flows from a sender to an ingress intermediate node, the ACKs of the flow may suffer a greater level of jitter and loss than the data packets. Will this affect the flow rate of the sender, and will this affect the steady-state objective of the flow-control mechanism to clock the sender's rate at the level of maximum availability of the sender's data path?

The most immediate effect of ACK loss is to cause the data-transfer rate to slow down by a one-segment size per RTT for each lost ACK, which has an impact on the cumulative flow throughput rate. The more subtle effect is caused by jitter imposed on the ACK sequence in slow-start mode. Because ACK packets drive the sender's data-transmission clock, jitter in the ACK sequence creates consequent jitter in the sender's behavior, which in turn increases the bursting nature of the sender. In slow-start mode, this jitter can be problematic, because the induced sender jitter can cause an overload of the interior queuing structure. The *steady state* of slow-start mode (if *steady state* can be used to describe exponential growth in a flow rate) already imposes load on the network's internal queuing structures by sending two data segments in the same time frame where a single data segment was delivered in the previous RTT interval. This sending overload rate is smoothed by interior queues on the critical intermediate-node hops. ACK jitter increases this burst level, which consequently increases the burst level of queuing on the interior queues. The risk here is that this induced ACK jitter will cause packet discard, which in turn causes premature signaling of rate saturation, switching the sender from slow-start mode to a more conservative level of rate growth via congestion-avoidance management and, in the worst case, may cause a reinitiated slow-start with an initial single segment per RTT.

Admittedly, this is a second-order result, and the effects of non-QoS reverse flow ACKs on a QoS data flow can be quite subtle. From a technical perspective, TCP sessions should reflect the precedence level of the incoming data stream in the reverse ACK flow to achieve the best possible overall flow rate. When only a subset of incoming data packets in the flow has the precedence field set, the correct behavior of the receiver is less obvious and is probably a productive subject for further investigation. However, this must be placed into a context of an overall QoS structure which, in the public Internet environment, probably will include a premium pricing element for generating precedence-level packets. Here, the receiver exercises an independent decision about whether to exercise precedence levels in response to the sender, and the consequent possibility of achieving less-than-optimal flow rates for this class of traffic must be recognized. Overriding this is the consideration that such behavior (to reflect precedence parameters in ACK packets) is a function of the end-system's TCP stack and any changes in the default behavior of the stack with respect to precedence.

QoS and IP Version 6

Also known as *IPng* or *IP Next Generation*, IPv6 provides a number of enhancements to the current IPv4 protocol specification. However, for all of the reasons used to rationalize an industry-wide migration to IPv6 (IP Version 6), integrated QoS is not one of the most compelling arguments. It is a common misperception that the IPv6 protocol specification [IETF1995c] somehow includes a magic knob that provides QoS support. Although the IPv6 protocol structure is significantly different from its predecessor (IPv4), the basic functional operation is still quite similar.

Two significant components of the IPv6 protocol may, in fact, provide a method to help deliver differentiated Classes of Service (CoS). The first component is a 4-bit *Priority field* in the IPv6 header, which is functionally equivalent to the IP precedence bits in the IPv4 protocol specification, with somewhat of an expanded scope. The *Priority* field can be used in a similar fashion as described earlier when IPv4 was discussed, in an effort to identify and discriminate traffic types based on contents of this field. The second component is a *Flow Label*, which was added to enable the labeling of packets that belong to particular traffic flows for which the sender might request special handling, such as nondefault QoS or real-time traffic. Figure 9.6 shows these IPv6 header fields.

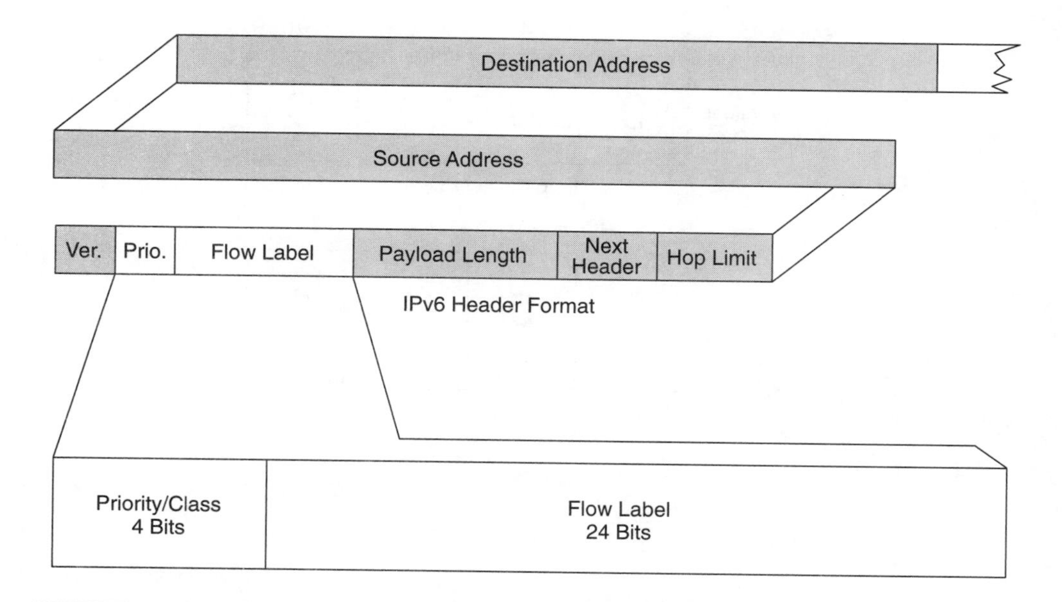

Figure 9.6: *The priority and flow label in the IPv6 header.*

TIP You can find more information on IPv6 by visiting the IETF IPng (IP — Next Generation) Working Group Web page at www.ietf.org/html .charters/ipngwg-charter.html as well as the IPv6 Web page hosted by Sun Microsystems, located at playground.sun.com/pub/ipng/html/ipng-main.html You can find interesting information related to the experimental deployment of IPv6 on the 6Bone Web site at www.6bone.net.

The IPv6 Priority Field

The 4-bit *Priority* field in the IPv6 header was designed to allow the traffic source to identify the desired delivery priority of its packets, as compared with other packets from the same traffic source. The 4 bits in the *Priority* filed provide a range of decimal values between 0 (binary value 0000) and 15 (binary value 1111), as shown in Figure 9.7. The *Priority* values are divided into two ranges. Values 0 through 7 specify the priority of traffic for which the source is providing congestion control—in other words, traffic, such as TCP, that backs off and responds gracefully to congestion and packet loss. Values 8 through 15 specify the priority of traffic that does not respond to congestion situations, such as real-time traffic being sent at a constant rate. For congestion-controlled traffic, RFC1883 [IETF1995c] recommends the *Priority* values listed in Table 9.3 for particular application categories.

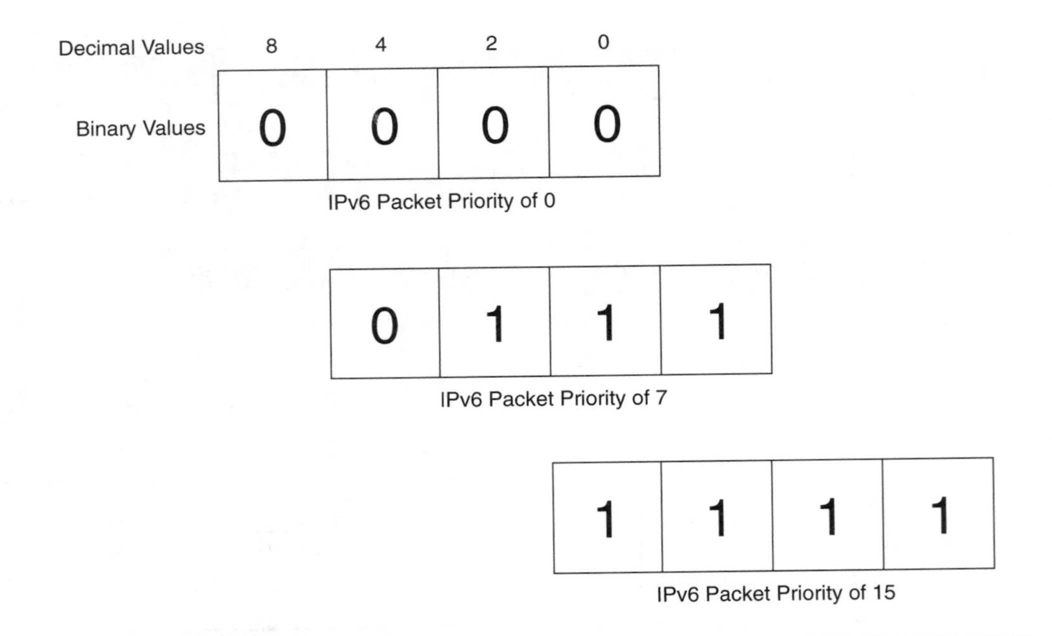

Figure 9.7: *The IPv6 Priority field.*

Table 9.3: *Priority Values for Application Categories*

Value	*Category*
0	Uncharacterized traffic
1	Filler traffic (e.g., NNTP)
2	Unattended data transfer (e.g., SMTP)
3	Reserved
4	Attended bulk transfer (e.g., FTP, NFS)
5	Reserved
6	Interactive traffic (e.g., Telnet, X)
7	Internet control traffic (e.g., routing protocols, SNMP)

For traffic that does not provide congestion control, the lowest *Priority* value, 8, should be used for packets the sender is most willing to discard in the face of congestion. Likewise, the highest *Priority* value, 15, should be used for packets the sender is least willing to discard.

It is not clear what immediate benefit the separation of traffic types that do or do not respond to congestion control (TCP versus non-TCP) provides in this priority scheme. However, it may very well prove beneficial as more advanced packet-drop algorithms are developed.

Proposed Modifications to the IPv6 Specification

Of course, technology often changes in midstream, and IPv6 certainly is no exception. A set of proposed modifications to the base IPv6 protocol specification has been proposed [ID1997a3] that includes changing the semantics of the IPv6 *Priority* field as originally set forth in RFC1883. This and other modifications were submitted to the IETF IPng Working Group in July 1997 and initially discussed at the IETF working group meetings in Munich, Germany, in mid-August 1997. Currently, it is unclear whether these semantics will be adopted, but it is worthwhile to outline them here, because there is a very good chance they will be advanced as standards track modifications to the base specification.

As illustrated in Figure 9.8, the name of the 4-bit field previously known in RFC1883 as the *Priority* field has been changed to the *Class* field. The placement of this field has not been changed in the IPv6 header, but the internal semantics have been changed significantly. The *Class* field has been modified to include two subfields: the *Delay Sensitivity Flag* subfield (or the *D-bit*) and the *Priority* subfield. The *D-bit* is a single bit that identifies packets that are delay sensitive and may require special handling by intermediate routers. This special handling might include special forwarding mechanics (for example, forwarding packets with the *D-bit* set along a lower-latency path), whereas packets without this bit set are forwarded along another, higher-latency path.

The 3-bit *Priority* subfield is similar in context and function to the IP precedence bit field in the IPv4 header and indicates relative priority of the packet. The higher the value in the field, the higher the priority. These modifications were made, in part, to facilitate a

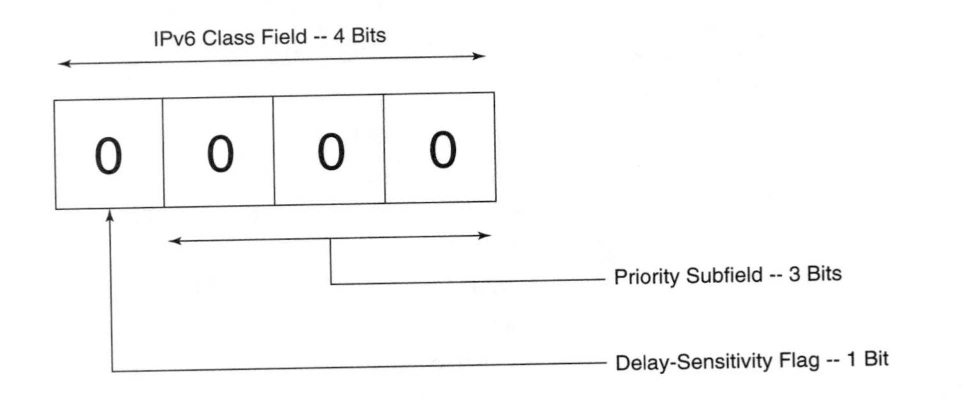

Figure 9.8: *The IPv6 Class field.*

cleaner mapping of the values contained in an IPv4 IP Precedence bit field into the IPv6 *Priority* field in an effort to simplify translation and transition schemes.

IPv6 Priority/Class and Differential Packet Drop

In a manner similar to that described earlier with regard to the IPv4 IP precedence field, the IPv6 *Priority* field setting can be used to identify and discriminate traffic based on these values as packets travel through the network. The basis of this CoS model is that during times of congestion, a prejudicial method of dropping packets is used, based on the content of the *Priority* field. The lower the value of the contents of the *Priority* field, the higher chances are that the packets will be dropped. The higher the values in the *Priority* field, the less susceptible the packets are to being dropped. Of course, if there is no congestion, all packets, regardless of their *Priority* designation, are happily delivered with equity. An enhanced congestion-avoidance mechanism, such as the eRED (Enhanced RED) algorithm described earlier [FKSS1997], could be used on a hop-by-hop basis through the network to provide the prejudicial method of packet drop.

There are clearly two benefits to using a packet-drop strategy such as eRED. The first benefit is that the basic underlying RED [Floyd1993] functionality provides a method to avoid global synchronization, where several hundred (or possibly even several hundred thousand) TCP sessions experience packet loss at roughly the same time, react by backing off at roughly the same time, and then begin to ramp up at roughly the same time because of buffer overflow at some point in the network. This congestion-avoidance mechanism is beneficial in a way that is more global in nature, providing a method to ensure the overall health of the network.

The second benefit is a byproduct of the first and is one of the central themes in providing for differentiated CoS. The eRED traffic-drop strategy provides for the prejudicial distinction of packets to be dropped in times of congestion.

Of course, it also would be wise to implement a combined policing and admission-control mechanism at the network ingress point to override the *Priority* field values that were set initially by the downstream hosts. Otherwise, allowing the downstream nodes to arbitrarily set the *Priority* values could result in an unrealistically large volume of high-priority traffic, which would negate the intrinsic value that eRED attempts to provide. The policing and admission control can use one of two possible mechanisms or a combination of both. One possible mechanism is simply to match an administratively configured criteria set, such as source or destination IP address, TCP or UDP port, or a combination of each, and when a match is found, the *Priority* bits are set to a predetermined value.

The second mechanism provides a more granular level of control. This mechanism entails implementing a token bucket, or perhaps multiple token buckets, on a router interface with predefined bit-rate thresholds, coupled with the source and/or destination criteria mentioned earlier. When traffic matching the defined criteria exceeds the configured bit-rate threshold, the *Priority* value of nonconforming packets can be set to a lower, predefined value. Packets from flows that remain below the bit-rate threshold are set with a higher predefined *Priority* value.

Using either of these mechanisms provides the necessary control to contribute predictive network behavior while allowing a method to discriminate traffic, effectively delivering differentiated classes of service.

The IPv6 Flow Label

The *Flow Label* in the IPv6 header is a 24-bit value designed to uniquely identify any traffic flow so that intermediate nodes can use this label to identify flows for special handling —for example, with a resource-reservation protocol, possibly RSVPv6 or a similar protocol. The *Flow Label* is assigned to a flow by the flow's source node or the flow originator. RFC1883 requires that hosts or routers that do not support the functions of the *Flow Label* field be required to set the field to 0 when originating a packet, pass the field on unchanged when forwarding a packet, and ignore the field when receiving a packet. It also specifies that flow labels must be chosen randomly and uniformly, ranging from hexadecimal 0x000001 to 0xffffff, as depicted in Figure 9.9. Additionally, all packets belonging to the same flow must be sent with the same source address, destination address, *Priority* value, and *Flow Label*.

Currently, aside from the suggestion mentioned in RFC1883, there are still no clearly articulated methods for using the *Flow Label* in the IPv6 header, apart from using it with a resource-reservation protocol such as RSVP. However, at least one recommendation has been published that provides a generalized proposal for using the *Flow Label* in the IPv6 header. RFC1809 [IETF1995d] suggests that *Flow Labels* may be used on individual sessions in which the flow originator may not have specified a *Flow Label*, and an upstream service provider may want to impose a *Flow Label* on specific flows as they enter their network, in an effort to differentiate these flows from others, possibly for preferential treatment. Currently, it is unclear what intrinsic value the *Flow Label* actually provides, because recently devised methods are available to build flow state in routers without the need for a special label in each packet.

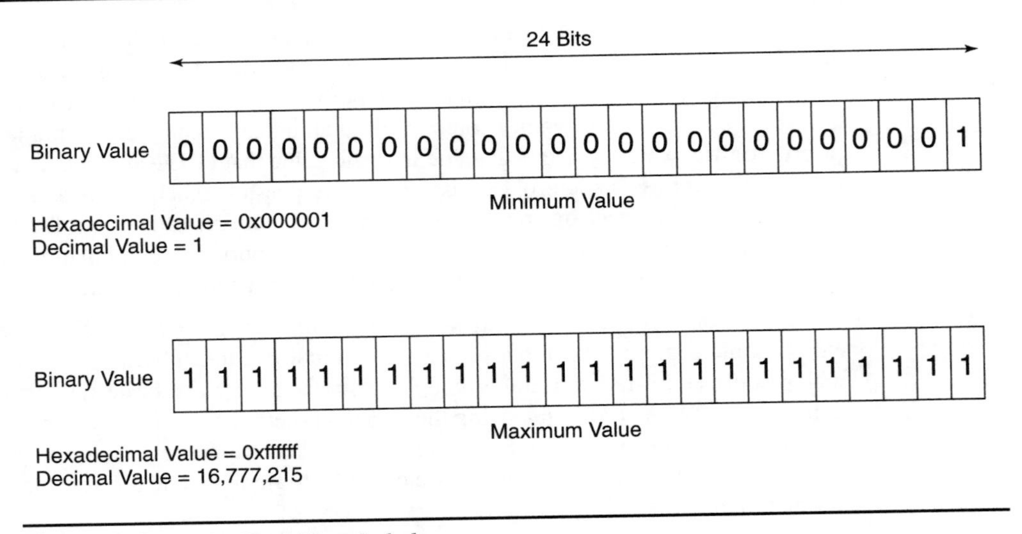

Figure 9.9: *The IPv6 Flow Label.*

IPv6 QoS Observations

There is continuing dissension in the networking community about the need for IPv6, and when closely examining the issues in these opposing viewpoints, you might be hard pressed to find a compelling reason to pervasively deploy IPv6, at least in the short term. In fact, most of the enhancements designed into the IPv6 protocol specification have been retrofitted for IPv4. The utility gained by migrating to IPv6 in an effort to benefit from the dynamic neighbor-discovery and address-autoconfiguration features, for example, might be negated by the availability and subsequent wide-scale deployment of DHCP (Dynamic Host Configuration Protocol) for IPv4 [IETF1997b].

In a similar vein, the IPv6 *Priority* field does not offer any substantial improvement over the utility of the IP precedence field in the IPv4 header. Granted, the IPv6 *Priority* field offers eight additional levels of distinction (16 possible values) than found in the IPv4 IP precedence field (8 possible values). However, this does not prove to be of considerable benefit. Conventional thinking suggests that because no one has yet to commercially (or otherwise) implement even a simple two-level class of distinction for differentiating service classes, increasing the possible number of service classes is not a compelling reason to prefer IPv6 over IPv4—at least not for the foreseeable future. A rough consensus of polled service providers indicates that if they deployed a service to provide differentiated CoS, they would not be interested in offering any more than three levels for matters of configuration simplicity and management complexity.

The utility of the IPv6 *Flow Label*, on the other hand, may prove to be quite useful. However, the only immediately recognizable use for the *Flow Label* is in conjunction with a resource-reservation protocol, such as RSVP for IPv6. The *Flow Label* possibly could be used to associate a particular flow with a specific reservation. The presence of a *Flow Label* in data packets also may help in expediting traffic through intermediate nodes that have previously established path and reservation states for a particular set of flows. Aside from using the IPv6 *Flow Label* with RSVP, its benefit is not immediately determinable.

All in all, IPv6 does not offer any substantial QoS benefits above and beyond what already is achievable with IPv4.

chapter 10

QoS: Final Thoughts and Observations

Quality of Service continues to remain a largely desirable property, a far-reaching requirement, and an ever-increasing topic of discussion with respect to the Internet today.

There's No Such Thing as a Free Lunch

A number of dichotomies exist within the Internet that tend to dominate efforts to engineer possible solutions to the quality of service requirement. Is this addressing an economic or a technical problem? Is QoS an economic or technical solution? Is QoS the imposition of state onto a stateless network? By attempting to differentiate degradation, do you increase total degradation of service? Is QoS a mechanism imposed by the network on a customer-by-customer basis or activated by the user on a per-transaction basis? Is the need to provide QoS the result of a network service provider that wants to oversell a limited network resource, and are some users being forced to compensate for a poor network model? Is QoS a product of marketing or a consumer-driven approach to the marketing problem of oversubscription? This final chapter attempts to tackle some of these questions by making some observations of the QoS landscape.

So far, QoS has been viewed as a wide-ranging solution set against a very broad problem area. This fact often can be considered a liability. Ongoing efforts to provide "perfect" solutions have illustrated that attempts to solve all possible problems result in technologies that are far too complex, have poor scaling properties, or simply do not integrate well into the diversity of the Internet. By the same token, and by close examination of the issues and technologies available, some very clever mechanisms are revealed under close scrutiny. Determining the usefulness of these mechanisms is perhaps the most challenging aspect in assessing the merit of a particular QoS approach. So, as we make these observations, we'll try to assess their practical utility as well.

One basic observation will be revisited here before we can begin to draw observations on how to practically implement a QoS methodology, and several criteria need to be quantified before launching into a dissertation on which one might be appropriate and in what environment. One conclusion is that when a network is under severe stress, the network operator can increase the available bandwidth (and scale up a simple switching environment to match the additional bandwidth) or leave the base capacity levels constant and attempt to ration access to the bandwidth according to some predetermined policy framework.

A very supportable second conclusion is the fact that bandwidth is not unlimited in every location where there is a need for data. Within this framework, the need to differentiate services becomes a technical argument, not necessarily one of introducing new revenue-generating products and services. Although the latter certainly may be a byproduct of the former, unfortunately, new business services rarely are driven by technical requirements.

A Matter of Marketing?

The current model of the Internet marketplace is an undifferentiated one. Every subscriber sees a constant best-effort Internet where, at any point within the network, packets are treated uniformly. No distinction is made in this model for the source or destination of the packet, no distinction is made in relation to any other packet header field, and no distinction is made in regard to any attempt to impose a state-based contextual "memory" within the network. The base best-effort model is remarkably simple. A single routing generated topology is imposed on the network, which is used by the switches to implement simple destination-based, hop-by-hop forwarding. At each hop, the packet may be queued in the next hop's FIFO (first in, first out) buffer, or if the buffer is full, the packet is discarded. If queuing delay or buffer exhaustion occurs, discarded packets are affected equally. Thus, although the Internet service model is highly variable across a large network, the elements of the service model are imposed uniformly on all traffic flows.

The marketing question is whether this uniform variable-service model creates unacceptable uncertainties in transaction times, and if this is the case, whether there is a secondary market for a more consistent service model within the same environment. Of course, the phrase "more consistent" is somewhat of a misnomer, because the clients of such a distinguished service doubtless would want to avail themselves of superior service levels at those times when the resources are available to other clients of the service. Their expectation would be that when the service levels degrade for other clients, their service levels are maintained at a *normal* or contracted level. The bottom line for this approach has a number of facets: Can a base-service QoS level be specified for general connectivity within an Internet environment, or is a QoS contract a comparative contract in which the service level for QoS-differentiated clients is specified as a differential from traditional best-effort services?

Is QoS Financially Viable?

The second facet of the QoS marketing service is whether they will provide a financial return to the operator and yield results for the client that can translate into justification of the presumably increased subscription fee. In very general terms, it is reasonable to state that QoS services may cause the network to operate at a lower level of overall efficiency within periods of peak load, and therefore the QoS service must operate at a pricing premium, which can offset the reduced base revenue stream because of the lower peak carriage capacity for such base normal service traffic. Working through the levels of service and price differential is a critical exercise in marketing this service.

The challenge here is that although the price differential will be fixed within the fee schedule, the service differential will be extremely difficult to determine from the customer's perspective. In an uncongested network, there will be no visible service differential (unless the operator undertakes traffic degradation for non-QoS traffic during unloaded periods, which in a competitive market appears to be a very short-sighted move). As the network load increases, the differential will become greater as the QoS traffic consumes a proportionately larger share of the congested resources. Of course, this is not an infinite resource, and at some load point, the QoS traffic will congest within the QoS category. If this happens, the service differential will then start to decline.

Subscription versus Packet-Option Models

Is QoS a subscription service or a user-specified per-packet option? From a marketing perspective, the subscription model offers many advantages, including some stability of the QoS revenue stream, some capability to plan the QoS traffic levels and consequent engineering load, and a resultant capability from the engineering perspective to be able to deliver a stable QoS service. Packet-option models create a more variable service environment with highly dynamic QoS loads that are visible only at periods of intense network use. A per-packet option also entails extensive packet-level accounting and verification at the entry to the network, because the consumer would anticipate that any fee premium would apply only to packets marked with precedence directives.

Effect on Non-QoS Clients

The other aspect to the marketing service is the negative message to non-QoS clients. When the network biases available resources toward QoS traffic, non-QoS traffic receives lower resource levels, which in turn leads to an average lower-service profile. Presumably, the fee would remain constant, so the resultant situation is constant pricing with declining expectation of service levels. In a competitive market, this is not a scenario that results in excellent levels of customer retention. The message this observation expresses is that although it is possible to implement traffic precedence within the network, the marketing approach should attempt to limit the negative impact of QoS on the network by deploying entry-level traffic profilers, creating the resultant environment of bounded levels of QoS network preemption in an attempt to preserve average levels of service performance for non-QoS customers.

Looking at QoS solely from the perspective of the transaction between the service provider and the consumer of the service and taking the approach that QoS is a matter of marketing, however, tends to ignore the larger issue of market economics, which nevertheless are very critical here. So is QoS a matter of market and supply-side economics?

A Matter of Economics?

Some of the more profound and intriguing aspects of examining what is possible in the realm of QoS is that it can, in some regard, be considered a situation that simply boils down to a set of compromises. Although some of the compromises are economic and some are technical, they are related inextricably in the grand scheme of things. If bandwidth were truly unlimited and access to installed fiber was readily available for a pittance, for example, QoS probably would not be an issue at all, except perhaps as an interesting research project to examine the possibility of changing the speed-of-light propagation delay characteristics on transcontinental links. In an ideal world, you also might choose to limit access to the network for certain types of traffic for a variety of reasons. Admission control appears to be an eternal desire, regardless of the economic or technical paradigm.

Of course, this is not the case, and bandwidth on a global scale is not as cheap or readily available as we would like. This presents a delicate balance among network engineering, network architectural design, and scales of economy, some of which are still not well understood in the telecommunications industry. The brokering of transcontinental bandwidth is a complex and convoluted game. The players are global telecommunications industry giants who have been playing the game without consideration of the traffic content, whether it be voice or data, but only within the realm of capacity. The process was crafted to reflect the very high investment levels required for such projects and the high risks of such investment, as well as the relatively slow growth of the traditional consumer of the product—the voice market. The industry geared itself to an artificial constraint of supply to ensure stability of pricing, which in turn was intended to ensure that the return on investment remained high (those familiar with the diamond industry no doubt will see some parallels in this situation). Thus, international cable systems are geared to ensuring that at any stage, the supply of capacity onto the market just matches the current level of demand, so that wholesale pricing does not crash through dumping and the investment profile remains adequately attractive for the associated investment risk.

Time Brings Change

Now that recent data demands are changing the demand model from a well-tempered demand growth to an "all you can eat" model of capacity requirement, the balances are changing. Historically, well-understood forward capacity plans are being scrapped as the data networks ravenously chew through all available inventory. The same holds true for Local Exchange Carriers (LECs) in a competitive domestic telecommunications market, who are just now beginning to understand that architecting "data-friendly" networks deviates somewhat from their traditional method of architecting circuit-switched voice networks. This produces an interesting paradigm.

Traditionally, Internet Service Providers (ISPs) simply have bought or leased circuits from the telco (the local telecommunications entity or telephone company) with which they construct their networks. This worked reasonably well in the earlier days of the Internet, when traffic volumes were relatively low and high-speed circuits were relatively low-speed by today's standards. The circuit orders were provisioned from the margins of the oversupply of a vastly greater voice network, where the supply model was one of advance provisioning up to two decades in advance of consumption. The supply of carriage for data leases can be seen as reducing the oversupply margin by some small number of months two decades hence. In the intervening period, the telco has a revenue stream for otherwise idle capacity.

A paradigm shift occurred over the course of the past 10 years where some of the telcos were getting into the ISP business and competing directly with the traditional ISP businesses to which they also are selling capacity. Coupled with the fact that the number of ISPs has grown phenomenally during the same time frame, as has the demand for bandwidth, capacity is not readily available for a variety of reasons. One reason is the sheer growth in bandwidth consumption and a staggering lag in provisioning new capacity. The provisioning systems and their associated capital investment structures still are well entrenched within a traditional slow linear growth model. The capital demands of this recent explosive wave of data expansion will take some time, because a radical change to the capital investment programs of the capacity providers (currently the telcos) will not occur quickly.

Another significant reason for this shift is the telco's emerging appreciation of an apparent conflict of interest. The traditional high barriers for entry into the voice market has been eroded by both deregulation and technology advance, so that telcos may perceive their current ownership of cable as the last bulwark of protection in their historical voice revenues. An additional factor, is that the telcos are now becoming active players in the "value-added" data business, either through direct business development or by acquisition, and now are somewhat reluctant to sell significant levels of capacity outside their immediate in-house interests in consideration of their own competitive interests in a deregulated market.

The Cost of Additional Capacity

The potential outcome of these factors is an artificial environment where, regardless of the levels of installed inventory, capacity is not readily available or is tariffed at a premium. Under these circumstances, the non-telco ISP must decide whether to pay such tariffs for additional capacity, find alternative methods for connectivity, or simply find methods to mitigate network degradation because of the lack of adequate network resources. This is the basis of the observation that, even when an abundance of bandwidth seems to exist, there is still an economic driver for QoS services that is fueled by a lack of bandwidth availability. It is possible that this will become an increasingly important regulatory issue as the market outcomes from the current pass of deregulation of this industry are examined, when the question will be posed as to whether the deployment of QoS is an acceptable outcome of the current market-based methods of bandwidth distribution to the telecommunications industry.

Of course, in some geographic locations, bandwidth cannot be purchased for any price because it simply does not exist. In this case, the deployment of QoS (in response to bandwidth scarcity) is based on local infrastructure investment conditions instead of an artifact of the current state of the industry.

We find ourselves asking some familiar questions—most of them technical, but surprisingly enough, some of them economic. Is the desire to implement QoS introduced by fundamental economic imbalances in the evolution of the telecommunications industry? There are far too many facets of this question to undertake a meaningful economic analysis of the issues involved here. However, one might suggest that the economic issues of QoS are somewhat orthogonal to the technical issues of QoS, because regardless of the reasons, differentiation is a bona fide outcome; therefore, QoS is indeed a technical matter.

A Matter of Technology?

This section approaches the technology viewpoint by using a top-down perspective. The first issue is one for which the most common denominator must be identified, because this is the place in which QoS is most easily implemented and least likely to produce unpredictable or undesirable effects.

The Best Environment for QoS

In the global Internet, this is clearly the IP protocol suite, because a single link-layer media will never be used pervasively end-to-end across all possible paths. What about the suggestion that it is certainly possible to construct a smaller network of a pervasive link-layer technology, such as ATM? Although this is certainly true in smaller private networks, and perhaps in smaller peripheral networks in the Internet, it currently is rarely the case that all end-systems are ATM-attached, and this does not appear to be a likely outcome in the coming years. In terms of implementing visibly differentiated services based on a quality metric, using ATM only on parts of the end-to-end path is not a viable answer. The ATM subpath is not aware of the complete network layer path, and it does not participate in the network or transport layer protocol end-to-end signaling.

The simplistic answer to this conundrum is to dispense with TCP/IP and run native cell-based applications from ATM-attached end-systems. This is certainly not a realistic approach in the Internet, though, and chances are that it is not very realistic in a smaller corporate network. Very little application support exists for native ATM. Of course, in theory, the same could have been said of frame relay transport technologies in the recent past and undoubtedly will be claimed of forthcoming transport technologies in the future. In general, transport-level technologies are similar to viewing the world through plumber's glasses: Every communications issue is seen in terms of point-to-point bit pipes. Each wave of transport technology attempts to add more features to the shape of the pipe, but the underlying architecture is a constant perception of the communications world as a set of one-on-one conversations, with each conversation supported by a form of singular communications channel.

One of the major enduring aspects of the communications industry is that no such thing as a ubiquitous single transport technology exists. Hence, there is an enduring need for an internetworking end-to-end transport technology that can straddle a heterogeneous transport substrate. Equally, there is a need for an internetworking technology that can allow differing models of communications, including fragmentary transfer, unidirectional data movement, multicast traffic, and adaptive data-flow management.

This is not to say that ATM itself, or any other link-layer technology for that matter, is not an appropriate technology to install into a network. Surely, ATM offers high-speed transport services, as well as the convenience of virtual circuits. However, what is perhaps more appropriate to consider is that any particular link-layer technology is not effective as far as providing QoS for reasons that have been discussed thus far.

The second technology issue is determining how, within the constructs of IP, QoS can be provided.

Providing QoS

Before looking in detail at the way in which QoS can be provided, the underlying model of the network itself is relevant here. To quote a work in progress from the Internet Research Task Force, "The advantages of [the Internet Protocol's] connectionless design, flexibility and robustness, have been amply demonstrated. However, these advantages are not without cost: careful design is required to provide good service under heavy load" [IRTFa]. Careful design is not exclusively the domain of the end-system's protocol stack, although good end-system stacks are of significant benefit. Careful design also includes consideration of the mechanisms within the routers that are intended to avoid congestion collapse. Differentiation of services places further demands on this design, because in attempting to allocate additional resources to certain classes of traffic, it is essential to ensure that the use of resources remains efficient and that no class of traffic is totally starved of resources to the extent that it suffers throughput and efficiency collapse.

IRTF Overview

The Internet Research Task Force (IRTF) is composed of a number of focused, long-term and small research groups. These groups work on topics related to Internet protocols, applications, architecture, and technology. Participation is by individual contributors instead of by representatives of organizations. The IRTF's mission is to promote research that further enables the evolution of the future Internet.The IRTF Research Group's guidelines and procedures are described more fully in RFC 2014 (BCP 8) [IETF1996f].

The IRTF is managed by the IRTF Chair in consultation with the Internet Research Steering Group (IRSG). The IRSG membership includes the IRTF Chair, the chairs of the various research groups, and possibly other individuals ("members at large") from the research community.

The IRTF Chair is appointed by the Internet Architecture Board (IAB), the research group chairs are appointed as part of the formation of research groups, and the IRSG members at large are chosen by the IRTF Chair in consultation with the rest of the IRSG and on approval of the IAB. In addition to managing the research groups, the IRSG may from time to time hold topical workshops focusing on research areas of importance to the evolution of the Internet or more general workshops to discuss research priorities from an Internet perspective, for example.

You can find more information on the IRTF at the IRTF Web site, located at www.irtf.org.

QoS and the Internet Router

A router has a very limited set of actions at its disposal after receipt of an IP datagram. It can select an outbound interface and then queue the datagram on the associated outbound queue or discard the datagram. The router subsequently selects a datagram to transmit from the queue. The router can perform only four fundamental actions: outbound interface selection, queue management, interface scheduling, and discard selection.

Outbound Interface Selection—Routing

In a traditional unicast environment, the router uses the destination IP address of the datagram when determining which outbound interface to pass the packet to for transmission. One course of possible action here is to make a forwarding decision based on some QoS-related setting. Quality of Service routing can take a variety of factors into consideration, as discussed earlier. One action is to overlay a number of distinct topologies based on differing results of calculating end-to-end (or hop-by-hop) metrics. These metric calculations might include such factors as estimates of propagation delay or configured-link capacity. It certainly is theoretically possible to also use current idle capacity as a metric, although a very tight relationship is required between forwarding decisions and the subsequent need to redefine the routing protocol, so this renders this option an unstable choice, at least for the short-term future. In fact, QoS-routing technologies still are in the conceptual stages, and nothing appears to be viable in this area for the near-term, especially when considering this as a candidate routing mechanism for large-scale Internet environments. Alternatively, the forwarding decision can be based on an imposed state, where the flow identification of the packet is matched against a table of maintained state information. Such constructs are used within the RSVP protocol mechanisms.

At present, knowledge of the stability and scalability of QoS mechanisms directed at altering the choice of the outbound interface as the mechanism of service differentiation is still very scant. Indeed, given the succession of routing protocols that have been deployed in the Internet over the past decade, it also may be relevant to observe that knowledge of the stability and scalability of large-scale routing protocols is still somewhat inadequate. Adding further complexity in terms of overlaying multiple QoS-related logical topologies is an area that should be undertaken with extreme caution. As with a dancing circus bear, the fact that it can dance is simply amazing. Getting it to sing is possibly asking too much.

Queue Management

Packet queuing and packet discard are related concepts.

Conventionally, FIFO queues are used in routers. This preserves the ordering of packets within the router, so that sequential packets of the same flow remain in order along a particular end-to-end path within the network. Given the finite length of an output queue, the router performs packet discard once the queue is exhausted, creating what commonly is referred to as a *tail-drop* behavior. In this state, discard is undertaken as an alternative to queuing. If space is available on the queue, the packet is added to the tail end of the queue. Otherwise, the packet is discarded.

This basic queue-admission policy can be modified by setting up admission criteria in which packets may be discarded even though space is available on the queue. This queue-admission policy, Random Early Discard (RED) [Floyd1993], is a critical component of queue-management practices in which the intent is to signal TCP sessions of the imminent risk of tail-drop queue saturation by using packet drop.

In general, tail-drop queue management should be avoided. Tail-drop causes all subsequent packets to be discarded, which causes all TCP sessions crossing the exhausted queue to undertake congestion-avoidance measures, which induce a large-scale efficiency drop in overall network throughput. The RED approach is to alter queue admission to allow packet discard before the onset of queue exhaustion. Within flow-controlled end-to-end environments, this approach is readily interpreted by the sender as a signal to throttle back the rate expansion instead of extensive packet drop and retransmits.

QoS criteria can be admitted into the queue-admission policy through the selective behavior of RED. The technique is to weight the preference for packet discard to lower-precedence packets, which causes a rate-damp signal to be sent to lower-precedence streams (assuming that all packets within a stream have identical precedence) before the signal is sent to the higher-precedence streams.

The intrinsic behavior of RED is a relatively subtle form of service differentiation. Short-lived TCP flows and nonflow-controlled UDP streams are relatively immune from the effects of RED, in that while the packet drop may cause retransmission, the total flow rate is not substantially altered. This is either because the flow is only of very short duration in any case, or the end-to-end application protocol is not drop sensitive. Thus, while RED can promote more efficient use of network resources by well-behaved TCP flows, and while weighting of RED by some form of QoS precedence can allow some level of QoS flow differentiation within the network, the overall result of visible differentiation of service level is difficult to discern within today's traffic profiles.

The queue-admission policy can be altered further by modifying the basic FIFO operation to one in which QoS packets may be inserted at the head of the queue or in an intermediate position that maintains an overall queue ordering by some QoS precedence value. However, this particular model is best regarded not as a queue-management issue, by as a use of multiple queues, with the consequent issue of scheduling packets from each of the queues.

Uncontrolled UDP (User Datagram Protocol) flows are a relevant consideration here, and with the increasing deployment of UDP traffic flows under the *multimedia* umbrella, this issue will become critical in the near future. A long-lived UDP flow, operating at a rate in excess of an intermediate hop's resource capacity, places growth pressure on the associated output queue, ultimately causing queue saturation and subsequently a queue tail-drop condition that affects all flows across this path. Controlling the level of non-flow-controlled UDP can be a difficult ingress policy to effectively police. An alternative is to use distinct queuing disciplines for TCP and UDP traffic, attempting to minimize the interaction between flow-controlled and non-flow-controlled data.

Packet Scheduling and Queue Management

Because a number of ways exist to modify this single FIFO queue behavior by using multiple output queues on a single outbound interface, the router behavior can be seen from the perspective of scheduling packets from the set of associated output queues. One such method includes varying the queue lengths and then adopting a scheduling algorithm that queues packets in some form of preferential basis. This could done by weighted round-robin queue selection or some other form of queue-selection algorithm.

The capability of the QoS application is an outcome of the queue structure, the queue-admission criteria, and the packet-scheduling mechanism. The use of QoS weighting on the queues, where higher-precedence packets are scheduled at a higher priority, does introduce grossly visible differentiation of service levels under load. As the network load increases, the incremental congestion load is expressed in increased queue delay in queues that surround the network load point. This queuing delay is visible in terms of extended RTT estimates for traffic flows, which in turn reduces the TCP traffic rates. Both slow-start and congestion-avoidance TCP algorithms are RTT sensitive. By using QoS scheduling, the higher-precedence traffic consumes a greater share of the congested segment(s), attempting to preserve the RTT estimates of the associated traffic flows. The difference between a relative constant RTT and one that exhibits relatively high variability and higher average value is eminently visible, both for short and long TCP traffic flows. It also is highly visible to simple diagnostic probes such as PING and TRACEROUTE. It is admittedly a relatively coarse differentiator of traffic and one that can be used to cause high levels of differentiation even under relatively light levels of load. Of course, if QoS is tied to a financial premium, one of the key market attributes of QoS will need to be highly visible QoS differentiation. Weighting the queue scheduling is the mechanism that offers the greatest promise here.

QoS Deployment

For QoS to be functional, it appears to be necessary that all the nodes in a given path behave in a similar fashion with respect to QoS parameters, or at the very least, do not impose additional QoS penalties other than conventional best effort into the end-to-end traffic environment. The sender or network ingress point must be able to create some form of signal associated with the data that can be used by down-flow routers to potentially modify their default outbound interface selection, queuing behavior, and discard behavior.

The insidious issue here is attempting to exert "control at a distance." The objective in this QoS methodology is for an end-system to generate a packet that can trigger a differentiated handling of the packet by each node in the traffic path, so that the end-to-end behavior exhibits performance levels in line with the end-user's expectations and perhaps even a contracted fee structure.

This *control-at-a-distance* model can take the form of a "guarantee" between the user and the network. This guarantee is one in which, if the ingress traffic conforms to a certain profile, the egress traffic maintains that profile state, and the network does not distort the desired characteristics of the end-to-end traffic expected by the requester. To provide such absolute guarantees, the network must maintain a transitive state along a determined path, where the first router commits resources to honor the traffic profile and passes this commitment along to a neighboring router that is closer to the nominated destination and also capable of committing to honor the same traffic profile. This is done on a hop-by-hop basis along the transit path between the sender and receiver, and yet again from receiver to sender. This type of state maintenance is viable within small-scale networks, but in the heart of large-scale public networks, the cost of state maintenance is overwhelming. Because this is the mode of operation of RSVP, RSVP presents some serious scaling considerations and is inappropriate for deployment in large networks.

RSVP scaling considerations present another important point, however. RSVP's deployment constraints are not limited simply to the amount of resources it might consume on each network node as per-flow state maintenance is performed. It is easy to understand that as the number of discrete flows increases in the network, the more resources it will consume. Of course, this can be somewhat limited by defining how much of the network's resources are available to RSVP; everything in excess of this value is treated as best effort. What is more subtle, however, is that when all available RSVP resources are consumed, all further requests for QoS are rejected until RSVP-allocated resources are released. This is similar in functionality to how the telephone system works, where the network's response to a flow request is commitment or denial, and such a service does not prove to be a viable method to operate a data network where better-than-best-effort services always should be available.

The alternative to state maintenance and resource reservation schemes is the use of mechanisms for preferential allocation of resources, essentially creating varying levels of best-effort. Given the absence of end-to-end guarantees of traffic flows, this removes the criteria for absolute state maintenance, so that better-than-best-effort traffic with classes of distinction can be constructed inside larger networks. Currently, the most promising direction for such better-than-best-effort systems appears to lie within the area of modifying the queuing and discard algorithms. These mechanisms rely on an attribute value within the packet's header, so these queuing and discard preferences can be made at each intermediate node. First, the ISP's routers must be configured to handle packets based on their IP precedence level. There are three aspects to this: First, you need to consider the aspect of using the IP precedence field to determine the queuing behavior of the router, both in queuing the packet to the forwarding process and queuing the packet to the output interface. Second, consider using the IP precedence field to bias the packet-discard processes by selecting the lowest precedence packets to discard first. Third, consider using any priority scheme used at Layer 2 that should be mapped to a particular IP precedence value.

The cumulative behavior of such stateless, local-context algorithms can yield the capability of distinguished service levels and hold the promise of excellent scalability. You still can mix best-effort and better-than-best-effort nodes, but all nodes in the latter class should conform to the entire QoS selected profile or a compatible subset (an example of the principle that it is better to do nothing than to do damage).

Service Quality and the Host

Each of the mechanisms discussed so far rely on the network to implement distinctions of quality of service. The user also has the capability to implement distinguished service, particularly in relation to TCP traffic.

Using TCP buffers and window advertisements commensurate with the delay-bandwidth product of the end-to-end traffic path allows the end-system to effectively utilize available bandwidth. The data rate is based on the amount of data that can be loaded into the end-to-end path, divided by the propagation delay. Queuing behavior attempts to reduce the propagation delay (which is the sum of the signal propagation delay across all intermediate hops plus the queuing time) by reducing queuing delay within this equation. If the system buffer is less than the delay-bandwidth product, no more data can be sent until the signal of receipt is received. Therefore, the buffer size must be greater than two times the bandwidth-delay product to increase data-transfer performance, so the only limiting factor becomes the true bandwidth of the network and not inadequate buffering.

The use of improved implementations of TCP window-scaling options, along with very large buffers, can provide significant improvements in end-to-end performance without any changes to the underlying behavior of the network. This is especially true in some cases where very long propagation delay is evident, such as in geostationary satellite paths.

Even within such a model, some amount of queuing is inevitable. A TCP session can be thought of as a quasiconstant delay loop. When a sender commences in slow-start mode, it transmits one packet, and upon receipt of the acknowledgment (following a delay of one RTT in time), immediately transmits two packets. This cycle of immediately transmitting two data packets upon receipt of each ACK packet continues through the slow-start mode. If, at any point in the traffic path, there are slower links than at the sender's end of the traffic path or there is the presence of other traffic, queuing of the second packet occurs. The steady state of slow start is an exponential increase of volume, placed into the end-to-end path using ACK pacing to double the data in each ACK "slot." The resulting behavior of this is naturally bursty traffic. At each RTT interval, the packet train contains successive trains of 2, 4, 8, and so on packets, where the pacing of the doubling is based on the bit rate of the slowest link in the end-to-end traffic path.

Within the network, this sequencing of packets at a rate of twice the previous rate within each RTT interval is smoothed out to the available bandwidth of the end-to-end path by the use of queuing within the routers. This queuing requirement can grow to half the delay-bandwidth product. For one-half of this RTT delay interval, the queuing requirement is to store one segment for every segment it can transmit. This queuing burst can be mitigated by the sender implementing bandwidth estimates using sender packet back-off, where the initial RTT is used to pace the emission of the second of the two packets, at one-half the RTT, for the second iteration at one-fourth the RTT, and so on. Alternatively, ACK pacing can be done at the receiver or in the interior of the network (although interior ACK pacing may require symmetric paths to allow automatic initiation of ACK spacing by the router). The slow-start phase of doubling the data rate per RTT continues to the onset of data loss; thereafter, further increases of data rate are linear. This next phase, the congestion-avoidance algorithm, instead of doubling the amount of data in the pipe each RTT, uses an algorithm intended to undertake a rate increase of one MTU per RTT.

How can the host optimize its behavior in such an environment? Consider these solutions:

Use a good TCP protocol stack. Much of the performance pathologies that exist in the network today are not necessarily the byproduct of oversubscribed networks and consequent congestion. Many of these performance pathologies exist because of poor implementations of TCP flow-control algorithms, inadequate buffers within the receiver, poor use of MTU discovery (if at all), and imprecise use of protocol-required timers. It is unclear whether network ingress-imposed QoS structures will adequately compensate for such implementation deficiencies, but the overall observation is that attempting to address the symptoms is not the same as curing the disease.

A good protocol stack can be made even better in the right environment, and the following suggestions are a combination of measures that are well studied, are known to improve performance, and appear to be highly productive areas of further research and investigation.

Implement a TCP Selective Acknowledgment (SACK) [IETF1996c] mechanism to filter out noise from bandwidth-saturation signals. SACK, combined with a selective repeat-transmission policy, can help overcome the limitation that traditional TCP experiences when a sender can learn only about a single lost packet per RTT.

Implement larger buffers with TCP window-scaling options. The TCP flow algorithm attempts to work at a data rate that is the minimum of the delay bandwidth product of the end-to-end network path and the available buffer space of the sender. Larger buffers at the sender assist the sender in adapting to a wider diversity of network paths more efficiently by permitting a larger volume of traffic to be placed in flight across the end-to-end path.

Use a higher initial TCP slow-start rate than the current 1 MSS (Maximum Segment Size) per RTT. A sample size that appears feasible is an initial burst of 4 MSS segments. The assumption is that there will be adequate queuing capability to manage this initial packet burst, while the provision to back off the send window to 1 MSS segment should remain to allow stale operation if the initial choice was too large for the path. The result of a successful start with a transmission window of 4 MSS units is that the initial data rate through the slow-start algorithm will move four times the data in the same time interval!

Implement sender data pacing or receiver ACK spacing so that the burst nature of slow-start increase is avoided. This is perhaps a more controversial recommendation. The intent of such pacing is to reduce the inherent burst nature of the slow-start TCP algorithm and, in so doing, relieve the queuing pressure placed on the network where the end-to-end path traverses a relatively slower hop. However, modern routers can use small, fast caches to detect and optimally switch packet trains, and packet pacing breaks apart such trains. The advantage of such an approach is to allow the network flow to quickly find the available end-to-end flow speed without receiving transient load signals that may confuse the availability calculation being performed.

All these actions have one feature in common: They can be deployed incrementally and individually, allowing end-systems to obtain better-than-best-effort service even in the absence of the network provider attempting to distinguish traffic by class distinction.

Also, because HTTP traffic typically is greater than 50 percent of all public IP network traffic, one additional point must be addressed.

Use HTTP Version 1.1 [IETF1997c]. HTTP 1.0 requires each client-server exchange to open a corresponding TCP connection. HTTP 1.1 is not bound by this restriction, though; it allows multiple client-server exchanges to be conducted over one (or more) TCP connections. In other words, it uses persistent TCP connections wherever possible. This behavior also negates the problem where TCP RSTs (Resets) are sent instead of FINs (Finish tags), which results in excessive TCP ACKs being generated.

The conclusion here is that if the performance of end-to-end TCP is the perceived problem, it is not necessarily the case that the most effective answer lies in adding service differentiation in the network. It is often the case that the greatest performance improvement can be made by improving the way in which the hosts and the network interact through the flow-control protocol.

Service Quality and the Network

The second part of this "good-housekeeping guide" list is intended to allow the network to play its role in working within reasonable operating parameters.

Of course, there is no substitute for proper network engineering in the first place. Network loads generally exhibit peak load conditions every day, and if the network cannot handle these loads, the consequent overload conditions created by bandwidth exhaustion, queue exhaustion, and switch saturation cannot be readily ameliorated by QoS measures. The underlying resource-starvation issue must be addressed if any level of service is to be delivered to the network's clients. Additionally, the stability of the routing environment is of paramount importance to ensure that the network platform behaves predictably. Therefore, two primary prerequisites exist for effective network management:

Adequate network resources to handle normal load conditions. This is both a need for adequate bandwidth to manage the imposed traffic without excessive queue build up and adequate queue sizes to cope with the inherent requirements for buffering under normal load. This is not entirely a bandwidth and queue issue, because a requirement also exists to deploy a switching capability commensurate with the packet loads each router faces. Because any imposition of QoS services imposes further load on the switching functions, this second requirement is somewhat more strict. The switching function is exercised most strenuously when the network is under peak load, so switching capability should be deployed so that it can comfortably manage the full extent of the peak network loads it will face.

Network stability. This refers to the stability of the underlying transport substrate and the routing system over which the network layers above this fabric. Without an environment where the routing system is allowed to converge (and then operate for an extended period without further need to recompute the internal forwarding tables), the network rapidly degenerates and degrades in such a way that no incremental introduction of QoS structures can salvage it.

Major performance and efficiency gains can be made by allowing the network to signal to the end-systems the likely onset of congestion conditions, so that the end-systems can take action to reduce the traffic rate well before the network is forced into queue tail-drop behavior. Three of the most effective steps a network operator may take to improve network efficiency and end-user flow performance follow:

Implement Random Early Detection (RED). This ensures that the initial congestion back-off signals are sent to the appropriate TCP senders (which are pushing hardest at the network) before comprehensive packet discard occurs. RED is statistically likely to signal those stacks that are operating with large transmission windows, and the effect of this discard mechanism is to signal a reduction in the transmit window size. This mechanism attempts to avoid first signaling those small-scale flows that are not causing the overall congestion problem.

Separate TCP and UDP queues within the router. One such mechanism is to implement class filters to bound the level of resources given to non-flow-controlled UDP traffic, allowing the flow-controlled queues to behave more predictably. The most direct way to implement this is to place UDP traffic and TCP traffic in different output queues and use a weighted scheduling algorithm to select packets from each queue according to a network-imposed policy constraint of relative resource allocation. This method allows the TCP end-system stacks to oscillate faster, in order to estimate the amount of total end-to-end traffic capacity, based on the behavior of the flow-controlled traffic passing through the queues. Although UDP does not provide this "network-clocking" function, it is assumed that the UDP application will be intelligent enough to understand how to pace itself. Of course, this may not be a valid assumption, but then again, the application may be fundamentally broken if it cannot do so. The outcome of this recommendation is to limit the extent of damage a non-network-clocked UDP flow can cause. By making UDP queues relatively short and TCP queues longer in the router, there is a greater probability that the TCP queues can behave in a way that attempts to avoid tail-drop congestion and therefore increases network-clocked throughput efficiency.

Traffic shaping, using a token bucket at the network edges, to reduce the burstiness of the data-traffic rates. This is somewhat controversial, because this may be the result of a fee contract and not really a service-quality issue, per se. The effect is a surrogate method of data-burst-rate limiting, using the queues at the periphery of the network to reduce the level of queue load in the more critical central interior of the network.

These actions alone have the potential to offer marked increases in the level of efficiency of bandwidth usage in terms of actual data transfers. Random Early Detection provides for the enhanced efficiency of bandwidth resource sharing, making best-effort more uniformly available to all flows at any given instant, whereas queue segregation attempts to limit the damage that high-volume real-time (externally clocked) flows may inflict on network-clocked reliable data flows.

QoS and the Network

It is unreasonable to believe that QoS deployment can fully compensate for poor network engineering or poor protocol implementations in hosts. The implication is that the following QoS measures have the greatest positive impact on congestion differentiation, while having a minimum negative impact on total network efficiency, if the network platform and the connected hosts already exhibit a relatively favorable level of service quality. In such a case, there is an additional consequential set of actions the network operator and the user can deploy to further direct the controlled differentiation of network resource availability. These actions are listed here:

Support of the IP precedence header field values to distinctively define service classes and an industry-wide standardization that defines the semantics of these field values. The semantics of the values need to implemented by a network-wide queue-management scheme that allows higher-precedence traffic to usurp network resources from lower-precedence traffic in some controlled fashion.

Weighted RED (WRED), with the weighting managed by IP precedence header field values. This allows the network to signal lower-precedence TCP traffic flows to back off their expanding window behavior gracefully, while higher-precedence traffic continues to flow unimpeded.

WFQ (Weighted Fair Queuing), with the weighting managed by IP precedence header field values, which allows higher-precedence traffic to be placed at the head of the output queues. This allows the higher-precedence traffic to have a better propagation time estimate, which leads to reduced RTT estimates. This result allows faster discovery of available transfer-path bandwidth.

These mechanisms constitute better-than-best-effort services, because there is no longer a single best-effort paradigm within the network. These measures also are independent of any dial-access trigger mechanism that may be used to complete the QoS picture. Obviously, the access mechanism can be through a network-admission implementation, which may combine other traffic-policy mechanisms with precedence settings, or it can be a simple acceptance of user-defined precedence values, allowing the differentiation function to be selected by the user rather than the network.

Deploying RSVP

One additional comment remains on the deployment by the user and the network of RSVP. RSVP could possibly be quite expensive, and often the results may be no better than a properly designed network. After implementing the mechanisms and principles discussed here, you can cut down queue length, reduce excessive extraneous congestion signaling, and control misbehaving flows so that propagation times more accurately reflect physical signal-hop propagation.

There is a tight coupling of interaction between the end-system behavior of flow-controlled protocols and the router's queuing behavior. The interaction of many such systems results in a system behavior that is somewhat similar to chaotic systems. When you further introduce non-flow-controlled UDP traffic, this results in a system in which predictive traffic management is virtually impossible. When a diverse collection of differentiation mechanisms is introduced, there is no doubt that not even an experienced protocol engineer, network architect, or network engineer would be able to determine how to make the network behave efficiently.

It is not all bad news, however. The model used by the Internet is one of distributed intelligence, where the functionality of the data flow is passed over the network fence to the end-user's platform. The interior of the network is reduced to its bare essentials of basic transmission elements and simple switches. The result is a network of unsurpassed cost efficiency, and it certainly is vastly different from the circuit-switching models of previous communications systems in which significant functionality and cost is preserved within the network in a centralized functionality model. The real long-term challenge is scaling this model efficiently in terms of switching, and in terms of consumer models, so that the cost of the transmission and switching elements is fairly distributed to the consumer. QoS is an intermediate step and, unfortunately, is perhaps a recognition of intermediate failure to converge on this path.

Choosing QoS

QoS is not a uniform requirement in any network. If networks are engineered to the point where there is never any congestion, the argument for QoS weakens considerably. This is not to say that the argument for service quality is any weaker; this is clearly not the case. However, differentiating services is somewhat of a non-sequitur. Network administrators will continue to be faced with the task of provisioning networks that meet customer demands for availability, speed, latency, and so on.

In fact, and as mentioned earlier, it has been suggested that this becomes nothing more than a pricing argument. There has been some indication that the initial hordes flocking to get connected to the Internet have been carried through on the margins of oversupply in, and subsequent build-out of, the traditional telephone system infrastructure. Now that Internet traffic has effectively chewed through that, there is no high-margin cross-subsidization agent for further build-out of Internet infrastructure, and prices will inevitably rise. However, service providers cannot raise prices uniformly in a competitive market. Accordingly, QoS may be the leverage mechanism for price escalation. QoS allows the service provider to undertake selective price escalation, offering differentiation within congestion periods as a mechanism for pricing increases. The current flat-rate pricing mechanisms that dominate much of the current Internet landscape do not readily provide the capability to rise to the financial challenge to install additional communications capacity necessary to ride the continued aggressive expansion of the Internet.

The alternative course of action is to increase the level of differentiation of delivered service and couple this with pricing premiums for access to such differentiation of services. The fervent hope is that this will allow network operators to break free of single-fee, single-quality level-market structures and allow the operator to market differentiated service levels at differentiated prices. It is perhaps an unspoken hope that the entry of premium pricing structures will correct the current problems of ever-decreasing quality levels associated with a continually decreasing unit-pricing model.

Again, things are not bad all over—it just pays to do your homework. Networking in the global Internet complicates matters tremendously, simply because of the diversity of administration. The Internet is the closest thing resembling true chaos on planet Earth. Network operators who want to implement QoS within a single administrative domain arguably will have much better success in providing differentiated services and QoS compared to anyone who attempts to provide similar services across multiple administrative domains, at least for the foreseeable future. However, the same technical principles still hold true, and similar considerations must be given to network performance, stability, scale, management, and control. Private networks that must purchase or lease wide-area network capacity from a telco or local common carrier also may face the same unavailability of wide-area capacity, and network administrators who choose to use public switched wide-area services may face even more insidious problems. Regardless of whether you are trying to implement QoS in a private network or within a segment of the global Internet, differentiation may come at a cost. There's no magic here. The cost may not be expressed in economic terms, but there are certainly other prices to pay that cannot be calculated in dollars and cents.

Caveat emptor.

afterword

QoS and Multiprotocol Corporate Networks

Although the majority of this book has dealt with TCP/IP networks and the global Internet, we cannot ignore the fact that thousands (if not hundreds of thousands) of private multiprotocol corporate networks still are deployed out there. Several pertinent issues need to be discussed as a postscript to these chapters, including possible methods of implementing QoS in a multiprotocol environment, and perhaps examining how the multiprotocol network paradigms are slowly changing.

First, we should clearly define what we mean by the term *multiprotocol*. Traditionally, when we speak of *multiprotocol traffic*, we are referring to multiple network-layer and transport-layer protocols, such as IPX, SNA, DECnet, AppleTalk, and so on. Some of these protocols are multifaceted, because they are architectural in nature and have multiple communication layers. Traditional IBM networking strategies (e.g., SNA, APPN) are principal examples of multifaceted multiprotocol traffic. These multiprotocol networks generally run several different protocols such as these—perhaps in addition to TCP/IP.

It is important to understand the historical context of how these multiprotocol networks came into existence and to understand where they might be going in the future.

A Brief Multiprotocol Historical Perspective

In a broad sense, the annals of networking history have somewhat of a convoluted, and sometimes conflicting, family tree. It is fairly clear, however, that IBM was one of the first companies to successfully build, develop, and deliver a networking environment in its Systems Network Architecture (SNA) platforms. Traditionally (and still to this day), large-scale SNA networks have provided network computing systems for everything from large-scale accounting applications to warehouse-inventory systems. In fact, when network computing was in its infancy, IBM was truly one of the only games in town.

The early IBM mainframes were quite large—monoliths, in fact, compared to today's computing platforms—and the initial investment and ongoing maintenance costs were equally substantial. Companies that invested in IBM mainframes (and other vendors' mainframe platforms as well) fully expected to use them for a given calculated lifetime. Companies that used these mainframes generally used a convoluted depreciation schedule that dictated how long the platform had to be used before they could consider moving to another technology platform without losing a substantial portion of their initial investment.

The consequence of this economic incentive is that we now have thousands of companies that have invested millions and millions of dollars into legacy IBM technology and intend to use these network systems until they have fully depreciated. This introduces a somewhat interesting paradigm, because of the challenges of attempting to make newer, more advanced technologies that interoperate with older, entrenched technologies. In fact, matters are complicated significantly when the older, entrenched technology platform vendors make "enhancements" to the base technology in an effort to modernize the older-platform characteristics and behavior.

This same paradigm exists with similar technologies, such as Digital Equipment Corporation's DECnet and LAT (Local-Area Transport) protocols, as well as more recent network protocols such as Novell Corporation's IPX (Internet Packet Exchange) protocol. Significant numbers of networks still use these older protocols, and this introduces significant complications with regard to providing any sort of structured QoS within a single multiprotocol network substrate.

Migrating to TCP/IP

In recent years, many companies that own legacy multiprotocol networks of this sort have expressed a desire to migrate these network applications to TCP/IP for ease of administration, management, and engineering maintenance. Generally, managing a two-protocol network is not just double the operational cost of running a single-protocol network; it often is much harder and much more expensive. Accordingly, the desire to operate a single-protocol system is not just for the sake of simplicity of the network, but is often an outcome of the objective to operate the network within tight bounds of cost-effectiveness parameters. However, it is not as simple as you might imagine.

As stated earlier, in many cases, economic factors are involved. Many corporations have invested millions of dollars in legacy technologies and equipment. Just because a simpler or more elegant solution has been introduced is not a compelling enough reason to immediately upgrade these legacy platforms. Often, an organization will look for better mousetraps—newer and improved ways of using existing legacy technologies. Of course, religious and political issues may tend to delay a migration, but these are not germane to this particular discussion. In any event, migrations to newer technologies do not happen overnight and in fact may take years. This sometimes results in a very unfortunate situation; a company may begin planning a migration to the latest and most current technology, but by the time it completes the migration, the technology is outdated.

What we have today in the corporate landscape is a seemingly random collection of applications and protocols which, when considered together as a set of network requirements, has become cumbersome, dysfunctional, chaotic, and unmanageable. This is perhaps the most compelling reason to consider QoS mechanisms in the corporate network—to maintain balance and functionality so that some element of overall management and policy determination can be used to allocate network resources to particular business functions and their associated applications. It is very unwise to leave dynamic resource allocation to the wiles of operational competition between various protocols as they interact on the same network. Of course, all the other reasons previously discussed also may be compelling reasons for QoS, but it is this issue of unmanageability that seems to be a recurring theme. This is due to several reasons, but predominantly because some of these different network protocols simply were not designed to be used with one another.

Routed versus Bridged Legacy Protocols

One of the more significant complications these older legacy protocols present is the fact that some of them cannot be routed. They can only be transparently or source-route bridged, so the entire network become a single MAC-layer broadcast domain or conceptually equivalent to a single local area network. The most insidious aspect of this point is that this significantly affects how many hosts can be accommodated in the network, because all participating hosts share the same broadcast domain. Bridging simply does not scale.

Problems with SNA and LAT

The two most notable legacy protocols that fall into this category are SNA and LAT. A couple of methods to encapsulate SNA traffic into TCP can bypass this problem; we'll examine them in a moment. However, LAT was designed to operate at Ethernet speeds on original Ethernet broadcast networks, and the protocol has a hard-coded time-out value of around 80 milliseconds and limited resiliency against packet loss.

Here is a specific example of a relatively commonly deployed protocol that has assumed a number of performance parameters within the original protocol design that were commonly seen in small-scale Ethernet networks in the late 1980s. As the network environment became far more diverse, it became a significant challenge to explicitly recreate these required conditions for LAT functionality in a wide-area multiprotocol environment. A LAT implementation in the wide area, in this case, might suffer from poor service quality, and at times, the parameters of the wide-area network are so far outside the original protocol design that the protocol simply does not function at all. Nonetheless, it is surprising how many networks still try to implement LAT services in a wide-area network on low-speed, high-latency links where the 80-millisecond network transit time simply is unachievable.

How Will Different Protocols Interact?

Another significant aspect to consider when mixing protocols in a network is how they will interact. In a multiprotocol environment, it is not unusual to route protocols that can be routed and to bridge protocols that cannot be routed. This concept is reflected in the seminal engineering credo, "Route when you can, bridge if you must." However, although these protocols do not explicitly communicate with one another, one misbehaving protocol may very well have an adverse impact on another protocol. In a multiprotocol network with limited amounts of bandwidth and several hundred IPX SAP (Service Advertisement Protocol) broadcast advertisements, for example, it is not unusual for SAP broadcast flooding to interfere with more sensitive protocol applications that reside in the network, such as established SNA sessions. Unfortunately, this adverse interaction of protocols is a very common occurrence. Basically, three types of protocol families can badly distort a network:

A gratuitous, broadcast-rich protocol that cannot be routed. Such protocols appear to have proliferated within the "plug-and-play" model of networking, which attempts to reduce the complexity of host configuration by making self-configuration and resource discovery a byproduct of comprehensive broadcasting. Although this is intrinsically not a scaleable approach, it has not stopped folks from continuing to attempt to deploy these types of technologies in large-scale, corporate environments.

A lightweight real-time protocol. Such protocols typically have no concept of adjustment to network congestion and simply can saturate the network with externally clocked data sources. These applications can work well over small-scale, high-speed local networks in which the underlying bandwidth resources are not under heavy contention. But again, such protocols do not readily scale in larger and more diverse corporate network environments. We can expect to see more of these protocols with the onset of widespread multimedia applications.

Real-time broadcast protocols. These appear to combine the performance attributes of gratuitous advertisements (an inherent characteristic with broadcast-intensive protocols) with external data-clocking attributes (an inherent characteristic with real-time applications). It may come as no surprise to suggest that plug-and-play multimedia "solutions" may well be the most common applications that would rely upon such fundamentally broken protocol-level functionality.

It becomes apparent how QoS might be desired in cases such as this, using QoS structures simply to provide priority to one protocol over another. In the example here, it may be highly desirable to prioritize SNA traffic over all other protocols in the network, especially if the company's core business is impacted by the inability to conduct critical SNA-based business transactions.

Network Abusiveness: Bridging versus Routing

It is important to recognize the intrinsic difference between how traffic is forwarded when it is bridged, as opposed to how it is forwarded when it is routed. When bridging, decisions on how traffic is forwarded are determined by information contained in the link-layer frame. When routing, the intermediate routers can look into information contained in higher layers of the protocol stack information (namely, the network layer) to make more intelligent decisions on how to forward traffic. Routing allows you to compartmentalize information about the network topology so that excessive or irrelevant information is not flooded onto every sublet in the network. When bridging, MAC (Media Access Control) information about every host in the bridging domain also must be readily available to every end-system; this consumes an inordinate amount of network bandwidth resources.

The major issue with wide-scale bridging is that it takes a non-routed LAN into areas in which the protocol designers never contemplated. Operational parameters relevant to protocol design will change in a wide-area bridged network, especially the number of attached devices and the end-to-end propagation time. More fundamentally, the packet-delivery concept is changed so that the reliability of packet transmission is reduced substantially.

In a local Ethernet broadcast configuration, for example, the sender's protocol stack can make the assumption that if the transmission attempt does not generate a collision indication, there is a high probability that the receiver has received the packet into its input buffer. The probability of the receiver then silently discarding the packet is relatively low, so that in this environment, if the sender's protocol stack receives a signal from the Ethernet driver of successful transmission, the sender can assume a high probability that the packet has been received. In this environment, undetected packet loss within the network is a rare occurrence, and the protocol designer can afford to make loss recovery a more lengthy process as an extraordinary event.

In the wide-area network, this is clearly not the case, and the number of events that can cause packet drop (without direct real-time notification to the sender) increases dramatically. Protocols that assume that the Local Area Network is indeed local make the fatal assumption that broadcast is easy, packet delivery reliability is high, and end-to-end propagation time is short. Wide-area bridging abuses all three assumptions and seriously degenerates the robustness of the original protocol design. QoS approaches can attempt to ameliorate this by performing service differentiation within the combined bridge/router units with a bias toward bridging, although this may not conform to other resource-allocation objectives for the network.

QoS and Multiprotocol Traffic

Providing QoS to multiprotocol traffic is significantly different from most of the methods already discussed for providing QoS to TCP/IP traffic. Although TCP provides some fundamental congestion-control mechanisms and provides service-quality fields within the packet headers that can interact with the network, most of the other protocols do not possess similar robust properties. No Precedence field exists in an IPX packet, for example, and more advanced mechanisms such as RSVP are geared toward IP traffic only. This leaves queuing mechanisms as the only feasible options for implementing QoS in a multiprotocol network, and this certainly has its limitations.

Effect of Different Queuing Strategies

As mentioned earlier, several queuing strategies provide preferential treatment to certain types of traffic. Priority queuing can give absolute queuing preference to specific traffic types to the detriment of lesser prioritized traffic. Class Based Queuing (CBQ) allows the network administrator to define multiple queues, each of a specified depth, and classify traffic so that specific traffic types (classes) have specific dedicated queues. Traffic relegated to larger queues gets larger portions of network resources than does traffic relegated to smaller queues, providing behavior consistent with the desire to prioritize some traffic classes over others. WFQ (Weighted Fair Queuing) similarly uses multiple queues for similarly classified types of traffic. However, it disallows traffic relegated to any one queue from dominating network resources, introducing a fairness concept to the resource-allocation scheme.

Each of these queuing schemes will work with multiprotocol traffic, because the queue does not make a distinction regarding what the packet it is given contains. It simply accepts the packet and queues it for transmission. The task of distinguishing the traffic type (what protocol the packet contains) is left to the classification process, which can be used with any of these queuing schemes. Recall that default queuing behavior is single-queue FIFO (First-In, First-Out). Although a simple priority queuing may also consist of a singular queue, there must be a traffic classifier that identifies the incoming traffic and performs packet reordering, moving higher-preference packets to the front of the queue. Similarly, a traffic classifier is used to place packets into separate queues in which CBQ and WFQ is used.

Of course, all the caveats mentioned earlier still apply here, so we will not repeat them. However, it is important to understand that these queuing schemes have limited applicability. At slower link speeds, of course, traffic coming into the router is at a slower rate, so the CPU has more time to perform packet classification, reordering, and queue management. However, at higher speeds, the CPU may become burdened to the point where these tasks become a liability and performance degradation results.

Using Layer 2 Switching for Traffic Engineering

At least one additional option is available for providing differentiated services in a multiprotocol network, but it requires the use of a switched wide-area network (such as Frame Relay or ATM). This particular method consists of directing traffic belonging to one protocol family across a distinctly different traffic path, or virtual circuit, than traffic belonging to other protocol families. If an organization specifically wants to ensure that its IPX traffic does not interfere with its TCP/IP traffic between Chicago and Los Angeles, for example, it simply could provision two PVCs between the routers at each location and route TCP/IP traffic across one PVC and IPX across the other. This task is relatively simple enough to accomplish. In fact, it is only a matter of enabling IPX routing on one PVC, and alternatively, only enabling IP routing on the other.

Also, the PVCs could each have different CIRs (Committed Information Rates), so that drop preferences in the frame relay network can be characteristically different for each type of traffic. Granted, some downsides to this approach are obvious—namely, that there is no redundant back-up path for either protocol in the event of PVC failure, and there are certainly economic considerations in provisioning multiple PVCs for individual protocol applications. Having said that, this method has proven to work quite well.

The capability that is readily feasible in either approach is to create a differentiated service environment based on protocol bias using precedence-based mechanisms to allocate network resources to various supported protocols within the network. Of course, the ultimate objective is to provision the routers with a set of generic QoS actions and remotely trigger those actions by setting protocol-specific header field values on an application-specific basis. This is a compelling requirement for providing QoS fields within the packet headers of various protocols—something that has not been a widespread feature of such protocols to date. In the absence of such facilities, the pragmatic conclusion is that service differentiation is feasible within multiprotocol networks on a protocol-by-protocol basis, but finer levels of granularity are limited to protocols that have readily recognizable "hooks" that can be used by the routers to treat specific applications, protocols, and user traffic with preference.

SNA CoS: A Contradiction in Terms?

SNA—or, more appropriately, IBM networking—does indeed provide Classes of Services (CoS) distinctions. However, it is imperative to understand how these classes work, when they are relevant, and their potential applications in a corporate network. First, you should briefly review the SNA architectural technology so that you have a basic understanding of what SNA provides as far as QoS is concerned. This is in no way intended to be an all-inclusive overview of SNA. On the contrary—entire volumes of textbooks have been written on the topic, and we cannot hope to provide a synopsis of the entire SNA architecture here. Here, we are simply illustrating the fact that IBM Class of Service distinctions may not provide the desired QoS functionality in your network.

The Legacy SNA Network

Traditional IBM networking technologies actually consist of two similar yet separate architectures, each which has its roots in a common origin. The first architecture commonly is referred to as *legacy SNA*, and the other, more recent, architecture is called *APPN* (Advanced Peer-to-Peer Networking). In a legacy SNA network, the mainframe is the central control entity in the network. The mainframe runs ACF/VTAM (Advanced Communications Facility/Virtual Telecommunications Access Method), which controls the activation and deactivation of network resources, as well as establishes all sessions to network clients. In this model, there is no client-to-client or peer-to-peer communications. All communication is solely between the mainframe and the clients, via devices called *front-end processors* (FEPs) and cluster controllers.

Four basic entities exist in the legacy SNA environment:

Hosts (the mainframe)

Communications controllers (FEPs)

Establishment controllers (cluster controllers)

Terminals (clients)

The mainframe (or host) performs all the computational tasks, and the FEP performs all the traffic routing for data in the network. Cluster controllers control the I/O operations of network-attached devices, and the terminals provide the user interface to the network.

SNA does provide the capability for Classes of Services (CoS) parameters to be defined for each session, either upon session establishment or manually by the user when logging into the network. The CoS parameters include virtual path information, called VRNs (virtual route numbers) and transmission priorities, called TPRIs (transmission priorities). Characteristics specified by a VRN profile include considerations such as response time, security level, and availability of the path. The TPRI granularity allows three priorities for each virtual route: 0, 1, and 2, where 0 is the lowest transmission priority and 2 is the highest.

All data paths in the SNA network are statically configured. Each of these CoS templates also must be manually configured and uniformly synchronized between the FEP and the mainframe before they can be used effectively. A CoS template can be configured statically for each client node, or it can be assigned based on client specification during network login. There are no real dynamics in traditional legacy SNA networks to speak of —everything is statically preconfigured.

APPN

APPN represents IBM's second-generation SNA architecture. Significant modifications were introduced with APPN, but it is not within the scope of this overview to outline all of them. It is relevant, however, to realize that APPN does introduce some radical modifications to the older legacy SNA architecture, primarily dynamic routing for data and peer-to-peer communications, neither of which was available in the older architecture. Each of these has a bearing on how APPN changed the basic SNA CoS service schemes.

One of the more important modifications is the fact that APPN provides dynamic routing similar to other link-state routing protocols. Network nodes within the APPN network domain maintain a topology database that contains information used for calculating paths with a particular CoS profile. However, like its predecessor, the CoS values are explicitly defined—only the granularity has been increased.

A few salient points need to be reiterated here. One pertinent point is that SNA communications occur at the data-link layer. Therefore, SNA traffic must be bridged in a network; it cannot be routed. APPN does provide routing capabilities, but these haven't seemed to have integrated well into the traditional multiprotocol network, for reasons that remain to be seen. Perhaps the complexity and migration path for APPN is considered too risky, but this is purely speculation.

RSRB and DLSw Alternatives

For whatever reason, SNA still enjoys an amazingly large installed base, especially in the campus network. In the wide-area network, SNA has proven to be problematic, especially in a multiprotocol network. This is because SNA traffic cannot be natively Source-Route Bridged across the wide-area network; it must be encapsulated in a network or transport-layer protocol and transported in this fashion to the appropriate edge router peers. Traditionally, this task has been accomplished by using Remote Source-Route Bridging (RSRB), which uses TCP as a transport encapsulation for the SNA traffic. A TCP session is established between two edge routers, SRB traffic is encapsulated on one end of the wide-area network, the traffic is unencapsulated at the remote end, and it then is placed into the remote system's SRB domain.

A more recent alternative to RSRB is Data Link Switching (DLSw), which is similar to RSRB. The primary difference between RSRB and DLSw is that DLSw provides local termination of link-layer acknowledgments and keep-alive messages. These Data Link Control (DLC) messages are handled on an end-to-end or peer-to-peer basis with RSRB, whereas they are terminated locally with DLSw. This provides quicker acknowledgment of these local messages and considerably lessens the possibility of an error condition due to an acknowledgment time-out.

The inherent problem in this scheme is similar to what was discussed in the chapter on ATM regarding the "distortion effect." Assuming that the underlying CoS distinctions within the SNA network are configured properly, fully functional, and appropriately engineered, there is still a possibility that SNA CoS distinctions for a session that traverses the wide-area network encapsulated in TCP may experience session degradation, or worse, session disconnect, when there is instability of congestion in the TCP/IP network.

SNA CoS distinctions alone are insufficient in a multiprotocol network. The addition of several mechanisms is necessary to ensure that TCP-encapsulated SNA sessions are given preferential treatment at the TCP layer. This includes things already discussed, such as congestion control, congestion avoidance, preferential packet-drop schemes, and perhaps a queue-management mechanism to preferentially schedule these packets for transmission.

Interestingly enough, many organizations have expressed a desire to migrate their legacy SNA networks to native TCP/IP and, in fact, many have done so already or are in the process of migrating to native TCP/IP. This does not necessarily mean that they are changing out hardware platforms in favor of more contemporary client-server platforms, such as UNIX or Microsoft Windows NT. It simply means that these organizations are beginning to integrate the TCP/IP stack into the traditional IBM networking hardware platforms. This is a major paradigm shift and reinforces the suspicion that many multiprotocol networks will take similar actions over the course of time, abandoning the complexity and aggravation of managing multiprotocol networks in favor of a single network protocol: TCP/IP.

❖ ❖ ❖

Economists often describe economic history in terms of cycles of boom and bust. Although we have not yet seen networking decline, let alone bust, we are witnessing some other cyclical trends of networking. The initial deployment of computer networks generally was dominated by single-vendor Information Technology (IT) environments and the associated exclusive use of the vendor's proprietary networking technologies. Multivendor IT environments, like multiprotocol computer networks, were a gradual refinement to this original model and were heralded by broadcast-based local area networks and wide-area bridges. Multiprotocol routers soon appeared, and many networks appeared to operate with a chaotic mix of TCP, SNA, IPX, DECnet, AppleTalk, and X.25 transport protocols, with an equally chaotic mix of PAD (Protocol Access Device), LAT (Local-Area Transport), and Telnet support for remote access.

We are now witnessing the decline of the aggressively heterogeneous networking environment. Such a plethora of protocols is highly expensive in terms of operational management, and the stability of performance is often a challenge. Attempting to efficiently refine this performance stability paradigm into differentiated quality levels in larger wide-area network becomes a tedious and sometime frustrating exercise when so many variables exist. The response to these factors has been a recognition of the value of using a smaller set of protocols as the foundation of the network and layering the application set based on this protocol choice. This is a cyclical return to a network that parallels the single vendor network of earlier days—although it is interesting to highlight that in the previous incarnation, the choice of the mainframe dictated the choice of the network, whereas today, the network protocol dictates the approach taken by the computer hardware and software vendors.

From a network-engineering perspective, this return to a single-network protocol suite is viewed with some relief, given that it is far easier to create a cost-effective network with a single protocol at its base than it is to attempt to balance the often conflicting requirements of multiple protocols. In this environment, the best answer to the requirement for QoS is due attention to the fundamentals of *service quality*, and like so many other aspects of technology, simplicity is the key to achieving it.

Glossary

AAL *ATM Adaptation Layer*. A connection of protocols that takes data traffic and frames it into a sequence of 48-byte payloads for transmission over an ATM (Asynchronous Transfer Mode) network. Currently, four AAL types are defined that support various service categories. *AAL1* supports constant bit-rate connection-oriented traffic. *AAL2* supports time-dependant variable bit-rate traffic. *AAL3/4* supports connectionless and connection-oriented variable bit-rate traffic. *AAL5* supports connection-oriented variable bit-rate traffic.

ABR *Available Bit Rate*. One of the service categories defined by the ATM Forum. ABR supports variable bit-rate traffic with flow control. The ABR service category supports a minimum guaranteed transmission rate and peak data rates.

ACF/VTAM *Advanced Communications Facility/Virtual Telecommunications Access Method*. In traditional legacy SNA networks, ACF/VTAM on the IBM mainframe is responsible for session establishment and activation of network resources.

ACK *Acknowledgment*. A message that indicates the reception of a transmitted packet.

ANSI *American National Standards Institute*. One of the American technology standards organizations.

API *Application Programming Interface*. A defined interface between an application and a software service module or operating system component. Conventionally, an API is defined as a subroutine library with a common definition set that extends across multiple computer platforms and operating systems.

APPN *Advanced Peer-to-Peer Networking*. APPN represents IBM's second-generation SNA networking architecture, which accommodates peer-to-peer communications, directory services, and dynamic data routing between SNA subdomains.

ARP *Address Resolution Protocol*. The discovery protocol used by host computer systems to establish the correct mapping of Internet layer addresses, also known as IP addresses, to Media Access Control (MAC) layer addresses.

ARPA *Advanced Research and Projects Agency.* A U.S. federal research funding agency credited with initially deploying the network now known as the Internet. The agency was referred to as DARPA (Defense Advanced Research and Projects Agency) in the past, indicating its administrative position as an agency of the U.S. Department of Defense.

AS *Autonomous System.* The term used to describe a collection of networks administered by a common network-management organization. The most common use of this term is in interdomain routing, where an Autonomous System is used to describe a self-connected set of networks that share a common external policy with respect to connectivity, or in other words, networks that generally are operated within the same administrative domain.

ASIC *Application Specific Integrated Circuit.* An integrated circuit that is an implementation of a specific software application or algorithm within a silicon engine.

ATM *Asynchronous Transfer Mode.* A data-framing and transmission architecture that features fixed-length data cells of 53 bytes, consisting of a fixed format of a 5-byte cell header and a 48-byte cell payload. The small cell size is intended to support high-speed switching of multiple traffic types. The architecture is asynchronous, so there is no requirement for clock control of the switching and transmission.

B-Channel *Bearer Channel.* Traditionally refers to a single, full-duplex physical ISDN (Integrated Services Digital Network) interface that operates at 64 Kbps.

BECN *Backward Explicit Congestion Notification.* A notification signal passed to the originator of traffic indicating that the path to the destination exceeds a threshold load level. This signal is defined explicitly in the Frame Relay frame header.

BGP *Border Gateway Protocol.* An Internet routing protocol used to pass routing information between different administrative routing domains or ASs (Autonomous Systems). The BGP routing protocol does not pass explicit topology information. Instead, it passes a summary of reachability between ASs. BGP is most commonly deployed as an inter-AS routing protocol.

border router Generally describes routers on the edge of an AS (Autonomous System). Uses BGP to exchange routing information with another administrative routing domain. However, this term also can describe any router that sits on the edge of a routing subarea, such as an OSPF (Open Shortest Path First) area border router. See also *edge device*.

BRI *Basic Rate Interface.* A user interface to an ISDN (Integrated Services Digital Network) that consists of two 64-Kbps data channels (B-Channels) and one 16-Kbps signaling channel (D-channel) sharing a common physical access circuit.

bridging The process of forwarding traffic based on address information contained at the data-link framing layer. Bridging allows a device to flood, forward, or filter frames based on the MAC (Media Access Control) address. Contrast with *routing*.

CBQ *Class Based Queuing.* A queuing methodology by which traffic is classified into separate classes and queued according to its assigned class in an effort to provide differential forwarding behavior for certain types of network traffic.

CBR *Constant Bit Rate.* An ATM service category that corresponds to a constant bandwidth allocation for a traffic flow. The CLP (Cell Loss Priority) bit is set to 0 in all cells to ensure that they are not discard eligible in the event of switch congestion. The service supports circuit emulation as well as continuous bitstream traffic sources (such as uncompressed voice or video signals).

CDV *Cell Delay Variation.* An ATM QoS (Quality of Service) parameter that measures the variation in transit time of a cell over a Virtual Connection (VC). For service classes that are jitter sensitive, this is a critical service parameter.

CHAP *Challenge Handshake Authentication Protocol.* An authentication mechanism for PPP (Point-to-Point Protocol) connections that encrypts the user password.

CIDR *Classless Inter Domain Routing.* An Internet routing paradigm that passes both the network prefix and a mask of significant bits in the prefix within the routing exchange. This supercedes the earlier paradigm of *classful* routing, where the mask of significant bits is inferred by the value of the prefix (where Class A network prefixes infer a mask of 8 bits, Class B network prefixes infer a mask of 16 bits, and Class C network prefixes infer a mask of 24 bits). CIDR commonly is used to denote an Internet environment in which no implicit assumption exists of the Class A, B, and C network addresses. BGP (Border Gateway Protocol) version 4 is used as the de facto method of providing CIDR support in the Internet today.

CIR *Committed Information Rate.* A Frame Relay term describing a minimum access rate at which the service provider commits to provide the customer for any given Permanent Virtual Circuit (PVC).

CLP *Cell Loss Priority.* A single-bit field in the ATM cell header to indicate the discard priority. A CLP value of 1 indicates that an ATM switch can discard this cell in a congestion condition.

CLR *Cell Loss Ratio.* An ATM QoS metric defined as the ratio of lost cells to the number of transmitted cells.

CoS *Class of Service* or *Classes of Services.* A categorical method of classifying traffic into separate classes to provide differentiated service to each class within the network.

CPE *Customer Premise Equipment.* The equipment deployed on the customer's site when the customer subscribes (or simply connects) to a carrier's service.

CPU *Central Processing Unit.* The arithmetic, logic, and control unit of a computer that executes instructions.

CSU/DSU *Channel Service Unit/Data Service Unit.* A Customer Premise Equipment (CPE) device that provides the telephony interface for circuit data services, including the physical framing, clocking, and channelization of the circuit.

CTD *Cell Transfer Delay.* An ATM QoS metric that measures the transit time for a cell to traverse a Virtual Connection (VC). The time is measured from source UNI (User-to-User Interface) to destination UNI.

D-Channel *Data Channel.* A full-duplex control and signaling channel on an ISDN BRI (Basic Rate Interface) or PRI (Primary Rate Interface). The D-Channel is 16 Kbps on an ISDN BRI and 64 Kbps on a PRI.

DCE *Data Communications Equipment.* A device on the network side of a User-to-Network Interface (UNI). Typically, this is the Customer Premise Equipment (CPE), such as a modem or Channel Service Unit/Data Service Unit (CSU/DSU).

DE *Discard Eligible.* A bit field defined within the Frame Relay header indicating that a frame can be discarded within the Frame Relay switch when the local queuing load exceeds a configured threshold.

DHCP *Dynamic Host Configuration Protocol.* A protocol that is beginning to be used quite pervasively on end-system computers to automatically obtain an IP (Internet Protocol) host address, subnet mask, and local gateway information. A DHCP server dynamically supplies this information in response to end-system broadcast requests.

Dijkstra algorithm Also commonly referred as SPF (Shortest Path First). The Dijkstra algorithm is a single-source, shortest-path algorithm that computes all shortest paths from a single point of reference based on a collection of link metrics. This algorithm is used to compute path preferences in both OSPF (Open Shortest Path First) and IS-IS (Intermediate System to Intermediate System). See also *SPF.*

DLC *Data Link Control.* Refers to IBM data-link layer support, which supports various types of media, including mainframe channels, SDLC (Synchronous Data Link Control), X.25, and token ring.

DLCI *Data Link Connection Identifier.* A numerical identifier given to the local end of a Frame Relay Virtual Circuit (VC). The local nature of the DLCI is that it spans only the distance between the first-hop Frame Relay switch and the router, whereas a VC spans the entire distance of an end-to-end connection between two routers that use the Frame Relay network for link-layer connectivity.

DLSw *Data Link Switching.* Provides a standards-based method for forwarding SNA (Systems Network Architecture) traffic over TCP/IP (Transmission Control Protocol/Internet Protocol) networks using encapsulation. DLSw provides enhancements to traditional RSRB (Remote Source-Route Bridging) encapsulation by eliminating hop-count limitations, removes unnecessary broadcasts and acknowledgments, and provides flow-control.

DS0 *Digital Signal Level 0.* A circuit-framing specification for transmitting digital signals over a single channel at 64 Kbps on a T1 facility.

DS1 *Digital Signal Level 1.* A circuit-framing specification for transmitting digital signals at 1.544 Mbps on a T1 facility in the United States, or at 2.108 Mbps on an E1 facility elsewhere.

DS3 *Digital Signal Level 3.* A circuit-framing specification for transmitting digital signals at 44.736 Mbps on a T3 facility.

DSBM *Designated Subnet Bandwidth Manager.* A device on a managed subnetwork that acts as the Subnet Bandwidth Manager (SBM) for subnetwork to which it is attached. This is done through a complicated election process specified in the SBM protocol specification. The SBM protocol is a proposal in the IETF (Internet Engineering Task Force) for handling resource reservations on shared and switched IEEE (The Institute of Electrical and Electronics Engineers) 802-style local area media. See also *SBM.*

DTE *Data Terminal Equipment.* A device on the user side of a User-to-Network Interface (UNI). Typically, this is a computer or a router.

E1 A WAN (Wide-Area Network) transmission circuit that carries data at a rate of 2.048 Mbps. Predominantly used outside the United States.

E3 A WAN transmission circuit that carries data at a rate of 34.368 Mbps. Predominantly used outside the United States.

edge device Any device on the edge or periphery of an administrative boundary. Traditionally used to describe an ATM-attached host or router that interfaces with an ATM network switch. See also *border router.*

end-system Any device that terminates an end-to-end communications relationship. Traditionally used to describe a host computer. However, may also include intermediate network nodes in situations where a particular end-to-end communications substrate relationship terminates on an intermediate device (e.g., a router and an ATM VC).

EPD *Early Packet Discard.* A congestion-avoidance mechanism generally found in ATM networks. EPD uses a method to preemptively drop entire AAL5 (ATM Adaptation Layer 5) frames instead of individual cells in an effort to anticipate congestion situations and make the most economic use of explicit signaling within the ATM network.

Ethernet A LAN (Local Area Network) specification invented by the Xerox Corporation and then jointly developed by Xerox, Intel, and Digital Equipment Corporation. Ethernet uses CSMA/CD (Carrier Sense Multiple Access/Collision Detection) and operates on various media types. It is similar to the IEEE 802.3 series of protocols.

FAQ *Frequently Asked Questions.* Compiled lists of the most frequent questions and their answers on a particular topic. An FAQ generally can be found in various formats, such as HTML (Hyper Text Markup Lanuage) Web pages, as well as traditional printed material.

FDDI *Fiber Distributed Data Interface.* A LAN standard defined in ANSI (American National Standards Institute) Standard X3T9.5 that operates at 100 Mbps, uses a token-passing technology, and uses fiber-optic cabling for physical connectivity. FDDI has a base transmission distance of up to 2 kilometers and uses a dual-ring architecture for redundancy.

FECN *Forward Explicit Congestion Notification.* A notification signal passed to the receiver of traffic indicating that the path to the originator exceeds a threshold load level. This signal is defined explicitly within the Frame Relay frame header.

FEP *Front-End Processor.* Typically, FEP describes the function of an IBM 3745, which provides a networking interface to the SNA (Systems Network Architecture) network for downstream nodes that have no knowledge of network data forwarding paths. The IBM 3745 FEP functions as an intermediary networking arbiter.

FIFO *First In, First Out.* FIFO queuing is a strict method of transmitting packets that are presented to a device for subsequent transmission. Packets are transmitted in the order in which they are received.

FIN *FINish flag*. Used in the TCP header to signal the end of TCP data.

FRAD *Frame Relay Access Device* or *Frame Relay Assembler/Disassembler*. A device that operates natively at the Frame Relay data-link layer and is less robust than multiprotocol routers (and in fact, usually does not provide network-layer routing). A FRAD simply frames and transmits traffic over a Frame Relay network, and a FRAD on the opposite side of a Frame Relay network unframes the traffic and places it on the local media.

FTP *File Transfer Protocol*. A bulk, TCP-based, transaction-oriented file transfer protocol used in TCP/IP networks, especially the Internet.

Gbps *Gigabits per second*. The data world avoided using the term *billion*, which invariably is interpreted as one thousand million or one million million, in favor of the term *giga* as one thousand million. Of course, some confusion between the telecommunications and data-storage worlds still exist as to whether a *giga* is really the value 10^9 or 2^{30}.

GCRA *Generic Cell Rate Algorithm*. A specification for implementing cell-rate conformance for ATM VBR (Variable Bit Rate) Virtual Connections (VC). The GCRA is an algorithm that uses traffic parameters to characterize traffic that is conformant to administratively defined admission criteria. The GCRA implementation commonly is referred to as a *leaky bucket*.

HDLC *High-Level Data Link Control*. A bit-oriented, synchronous data-link layer transport protocol developed by the ISO (International Standards Organization). HDLC provides an encapsulation mechanism for transporting data on synchronous serial links using framing characters and checksums. HDLC was derived from SDLC (Synchronous Data Link Control).

HSSI *High-Speed Serial Interface*. The networking standard for high-speed serial connections for wide-area networks (WANs), accommodating link speeds up to 52 Mbps.

HTML *Hyper Text Markup Language*. A simple hypertext document-formatting language used to format content that is presented in Web pages on the World Wide Web (WWW), which is read using one of the many popular Web browsers.

HTTP *Hyper Text Transfer Protocol*. A TCP-based application-layer protocol used for communicating between Web servers and Web clients, also known as Web browsers.

IAB *Internet Architecture Board*. A collection of individuals concerned with the ongoing architecture of the Internet. IAB members are appointed by the trustees of the Internet Society (ISOC). The IAB also appoints members to several other organizations, such as the IESG (Internet Engineering Steering Group).

iBGP *Internal BGP* or *Interior BGP*. A method to carry exterior routing information within the backbone of a single administrative routing domain, obviating the need to redistribute exterior routing into interior routing. iBGP is a unique implementation of BGP (Border Gateway Protocol) rather than a separate protocol unto itself.

ICMP *Internet Control Message Protocol*. A network-layer protocol that provides feedback on errors and other information specifically pertinent to IP packet handling.

I-D *Internet Draft*. A draft proposal in the IETF submitted as a collaborative effort by members of a particular working group or by individual contributors. I-Ds may or may not be subsequently published as IETF Requests for Comments (RFCs).

IEEE *The Institute of Electrical and Electronics Engineers*. A professional organization that develops communications and network standards—traditionally, link-layer LAN signaling standards.

IESG *Internet Engineering Steering Group*. IESG members are appointed by the Internet Architecture Board (IAB) and manage the operation of the IETF.

IETF *Internet Engineering Task Force*. An engineering and protocol standards body that develops and specifies protocols and Internet standards, generally in the network layer and above. These include routing, transport, application, and occasionally, session-layer protocols. The IETF works under the auspices of the Internet Society (ISOC).

Integrated Services In a broad sense, this term encompasses the transport of audio, video, real-time, and classical data traffic within a single network infrastructure. In a more narrow focus, it also refers to the Integrated Services architecture (the focus of the Integrated Services working group in the IETF), which consists of five key components: QoS requirements, resource-sharing requirements, allowances for packet dropping, provisions for usage feedback, and a resource-reservation model (RSVP).

Internet The global Internet. Commonly used as a reference for the loosely administered collection of interconnected networks around the globe that share a common addressing structure for the interchange of traffic.

intranet Generally used as a reference for the interior of a private network, either not connected to the global Internet or partitioned so that access to some network resources is limited to users within the administrative boundaries of the domain.

I/O *Input/Output*. The process of receiving and transmitting data, as opposed to the actual processing of the data.

IP *Internet Protocol*. The network-layer protocol in the TCP/IP stack used in the Internet. IP is a connectionless protocol that provides extensibility for host and subnetwork addressing, routing, security, fragmentation and reassembly, and as far as QoS is concerned, a method to differentiate packets with information carried in the IP packet header.

IP precedence A bit value that can be indicated in the IP packet header and used to designate the relative priority with which a particular packet should be handled.

IPng *IP Next Generation*. A vernacular reference to the follow-on technology for IP version 4, otherwise known as IP version 6 (IPv6).

IPv4 *Internet Protocol version 4*. The version of the Internet protocol that is widely used today. This version number is encoded in the first 4 bits of the IP packet header and is used to verify that the sender, receiver, and routers all agree on the precise format of the packet and the semantics of the formatted fields.

IPv6 *Internet Protocol version 6*. The version number of the IETF standardized next-generation Internet protocol (IPng) proposed as a successor to IPv4.

IPX *Internet Packet eXchange*. The predominant protocol used in NetWare networks. IPX was derived from XNS (Xerox Networking Services), a similar protocol developed by the Xerox Corporation.

IRTF *Internet Research Task Force*. Composed of a number of focused and long-term research groups, working on topics related to Internet protocols, applications, architecture, and technology. The chair of the IRTF is appointed by the Internet Architecture Board (IAB). The IRTF is described more fully in RFC2014.

ISDN *Integrated Services Digital Network*. An early adopted protocol model currently offered by many telephone companies for digital end-to-end connectivity for voice, video, and data.

IS-IS *Intermediate System to Intermediate System*. A link-state routing protocol for connectionless OSI (Open Systems Interconnection) networks, similar to OSPF (Open Shortest Path First). The protocol specification for IS-IS is documented in ISO 10589.

ISO *International Standards Organization*. The complete name for this body is the *International Organization for Standardization and International Electrotechnical Committee*. The members of this body are the national standards bodies, such as ANSI (American National Standards Institute) in the United States and BSI (the British Standards Institution) in the United Kingdom. The documents produced by this body are termed *International Standards*.

ISOC *Internet Society*. An international user society of Internet users and professionals that share a common interest in the development of the Internet.

ISP *Internet Service Provider*. A service provider that provides external transit for a client network or individual user, providing connectivity and associated services to access the Internet.

ISSLL *Integrated Services over Specific Link Layers*. An IETF working group that defines specifications and techniques needed to implement Internet Integrated Services capabilities within specific subnetwork technologies, such as ATM or IEEE 802.3z Gigabit Ethernet.

jitter The distortion of a signal as it is propagated through the network, where the signal varies from its original reference timing. In packet-switched networks, jitter is a distortion of the interpacket arrival times compared to the interpacket times of the original signal transmission. Also known as *delay variance*.

Kbps *Kilobits per second*. A measure of data-transfer speed. Some confusion exists as to whether this refers to a rate of 10^3 bits per second or 2^{10} bits per second. The telecommunications industry typically uses this term to refer to a rate of 10^3 bits per second.

L2TP *Layer 2 Tunneling Protocol*. A proposed mechanism whereby discrete virtual tunnels can be created for each dial-up client in the network, each of which may terminate at different points upstream from the access server. This allows individual dial-up clients to do interesting things, such as to use discrete addressing schemes and have their traffic forwarded, via the tunneling mechanisms, along completely different traffic paths. At the time of this writing, the L2TP protocol specification is still being developed with the IETF.

LAN *Local Area Network.* A local communications environment, typically constructed using privately operated wiring and communications facilities. The strict interpretation of this term is a broadcast media in which any connected host system can contact any other connected system without the need for explicit assistance of a routing protocol.

LANE *LAN Emulation.* A technique and ATM forum specification that defines how to provide LAN-based communications across an ATM subnetwork. LANE specifies the communications facilities that allow ATM to be interoperable with traditional LAN-based protocols, so that among other things, address resolution and broadcast services will function properly.

LAT *Local-Area Transport.* An older virtual terminal network protocol developed by Digital Equipment Corporation. LAT has been notorious for its inability to be routed in a network, as well as its insensitivity to induced latency.

Layer 1 Commonly used to describe the physical layer in the OSI (Open Systems Interconnection) reference model. Examples include the copper wiring or fiber-optic cabling that interconnects electronic devices.

Layer 2 Commonly used to describe the data-link layer in the OSI reference model. Examples include Ethernet and ATM (Asynchronous Transfer Mode).

Layer 3 Commonly used to describe the network layer in the OSI reference model. Examples include IP (Internet Protocol) and IPX (Internet Packet eXchange).

leaky bucket Generally, a traffic-shaping mechanism in which the input side of the shaping mechanism is an arbitrary size, and the output side of the mechanism is of a smaller, fixed size. This implementation has a smoothing effect on bursty traffic, because traffic is "leaked" into the network at a fixed rate. Contrast with *token bucket.*

LEC *Local Exchange Carrier.* Usually considered the local telephone company or any local telephony entity that provides telecommunications facilities within a local tariffing area. See also *RBOC.*

LIJ *Leaf Initiated Join.* A feature introduced in the ATM Forum Traffic Management 4.0 Specification in which any remote node in an ATM network can connect arbitrarily to a point-to-multipoint Virtual Connection (VC) without explicitly signaling the VC originator.

LIS *Logical IP Subnetwork.* An IP subnetwork in which all devices have a direct communication path to other devices sharing the same LIS, such as on a shared LAN or point-to-point circuit. In an NBMA (Non Broadcast Multi Access) ATM network where all devices are attached to the network via VCs, the LIS is a method by which attached devices can communicate at the IP layer so that the IP protocol believes all devices are connected directly to a local network media, although they are not.

LLC *Link Layer Control.* The higher of the two sublayers of the data-link layer defined by the IEEE. The LLC sublayer handles flow control, error correction, framing, and MAC-sublayer addressing. See also *MAC.*

LSA *Link State Advertisement.* A packet-forwarding link-state routing process to neighboring nodes that includes information concerning the local node, the link state of attached interfaces, or the topology of the network. LSAs are generated by link-state routing protocols such as OSPF (Open Shortest Path First) and IS-IS (Intermediate System to Intermediate System).

MAC *Media Access Control*. The lower of the two sublayers of the data-link layer defined by the IEEE. The MAC sublayer handles access to shared media—for example, Ethernet and token ring, and whether methods such as media contention or token passing are used. See also *LLC*.

MARS *Multicast Address Resolution Server*. A mechanism for supporting multicast in ATM (Asynchronous Transfer Mode) networks. The MARS serves a collection of nodes by proving a point-to-multipoint overlay for multicast traffic.

maxCTD *Maximum Cell Transfer Delay*. An ATM QoS metric that measures the transit time for a cell to traverse a VC (Virtual Connection). The time is measured from the source UNI (User-to-Network Interface) to the destination UNI.

Mbps *Megabits per second*. A unit of data transfer. The communications industry typically refers to a *mega* as the value 10^6, whereas the data-storage industry uses the same term to refer to the value 2^{20}.

MBS *Maximum Burst Size*. An ATM QoS metric describing the number of cells that may be transmitted at the peak rate while remaining within the Generic Cell Rate Algorithm (GCRA) threshold of the service contract.

MCR *Minimum Cell Rate*. An ATM service parameter related to the ATM Available Bit Rate (ABR) service. The allowed cell rate can vary between the Minimum Cell Rate (MCR) and the Peak Cell Rate (MCR) to remain in conformance with the service.

MIB *Management Information Base*. A database of network-management information used by the network-management protocol SNMP (Simple Network Management Protocol). Network-managed objects implement relevant MIBs to allow remote-management operations. See also *SNMP*.

MPLS *Multi Protocol Label Switching*. An emerging technology in which forwarding decisions are based on fixed-length labels inserted between the data-link and network layer headers to increase forwarding performance and path-selection flexibility.

MPOA *Multi Protocol Over ATM*. An ATM Forum standard specifying how multiple network-layer protocols can operate over an ATM substrate.

MSS *Maximum Segment Size*. A TCP option in the initial TCP SYN (Synchronize Sequence Numbers) three-way handshake that specifies the maximum size of a TCP packet that the remote end can send to the receiver. The resultant TCP data-packet size is normally 40 bytes larger than the MSS: 20 bytes of IP header and 20 bytes of TCP header.

MTU *Maximum Transmission Unit*. The maximum size of a data frame that can be carried across a data-link layer. Every host and router interface has an associated MTU related to the physical media to which the interface is connected, and an end-to-end network path has an associated MTU that is the minimum of the individual-hop MTUs within the path.

NANOG *North American Network Operators Group*. A group of Internet operators who share a mailing list. A subset of this group meets regularly in North America. The conversation on the mailing list ranges from the pertinent to the inane. The overall characterization of the group manages to remain as the conspicuous absence of suits and ties.

NAS *Network Access Server*. A modernized and "kinder, gentler" form of its precursor, the terminal server. In other words, a device used to terminate dial-up access to a network. Predominantly used for analog or digital dial-up PPP (Point-to-Point Protocol) access services.

NBMA *Non Broadcast Multi Access*. Describes a multiaccess network that does not support broadcasting or on which broadcasting is not feasible.

NetWare A Novell, Inc. network operating system still largely popular in the corporate enterprise. The use of NetWare is experiencing somewhat of a decline because of the popularity and critical success of TCP/IP. IPX (Internet Packet eXchange) is the principal protocol used in NetWare networks.

NGI *Next Generation Internet*. An obligatory inclusion in every current network research proposal. Also used as a reference for a U.S. government sponsored advanced Internet research initiative, called the Next Generation Internet Initiative, which is somewhat controversial.

NHOP *Next Hop*, as referenced as an object within the Integrated Services Architecture protocol specifications.

NHRP *Next Hop Resolution Protocol*. A protocol used by systems in an NBMA (Non Braodcast Multi Access) network to dynamically discover the MAC address of other connected systems.

NLRI *Network Layer Reachability Information*. Information carried within BGP updates that includes network-layer information about the routing-table entries and associated previous hops, annotated as prefixes (IP addresses).

NMS *Network Management System*. The distant dream of many a network operations manager: a computer system that understands the network so well that it can warn the operator of impending disaster (humor implied).

NNI *Network-to-Network Interface*. An ATM Forum standard that defines the interface between two ATM switches operated by the same public or private network operator. The term also is used within Frame Relay to define the interface between two Frame Relay switches in a common public or private network.

NNTP *Network News Transfer Protocol*. An application protocol used to support the transfer of network news (Usenet) within the Internet. The protocol is used for bulk news transfer and remote access from clients to a central server. NNTP uses TCP to support reliable transfer. This protocol is a point-to-point transfer protocol. Efforts to move to a reliable multicast structure for Usenet news are still an active area of protocol refinement and research.

NOC *Network Operations Center*. The people you try to ring when your network is down. Traditionally staffed 24 hours a day, 7 days a week, the NOC primarily logs network-problem reports and attempts to redirect responsibility for a particular network problem to the appropriate responsible party for resolution. The NOC is analogous to a Help Desk.

nrt-VBR *Non-Real-Time Variable Bit Rate.* One of two variable-bit rate ATM service categories in which timing information is not crucial. Generally used for delay-tolerant applications with bursty characteristics.

NSF *National Science Foundation.* A U.S. government agency that funds U.S. scientific research programs. This agency funded the operation of the academic and research NSFnet (a successor of the ARPAnet and a predecessor to the current commodity Internet) network from 1986 until 1995.

OSI *Open Systems Interconnection.* A network architecture developed under the auspices of ISO (International Standards Organization) throughout the 1980s as a standards-based technology suite to allow multivendor interoperability. Now primarily of historical interest.

OSPF *Open Shortest Path First.* An interior gateway routing protocol that uses a link-state protocol coupled with a Shortest Path First (SPF) path-selection algorithm. The OSPF protocol is widely deployed as an interior routing protocol within administratively discrete routing domains.

PAP *PPP Authentication Protocol.* A protocol that allows peers connected by a PPP link to authenticate each other using the simple exchange of a username and password.

PCR *Peak Cell Rate.* An ATM service parameter. PCR is the maximum value of the transmission rate of traffic on an Available Bit Rate (ABR) service category Virtual Connection (VC).

PHOP *Previous Hop*, as referenced as an object within the Integrated Services architecture protocol specifications.

PNNI *Private Network-to-Network Interface.* The ATM Forum specification for distribution of topology information among switches in an ATM network to allow the computation of end-to-end paths. The specification is based on similar link-state routing protocols. Otherwise known as the *ATM routing protocol.*

PPP *Point-to-Point Protocol.* A data-link framing protocol used to frame data packets on point-to-point links. PPP is a variant of the HDLC (High-Level Data Link Control) data-link framing protocol. The PPP specification also includes remote-end identification and authentication (PAP and CHAP), a link-control protocol (to establish, configure, and test the integrity of data transmitted on the link), and a family of network-control protocols specific to different network-layer protocols.

PRA *Primary Rate Access.* Commonly used as an off-hand reference for ISDN PRI network access.

PRI *Primary Rate Interface.* An ISDN user-interface specification. In North America, a PRI is a single 64-Kbps D-Channel used for signaling and 23 64-Kbps B-Channels used for voice or data (using a T1 access bearer). Elsewhere, the specification is two 64-Kbps D-Channels and 30 64-Kbps B-Channels (using an E1 access bearer).

PSTN *Public Switched Telephone Network.* A generic term referring to the public telephone network architecture.

PTSP *PNNI Topology State Packet*. A link-state advertisement distributed between adjacent ATM switches that contain node and topology information. Analogous to an OSPF Link State Advertisement (LSA).

PVC *Permanent Virtual Connection* or *Permanent Virtual Circuit*. An end-to-end Virtual Circuit (VC) that is established permanently.

QoS *Quality of Service*. Read this book and find out. Better yet, buy your own copy and then read it.

QoSR *Quality of Service Routing*. A dynamic routing protocol that has expanded its path-selection criteria to consider issues such as available bandwidth, link and end-to-end path utilization, node-resource consumption, delay and latency, and induced jitter.

RAPI *RSVP Application Programming Interface*. An RSVP-specific API that enables applications to interface explicitly with the RSVP (Resource ReSerVation Setup Protocol) resource-reservation process.

RBOC *Regional Bell Operating Company*. Specific to the United States. Basically, the terms *LEC* (Local Exchange Carrier) and *RBOC* are interchangeable. RBOCs were formed in 1984 with the breakup of AT&T. RBOCs handle local telephone service, while AT&T and other long-distance companies, such as MCI and Sprint, handle long-distance and international calling. The seven original RBOCs after the AT&T breakup were Bell Atlantic, Southwestern Bell (recently changed to SBC, which acquired Pacific Bell on April 1, 1996), NYNEX (recently merged with Bell Atlantic), Pacific Bell (bought by SBC), Bell South, Ameritech, and U.S. WEST. Independent telephone companies also exist, such as GTE, that cover particular areas of the United States. The current landscape in the United States is still evolving. The Telecommunications Deregulation Act of 1996 now allows both RBOCs and long-distance companies to sell local, long-distance, and international services.

RED *Random Early Detection*. A congestion-avoidance algorithm developed by Van Jacobson and Sally Floyd at the Lawrence Berkeley National Laboratories in the early 1990s. In a nutshell, when queue depth begins to fill on a router to a predetermined threshold, RED begins to randomly select packets from traffic flows that are discarded in an effort to implicitly signal the TCP senders to throttle back their transmission rate. The success of RED is dependent on the basic TCP behavior, where packet loss is an implicit feedback signal to the originator of a flow to slow down its transmission rate. The ultimate success of RED is that congestion collapse is avoided.

RFC *Request For Comments*. RFCs are documents produced by the IETF for the purpose of documenting IETF protocols, operational procedures, and similarly related technologies.

RIP *Routing Information Protocol*. RIP is a classful, distance-vector, hop-count-based, interior-routing protocol. RIP has been moved to a "historical" status within the IETF and is widely considered to have outlived its usefulness.

routing The process of calculating network topology and path information based on the network-layer information contained in packets. Contrast with *bridging*.

RSRB *Remote Source-Route Bridging*. A method for encapsulating SNA traffic into TCP for reliable transport, and the capability to be routed over a wide-area network (WAN).

RSVP *Resource ReSerVation Setup Protocol*. An IP-based protocol used for communicating application Quality of Service (QoS) requirements to intermediate transit nodes in a network. RSVP uses a *soft-state* mechanism to maintain path and reservation state in each node in the reservation path.

RTT *Round Trip Time*. The time required for data traffic to travel from its origin to its destination and back again.

rt-VBR *Real-Time Variable Bit Rate*. One of the two variable-bit rate ATM service categories in which timing information is indeed critical. Generally used for delay-intolerant applications with bursty transmission characteristics.

SAP *Service Advertisement Protocol*. A broadcast-based, Novell NetWare protocol used to advertise the availability of individual application services in a NetWare network.

SBM *Subnet Bandwidth Manager*. A proposal in the IETF for handling resource reservations on shared and switched IEEE 802-style local-area media. See also *DSBM*.

SCR *Sustained Cell Rate*. An ATM traffic parameter that specifies the average rate at which ATM cells may be transmitted over a given Virtual Connection (VC).

SDH *Synchronous Digital Hierarchy*. The European standard that defines a set of transmission and framing standards for transmitting optical signals over fiber-optic cabling. Similar to the SONET standards developed by Bellcore.

SDLC *Synchronous Data-Link Control*. A serial, bit-oriented, full-duplex, SNA data-link layer communications protocol. Precursor to several similar protocols, including HDLC (High-Level Data Link Control).

SECBR *Severely Errored Cell Block Rate*. An ATM error parameter used to measure the ratio of badly formatted cell blocks (or AAL frames) to blocks that have been received error-free.

SLA *Service Level Agreement*. Generally, a service contract between a network service provider and a subscriber guaranteeing a particular service's quality characteristics. SLAs vary from one provider to another and usually are concerned with network availability and data-delivery reliability. Violations of an SLA by a service provider may result in a prorated service rate for the next billing period for the subscriber as a compensation for the service provider not meeting the terms of the SLA, for example.

SMTP *Simple Mail Transfer Protocol*. The Internet standard protocol for transferring electronic mail.

SNA *Systems Network Architecture*. General reference for the large, complex network systems architecture developed by IBM in the 1970s.

SNMP *Simple Network Management Protocol*. A UDP (User Datagram Protocol)-based network-management protocol used predominantly in TCP/IP networks. SNMP can be used to monitor, poll, and control network devices. SNMP traditionally is used to manage device configurations, gather statistics, and monitor performance thresholds.

SONET *Synchronous Optical Network*. A high-speed synchronous network specification for transmitting optical signals over fiber-optic cable. Developed by Bellcore. SONET is the North American functional equivalent of the European SDH

(Synchronous Digital Hierarchy) optical standards. SONET transmission speeds range from 155 Mbps to 2.5 Gbps.

SPF *Shortest Path First*. Also commonly referred as the *Dijkstra algorithm*. SPF is a single-source, shortest-path algorithm that computes all shortest paths from a single point of reference based on a collection of link metrics. This algorithm is used to compute path preferences in both OSPF and IS-IS. See also *Dijkstra algorithm*.

SRB *Source-Route Bridging*. A method of bridging developed by IBM and used in token ring networks, where the entire route to the destination is determined prior to the transmission of the data. Contrast with *transparent bridging*.

SVC *Switched Virtual Connection* or *Switched Virtual Circuit*. A Virtual Circuit (VC) dynamically established in response to UNI (User-to-Network Interface) signaling and torn down in the same fashion.

SYN *SYNchronize sequence numbers flag*. Contained in the TCP header. A bit field in the TCP header used to negotiate TCP session establishment.

T1 A wide-area network (WAN) transmission circuit that carries DS1-formatted data at a rate of 1.544 Mbps. Predominantly used within the United States.

T3 A WAN transmission circuit that carries DS3-formatted data at a rate of 44.736 Mbps. Predominantly used within the United States.

TCP *Transmission Control Protocol*. TCP is a reliable, connection- and byte-oriented transport layer protocol within the TCP/IP protocol suite. TCP packetizes data into segments, provides for packet sequencing, and provides end-to-end flow control. TCP is used by many of the popular application-layer protocols, such as HTTP, Telnet, and FTP.

TDM *Time Division Multiplexing*. A multiplexing method popular in telephony networks. TDM works by combining several signal sources onto a single circuit, allowing each source to transmit during a specific timing interval.

Telnet A TCP-based terminal-emulation protocol used in TCP/IP networks predominantly for connecting to and logging into remote systems.

TLV *Type, Length, Value*. A standard IETF format for protocol packet formats, where individual fields are allocated to indicate the *type* and *length* of a particular packet, as determined by a specific *value* expressed in each field.

token bucket Generally, a traffic-shaping mechanism by which the capability to transmit packets from any given flow is controlled by the presence of tokens. For packets belonging to a specific flow to be transmitted, for example, a token must be available in the bucket. Otherwise, the packet is queued or dropped. This particular implementation controls the transmit rate and accommodates bursty traffic. Contrast with *leaky bucket*.

TOS *Type of Service*. A bit field in the IP packet header designed to contain values indicating how each packet should be handled in the network. This particular field has never been used much, though.

transparent bridging A method of bridging used in Ethernet and IEEE 802.3 networks by which frames are forwarded along one hop at a time, based on forwarding information at each hop. Transparent bridging gets its name from the fact that the bridges themselves are transparent to the end-systems. Contrast with *SRB* (Source-Route Bridging).

TTL *Time To Live*. A field in an IP packet header that indicates how long the packet is valid. The TTL value is decremented at each hop, and when the TTL equals 0, the packet no longer is considered valid, because it has exceeded its maximum hop count.

UBR *Unspecified Bit Rate*. An ATM service category used for best-effort traffic. The UBR service category provides no QoS controls, and all cells are marked with the Cell Loss Priority (CLP) bit set. This indicates that all cells may be dropped in the case of network congestion.

UDP *User Datagram Protocol*. A connectionless transport-layer protocol in the TCP/IP protocol suite. UDP is a simplistic protocol that does not provide for congestion management, packet loss notification feedback, or error correction; UDP assumes these will be handled by a higher-layer protocol.

UNI *User-to-Network Interface*. Commonly used to refer to the ATM Forum specification for ATM signaling between a user-based device, such as a router or similar end-system, and the ATM switch.

UPC *Usage Parameter Control*. A reference to the traffic policing done on ATM traffic at the ingress ATM switch. UPC is performed at the ATM UNI level and in conjunction with the GCRA (Generic Cell Rate Algorithm) implementation.

VBR *Variable Bit Rate*. An ATM service characterization for traffic that is bursty by nature or is variable in the average, peak, and minimum rates in which data is transmitted. There are two service categories for VBR traffic: Real-Time and Non-Real-Time VBR. See also *rt-VBR* and *nrt-VBR*.

VC *Virtual Connection* or *Virtual Circuit*. An end-to-end connection between two devices that spans a Layer 2 switching fabric (e.g., ATM or Frame Relay). A VC may be permanent (PVC) or temporary (SVC) and is wholly dependent on the implementation and architecture of the network. Contrast with *VP*.

VCI *Virtual Connection Identifier* or *Virtual Circuit Identifier*. A numeric identifier used to identify the local end of an ATM VC. The local nature of the VCI is that it spans only the distance between the first-hop ATM switch and the end-system (e.g., router), whereas a VC spans the entire distance of an end-to-end connection between two routers that use the ATM network for link-layer connectivity.

VLAN *Virtual Local Area Network* or *Virtual LAN*. A networking architecture that allows end-systems on topological disconnected subnetworks to appear to be connected on the same LAN. Predominantly used in reference to ATM networking. Similar in functionality to *bridging*.

VP *Virtual Path*. A connectivity path between two end-systems across an ATM switching fabric. Similar to a VC. However, a VP can carry several VCs within it. Contrast with *VC*.

VPDN *Virtual Private Dial Network*. A VPN tailored specifically for dial-up access. A more recent example of this is L2TP (Layer 2 Tunneling Protocol), where tunnels are created dynamically when subscribers dial into the network, and the subscriber's initial Layer 3 connectivity is terminated on an arbitrary tunnel end-point device that is predetermined by the network administrator.

VPI *Virtual Path Identifier.* A numeric identifier used to identify the local end of an ATM VP. The local nature of the VPI is that it spans only the distance between the first-hop ATM switch and the end-system (e.g., router), whereas a VP itself spans the entire distance of an end-to-end connection between two routers that use the ATM network for link-layer connectivity.

VPN *Virtual Private Network.* A network that can exist discretely on a physical infrastructure consisting of multiple VPNs, similar to the "ships in the night" paradigm. There are many ways to accomplish this, but the basic concept is that many individual, discrete networks may exist on the same infrastructure without knowledge of one another's existence.

WAN *Wide-Area Network.* A network environment where the elements of the network are located at significant distances from each other, and the communications facilities typically use carrier facilities rather than private wiring. Typically, the assistance of a routing protocol is required to support communications between two distant host systems on a WAN.

WDM *Wave Division Multiplexing.* A mechanism to allow multiple signals to be encoded into multiple wavelengths, so that the light signals can be transmitted on a single strand of fiber-optic cable.

WFQ *Weighted Fair Queuing.* A combination of two distinct concepts—fair queuing and preferential weighting. WFQ allows multiple queues to be defined for arbitrary traffic flows, so that no one flow can starve other, lesser flows of network resources. The weighting component in WFQ is that the administrator can create the queue size and also delegate what traffic is identified for a particular-sized queue.

WRED *Weighted Random Early Detection* or *Weighted RED.* A variant of the standard RED mechanism for routers, in which the threshold for packet discard varies according to the IP precedence level of the packet. The weighting is such that RED is activated at higher queue-threshold levels for higher-precedence packets.

WWW *World Wide Web.* A global collection of Web servers interconnected by the Internet that use the HTTP (Hyper Text Transfer Protocol).

X *X Windows Protocol.* A protocol developed in the 1980s to provide a common graphical user interface that is independent of the host computer architecture and includes a specification of a remote access mechanism to allowed distributed access to remote applications from a single workstation. The protocol specification covers the screen, keyboard, and mouse of the workstation.

Bibliography

[AF1995a] The ATM Forum Technical Committee, "BISDN Inter Carrier Interface (B-ICI) Specification Version 2.0 (Integrated)," af-bici-0013.003, December 1995.

[AF1995b] The ATM Forum Technical Committee, "LAN Emulation over ATM Version 1.0 Specification," af-lane-0021.000, January 1995.

[AF1996a] The ATM Forum Technical Committee, "Traffic Management Specification Version 4.0," af-tm-0056.000, April 1996.

[AF1996b] The ATM Forum Technical Committee, "Traffic Management Specification Version 4.0," af-tm-0056.000, April 1996, Section 2.4, "Flow Control Model and Service Model for the ABR Service Category."

[AF1996c] The ATM Forum Technical Committee, "Private Network-Network Interface Specification Version 1.0," af-pnni-0055.000, March 1996.

[AF1996d] The ATM Forum Technical Committee, "Traffic Management Specification Version 4.0," af-tm-0056.000, April 1996, "Normative Annex B: Measurement & Analysis of QoS Parameters."

[AF1997a] "ATM Service Categories: The Benefit to the User," Editor: Livio Lambarelli, CSELT, for the ATM Forum.
www.atmforum.com/atmforum/library/service_categories.html

[AF1997b] The ATM Forum Technical Committee, "Multi Protocol Over ATM (MPOA) Specification Version 1.0," af-mpoa-0087.000, July 1997.

[Cisco1995] "Internetworking Terms and Acronyms," September 1995, Text Part Number 78-1419-02, Cisco Systems, Inc.

[DARPA1980] RFC768, "User Datagram Protocol," J. Postel, August 1980.

[DARPA1981a] RFC793, "Transmission Control Protocol," DARPA Internet Program Protocol Specification, September 1981.

[DARPA1981b] RFC791, "Internet Protocol," DARPA Internet Program Protocol Specification, September 1981.

[DARPA1981c] RFC792, "Internet Control Message Protocol," J. Postel, DARPA Internet Program Protocol Specification, September 1981.

[DKS1990] "Analysis and Simulation of a Fair Queuing Algorithm," Internetwork: Research and Experience, Volume 1, Number 1, John Wiley & Sons, September 1990, pages 3–26.

[FKSS1997] "Understanding TCP Dynamics in an Integrated Services Internet," W. Feng, D. Kandlur, D. Saha, K. Shin, NOSSDAV '97, May 1997.

[Floyd1993] "Random Early Detection Gateways for Congestion Avoidance," S. Floyd, V. Jacobson, IEEE/ACM Transactions on Networking, v. 1, n. 4, August 1993.

[ID1996b] IETF Internet Draft, "Protocol Independent Multicast-Dense Mode (PIM-DM): Protocol Specification," draft-ietf-idmr-pim-dm-spec-04.ps, D. Estrin, D. Farinacci, A. Helmy, V. Jacobson, and L. Wei, September 1996.

[ID1996c] IETF Internet Draft, "Protocol Independent Multicast-Sparse Mode (PIM-SM): Motivation and Architecture," draft-ietf-idmr-pim-arch-04.ps, S. Deering, D. Estrin, D. Farinacci, V. Jacobson, C. Liu, and L. Wei, November, 1996.

[ID1996d] IETF Internet Draft, "Protocol Independent Multicast-Sparse Mode (PIM-SM): Protocol Specification," draft-ietf-idmr-pim-sm-spec-09.ps>, D. Estrin, D. Farinacci, A. Helmy, D. Thaler; S. Deering, M. Handley, V. Jacobson, C. Liu, P. Sharma, and L. Wei, October 1996.

[ID1996e] IETF Internet Draft, "Path Precedence Discovery," draft-huston-pprec-discov-00.txt, G. Huston, M. Rose, December 1996.

[ID1996f] IETF Internet Draft, "QoS Extension to OSPF," draft-zhang-qos-ospf-00.txt, Z. Zhang, C. Sanchez, B. Salkewicz, E. Crawley, June 1996.

[ID1996g] IETF Internet Draft, "BGP Route Calculation Support for RSVP, " draft-zappala-bgp-rsvp-01.txt, D. Zappala, June 1996.

[ID1996h] IETF Internet Draft, "RSRR: A Routing Interface for RSVP," draft-ietf-rsvp-routing-01.txt, D. Zappala, November 1996.

[ID1997e] IETF Internet Draft, "Partial Service Deployment in the Integrated Services Architecture," drafte-ietf-rsvp-partial-service-00.txt, L. Breslau, S. Shenker, April 1997.

[ID1997h] IETF Internet Draft, "Open Outsourcing Policy Service (OOPS) for RSVP," draft-ietf-rsvp-policy-oops-01.txt, S. Herzog, D. Pendarakis, R. Rajan, R. Guerin, April 1997.

[ID1997i] IETF Internet Draft, "RSVP Extensions for Policy Control," draft-ietf-rsvp-policy-ext-02.txt, S. Herzog, April 1997.

[ID1997j] IETF Internet Draft, "Providing Integrated Services over Low-Bitrate Links," draft-ietf-issll-isslow-02.txt, C. Bormann, May 1997.

[ID1997k] IETF Internet Draft, "Network Element Service Specification for Low Speed Networks," draft-ietf-issll-isslow-svcmap-00.txt, S. Jackowski, May 1997.

[ID1997l] IETF Internet Draft, "Flow Grouping for Reducing Reservation Requirements For Guaranteed Delay Service," draft-rampal-flow-delay-service-01.txt, S. Rampal, R. Guerin, July 1997.

[ID1997m] IETF Internet Draft, "A Framework for Integrated Services and RSVP over ATM,"draft-ietf-issll-atm-framework-00.txt, E. Crawley, L. Berger, S. Berson, F. Baker, M. Borden, J. Krawczyk, July 1997.

[ID1997n] IETF Internet Draft, "Issues for Integrated Services and RSVP over ATM," draft-ietf-issll-isatm-issues-00.txt, E. Crawley, L. Berger, S. Berson, F. Baker, M. Borden, J. Krawczyk, July 1997.

[ID1997o] IETF Internet Draft, "Interoperation of Controlled-Load and Guaranteed Services with ATM," draft-ietf-issll-atm-mapping-02.txt, M. Garrett, M. Borden, March 1997.

[ID1997p] IETF Internet Draft, "NBMA Next Hop Resolution Protocol (NHRP)," draft-ietf-rolc-nhrp-11.txt, D. Katz, D. Piscitello, B. Cole, J. Luciani, March 1997.

[ID1997q] IETF Internet Draft, "A Framework for Providing Integrated Services over Shared Media and Switched LAN Technologies," draft-ietf-issll-is802-framework-02.txt, A. Ghanwani, J. W. Pace, V. Srinivasan, May 1997.

[ID1997r] IETF Internet Draft, "SBM (Subnet Bandwidth Manager): A Proposal for Admission Control over IEEE 802-style networks," draft-ietf-issll-is802-sbm-04.txt, R. Yavatkar, D. Hoffman, Y. Bernet, F. Baker, July 1997.

[ID1997s] IETF Internet Draft, "Integrated Service Mappings on IEEE 802 Networks," draft-ietf-issll-is802-svc-mapping-00.txt, M. Seaman, A. Smith, E. Crawley, July 1997.

[ID1997t] IETF Internet Draft, "A Framework for QoS-based Routing in the Internet," draft-ietf-qosr-framework-01.txt, E. Crawley, R. Nair, B. Rajagopalan, H. Sandick, July 1997.

[ID1997u] IETF Internet Draft, "QoS Routing Mechanisms and OSPF Extensions," draft-guerin-qos-routing-ospf-01.txt, R. Guerin, S. Kamat, A. Orda, T. Przygienda, D. Williams, March 1997.

[ID1997v] IETF Internet Draft, "Protocol for Exchange of PoliCy Information (PEPCI)," J. Boyle, R. Cohen, L. Cunningham, D. Durham, A. Sastry, R. Yavatkar, July 1997.

[ID1997w] IETF Internet Draft, "A Framework for Multiprotocol Label Switching," draft-ietf-mpls-framework-00.txt, R. Callon, P. Doolan, N. Feldman, A. Fredette, G. Swallow, A. Viswanathan, May 1997.

[ID1997x] IETF Internet Draft, "Extended RSVP-Routing Interface," draft-guerin-ext-rsvp-routing-untf-00.txt, R. Guerin, S. Kamat, E. Rosen, July 1997.

[ID1997y] IETF Internet Draft, "QoS Path Management with RSVP," draft-guerin-qospath-mgmt-rsvp-00.txt, R. Guerin, S. Kamat, S. Herzog, March 1997.

[ID1997z] IETF Internet Draft, "Setting up Reservations on Explicit Paths using RSVP," draft-guerin-expl-path-rsvp-00.txt, R. Guerin, S. Kamat, E. Rosen, July 1997.

[ID1997a1] IETF Internet Draft, "Layer Two Tunneling Protocol 'L2TP,'" draft-ietf-pppext-l2tp-04.txt, K. Hamzeh, T. Kolar, M. Littlewood, G. Singh Pall, J. Taarud, A. Valencia, W. Verthein, June 1997.

[ID1997a2] IETF Internet Draft, "The Multi-Class Extension to Multi-Link PPP," draft-ietf-issll-isslow-mcml-02.txt, C. Bormann, May 1997.

[ID1997a3] IETF Internet Draft, "Internet Protocol, Version 6 (IPv6) Specification," draft-ietf-ipngwg-ipv6-spec-v2-00.txt, S. Deering, R. Hinden, July 1997.

[ID1997a4] IETF Internet Draft, "Soft State Switching: A Proposal to Extend RSVP for Switching RSVP Flows," draft-viswanathan-mpls-rsvp-00.txt, A. Viswanathan, V. Srinivasan, S. Blake, August 1997.

[ID1997a5] IETF Internet Draft, "Use of Label Switching With RSVP," draft-davie-mpls-rsvp-00.txt, B. Davie, Y. Rekhter, E. Rosen, May 1997.

[ID1997a6] IETF Internet Draft, "End-to-End Traffic Issues in IP/ATM Internetworks," draft-jagan-e2e-traf-mgmt-00.txt, S. Jagannath, S. Yin, August 1997.

[IEEE-1] "MAC Bridges," ISO/IEC 10038, ANSI/IEEE Std. 802.1D-1993.

[IEEE-2] "Supplement to MAC Bridges: Traffic Class Expediting and Dynamic Multicast Filtering," IEEE P802.1p/D6, May 1997.

[IEEE-3] "Carrier Sense Multiple Access with Collision Detection (CSMA/CD) Access Method and Physical Layer Specifications," ANSI/IEEE Std. 802.3-1985.

[IEEE-4] "Token-Ring Access Method and Physical Layer Specifications," ANSI/IEEE std. 802.5-1995.

[IEEE-5] "Draft Standard for Virtual Bridged Local Area Networks," IEEE P802.1Q/D6, May 1997.

[IETF1983] RFC854, "TELNET Protocol Specification," J. Postel, J. Reynolds, May 1983.

[IETF1985] RFC959, "File Transfer Protocol (FTP), " J. Postel, J. Reynolds, October 1985.

[IETF1986] RFC977, "Network News Transfer Protocol," B. Kantor, P. Lampsley, February 1986.

[IETF1988] RFC1058, "Routing Information Protocol," C. Hedrick, June 1988. See also: RFC1923, "RIPv1 Applicability Statement for Historic Status," J. Halpern, S. Bradner, March 1996.

[IETF1989a] RFC112, "Host Extensions for IP Multicast," S. Deering, August 1989.

[IETF1990a] RFC1157, "A Simple Network Management Protocol (SNMP)," J. Case, M. Fedor, M. Schoffstall, J. Davin, May 1990.

[IETF1990b] RFC1195, "Use of OSI IS-IS for Routing in TCP/IP and Dual Environments," R. Callon, December 1990.

[IETF1991a] RFC1241, "A Scheme for an Internet Encapsulation Protocol: Version 1," R. Woodburn, D. Mills, July 1991.

[IETF1992a] RFC1363, "A Proposed Flow Specification," C. Partridge, July 1992.

[IETF1992b] RFC1349, "Type of Service in the Internet Protocol Suite," P. Almquist, July 1992.

[IETF1993a] RFC1492, "An Access Control Protocol, Sometimes Called TACACS," C. Finseth, July 1993.

[IETF1993b] RFC1510, "The Kerberos Network Authentication Service (V5)," J. Kohl, C. Neuman, September 1993.

[IETF1993c] RFC1548, "The Point-to-Point Protocol (PPP)," W. Simpson, December 1993.

[IETF1994a] RFC1583, "OSPF Version 2," J. Moy, March 1994.

[IETF1994b] RFC1633, "Integrated Services in the Internet Architecture: An Overview," R. Braden, D. Clark, S. Shenker, June 1994.

[IETF1994c] RFC1577, "Classical IP and ARP over ATM," M. Laubach, January 1994.

[IETF1994d] RFC1701, "Generic Routing Encapsulation (GRE)," S. Hanks, T. Li, D. Farinacci, P. Traina, October 1994.

[IETF1995a] RFC1771, "A Border Gateway Protocol 4 (BGP-4)," Y. Rekhter, T. Li, March 1995.

[IETF1995b] RFC1812, "Requirements for IP Version 4 Routers," F. Baker, June 1995.

[IETF1995c] RFC1883, "Internet Protocol, Version 6 (IPv6)," S. Deering, R. Hinden, December 1995.

[IETF1995d] RFC1809, "Using the Flow Label Field in IPv6," C. Partridge, June 1995.

[IETF1996a] RFC1932, "IP over ATM: A Framework Document," R. Cole, D. Shur, C. Villamizar, April 1996.

[IETF1996b] RFC2022, "Support for Multicast over UNI3.0/3.1 based ATM Networks," G. Armitage, November 1996.

[IETF1996c] RFC2018, "TCP Selective Acknowledgment Options," M. Mathis, J. Mahdavi, S. Floyd, A. Romanow, October 1996.

[IETF1996d] RFC1994, "PPP Challenge Handshake Authentication Protocol (CHAP)," W. Simpson, August 1996.

[IETF1996e] RFC1990, "The PPP Multilink Protocol (MP)," K. Sklower, B. Lloyd, G. McGregor, D. Carr, T. Coradetti, August 1996.

[IETF1996f] RFC2014, "IRTF Research Group Guidelines and Procedures," A. Weinrib, J. Postel, October 1996.

[IETF1997a] RFC2113, "IP Router Alert Option," D. Katz, February 1997.

[IETF1997b] RFC2131, "Dynamic Host Configuration Protocol," R. Droms, March 1997.

[IETF1997c] RFC2068, "Hypertext Transfer Protocol—HTTP/1.1," R. Fielding, J. Gettys, J. Mogul, H. Frystyk, T. Berners-Lee, January 1997.

[IETF1997d] RFC2138, "Remote Authentication Dial In User Service (RADIUS)," C. Rigney, A. Rubens, W. Simpson, S. Willens, April 1997.

[IETF1997e] RFC2216, "Network Element Service Specification Template," S. Shenkar, J. Wroclawski, September 1997.

[IETF1997f] RFCRFC2205, "Resource ReSerVation Protocol (RSVP)—Version 1 Functional Specification," R. Braden, L. Zhang, S. Berson, S. Herzog, S. Jamin, September 1997.

[IETF1997g] RFC2215, "General Characterization Parameters for Integrated Service Network Elements," S. Shenker, J. Wroclawski, September, 1997.

[IETF1997h] RFC2211, "Specification of the Controlled-Load Network Element Service," J. Wroclawski, September 1997.

[IETF1997i] RFC2212, "Specification of Guaranteed Quality of Service," S. Shenker, C. Partridge, R. Guerin, September 1997.

[IETF1997j] RFC2210, "The Use of RSVP with IETF Integrated Services," J. Wroclawski, September 1997.

[IETF1997k] RFC2208, "Resource ReSerVation Protocol (RSVP) Version 1 Applicability Statement, Some Guidelines on Deployment," A. Mankin, F. Baker, R. Braden, S. Bradner, M. O'Dell, A. Romanow, A. Weinrib, L. Zhang, September 1997.

[INTSERVa] The IETF Integrated Services Working Group charter can be found at www.ietf.org/html.charters/intserv-charter.html.

[IRTFa] Internet Research Task Force draft, "Recommendations on Queue Management and Congestion Avoidance in the Internet," draft-irtf-e2e-queue-mgt-recs.ps, R. Braden, D. Clark, J. Crowcroft, B. Davie, D. Estrin, S. Floyd, V. Jacobson, G. Minshall, C. Partridge, L. Peterson, K. Ramakrishnan, S. Shenker, J. Wroclawski, L. Zhang, March 1997.

[Jacobson1988] V. Jacobson, "Congestion Avoidance and Control," Computer Communication Review, vol. 18, no. 4, pp. 314–329, August 1988.

[Kleinrock1976] "Queuing Systems, Vol. 2: Computer Applications," L. Kleinrock, John Wiley & Sons, 1976.

[MV1994] "Pricing Congestible Network Resources," Jeffrey K. MacKie-Mason, Hal. R. Varian, University of Michigan, July 1994.

[Nagle1985] RFC970, "On Packet Switches with Infinite Storage," J. Nagel, December 1985.

[Partridge1994a] "Gigabit Networking," by Craig Partridge, Addison-Wesley Publishing, 1994, pages 178, 179, 314, 315, ISBN 0-201-56333-9.

[Partridge1994b] "Gigabit Networking," by Craig Partridge, page 276, section 12.4, "Weighted Fair Queuing," Addison-Wesley Publishing, 1994, ISBN 0-201-56333-9.

[Partridge1994c] "Gigabit Networking," by Craig Partridge, page 20, Addison-Wesley Publishing, 1994, ISBN 0-201-56333-9.

[Perlman1992] "Interconnections: Bridges and Routers," by Radia Perlman, Addison-Wesley Publishing, 1992, ISBN 0-201-56332-0.

[Tannenbaum1988] "Computer Networks," 2nd Edition, A. Tannenbaum, pages 631–641, Prentice Hall, 1988.

[Zhang1990] "Oscillating Behavior of Network Traffic: A Case Study Simulation," L. Zhang, D. Clark, Internetwork: Research and Experience, Volume 1, Number 2, John Wiley & Sons, 1990, pages 101–112.

index

S

T

U

V

W

X